Expotecnologia 2018

AA006463

Research, Innovation and Development in Engineering

IOP Conference Series: Materials Science and Engineering
Volume 519

Cartagena, Colombia
31 October - 2 November 2018

Editor:

John R. Castro Suarez

ISBN: 978-1-5108-8718-3
ISSN: 1757-8981

Printed from e-media with permission by:

Curran Associates, Inc.
57 Morehouse Lane
Red Hook, NY 12571

Some format issues inherent in the e-media version may also appear in this print version.

This work is licensed under a Creative Commons Attribution 3.0 International Licence. Licence details: http://creativecommons.org/licenses/by/3.0/.

No changes have been made to the content of these proceedings. There may be changes to pagination and minor adjustments for aesthetics.

Printed by Curran Associates, Inc. (2019)

For permission requests, please contact the Institute of Physics at the address below.

Institute of Physics
Dirac House, Temple Back
Bristol BS1 6BE UK

Phone: 44 1 17 929 7481
Fax: 44 1 17 920 0979

techtracking@iop.org

Additional copies of this publication are available from:

Curran Associates, Inc.
57 Morehouse Lane
Red Hook, NY 12571 USA
Phone: 845-758-0400
Fax: 845-758-2633
Email: curran@proceedings.com
Web: www.proceedings.com

Table of contents

Volume 519

Expotecnología 2018 Research, Innovation and Development in Engineering

31 October to 2 November 2018, Cartagena, Colombia

Accepted papers received: 21 March 2019

Preface

Papers

IOP Conf. Series: Materials Science and Engineering **519** (2019) 011001 doi:10.1088/1757-899X/519/1/011001

Expotecnología 2018

Organized by

Fundación Tecnológica Antonio de Arévalo TECNAR

PROCESO DE RENOVACIÓN ACREDITACIÓN INSTITUCIONAL

FACULTAD DE DISEÑO E INGENIERÍA - TECNAR

CICTAR Centro de Investigación, Ciencia y Tecnología

Event sponsored by:

renata Colombia

BOLÍVAR SÍ INNOVA Territorio de conocimiento y transformación

República de Colombia COPNIA Consejo Profesional Nacional de Ingeniería

CAMACOL BOLIVAR

EPA Cartagena Establecimiento Público Ambiental

EUCLID GROUP TOXEMENT

CLOUDEX

Content from this work may be used under the terms of the Creative Commons Attribution 3.0 licence. Any further distribution of this work must maintain attribution to the author(s) and the title of the work, journal citation and DOI.
Published under licence by IOP Publishing Ltd

IOP Conf. Series: Materials Science and Engineering **519** (2019) 011001 doi:10.1088/1757-899X/519/1/011001

Expotecnología 2018 "*Research, Innovation and Development in Engineering*", Expotecnología 2018, took place from October 31 to November 2, 2018, in the beautiful city of Cartagena de Indias (Colombia), with an agenda that joins to the academic dynamics through the promotion of activities that promote culture and research commitment, transferring knowledge and technological innovation, and the management year after year to create spaces updating the educational and professional community of all areas of engineering in Colombia.

Expotecnología is a scientific-academic event that comes from the Faculty of Design and Engineering of the Technological Foundation Antonio de Arévalo - TECNAR, in order to provide teachers, students and other professionals, a specialized meeting that seeks, through the creation of academic networking, create a new relationship between science, technology, and society.

The conferences took place in an environment of socialization of experiences, research results or discussions around several themes focused on 4 topics, which have been strategically selected to approach from the academic, industrial and social point of view, the present and the future of our profession in our country.

The general topics discussed were:
New Technologies: Information systems, Software Engineering, Information technology, Internet of Things and Mobile Applications.
Petrochemical and Process Control: Material science, Instrumentation and Control Systems, Robotics and Automation, Industrial maintenance.
Electricity, Electronics and Telecommunications: Renewable energy, Energy and Power Systems, Digital Signal Processing, Wireless networks, Mobile Communications, Bioengineering.
Construction Summit: Modern Construction Systems, Computational Modeling, Structural analysis.

For Expotecnología 2018, almost 200 participants attended, representing 21 national and international universities, including Brazil, Jamaica, France, Cuba, Ecuador, Argentina, Costa Rica, Mexico and United States. 67 oral presentations took place on this occasion, of which 36 articles were selected for this volume of IOP Conference Series: Materials Science and Engineering, after a peer review process. Three international speakers and six national speakers were invited.

Invited speakers

Eufrasio Pérez Navío, Ph.D, Universidad de Jaén (España)
Eduardo Olguín Macaya, M.Sc, Universidad de San Sebastián (Chile)
Antonio Medina Rivilla, Ph.D, Universidad Nacional de Educación a Distancia (España)
Ferney Carvajalino Arrieta, EPA (Colombia)
Carlos Amor, COPNIA (Colombia)
Ana Velazco Barreto - Government of Bolivar (Colombia)
Indira Martin, Coastal Concession (Colombia)
Alfredo Meza, TOXEMENT (Colombia)
Fabián Arango Pineda, MER. Proycca Ingenie (Colombia)

Organizing Committee

Presidents

Libis Valdez Cervantes, M.Sc, Technological Foundation Antonio de Arévalo

Vice-Presidents

Mayerlis Vasquez Osorio, M.Sc, Technological Foundation Antonio de Arévalo

Coordinators
Luz Ahumada Torres, M.Sc, Technological Foundation Antonio de Arévalo
Yesid Tarriba Lezama, M.Sc, Technological Foundation Antonio de Arévalo

Scientific Committee

Co-Presidents
John R Castro Suarez, Ph.D, Technological Foundation Antonio de Arévalo
Luis Rojas Martinez, M.Sc, Technological Foundation Antonio de Arévalo

Finally, we want to thank all the authors for submitting their research papers to Expotecnología 2018. We express our gratitude to all the peer reviewers for their time and dedication in reviewing the articles assigned to be published in IOP.

John R. Castro Suarez, Ph.D
Editor of the Proceedings of EXPOTECNOLOGIA 2018

IOP Conf. Series: Materials Science and Engineering **519** (2019) 011002 doi:10.1088/1757-899X/519/1/011002

Peer review statement

All papers published in this volume of *IOP Conference Series: Materials Science and Engineering* have been peer reviewed through processes administered by the proceedings Editors. Reviews were conducted by expert referees to the professional and scientific standards expected of a proceedings journal published by IOP Publishing.

Content from this work may be used under the terms of the Creative Commons Attribution 3.0 licence. Any further distribution of this work must maintain attribution to the author(s) and the title of the work, journal citation and DOI.
Published under licence by IOP Publishing Ltd

IOP Conf. Series: Materials Science and Engineering **519** (2019) 012001 doi:10.1088/1757-899X/519/1/012001

Design and implementation of a level control at the bottom of the first column of a continuous binary distillation tower

M M Fitatá-Bojacá[1] and O García-Bedoya[1]

[1]University of Bogotá Jorge Tadeo Lozano, Faculty of Natural Sciences and Engineering, Department of Engineering, Career 4 Nº 22-61, Bogotá D. C., Colombia.

monicam.fitatab@utadeo.edu.co

Abstract. The laboratory of the university has a continuous distillation tower where students carry out laboratory practice, require that student manipulate valve at high temperatures generating risk at operation. A control system was carried out on the plant to maintain the level of 0.18 meters in the bottom of the first column. For its identification, the inspection of the tower was carried out to identify the components of each distillation column. These tests allowed obtaining data from the point of operation during the start-up of the tower. With this information collected, a model was defined through the mass and energy balance. Additionally, the tower was simulated under the Aspen Hysys software to obtain the steady state and, over this point the dynamic system was analyzed through the linearized model to design the controller and test robustness. The control strategy implemented, with LabVIEW software, was ON-OFF with hysteresis due to limitations in the instrumentation. As a result, it was found that the time stabilization of the tower decreased from three hours to one hour to concerning the manual level control and, a better regulation was observed in the temperatures of the top of first column 96°C and second column 72°C.

1. Introduction

Distillation is one of the most important unit operations in the chemical industry, consisting of separating streams with two or more components [1]. The University Jorge Tadeo Lozano of Bogotá has a distillation tower located in the Centro de Investigación en Procesos de Ingeniería (C.I.P.I); which presents stabilization times of up to three hours, which reduces the operational experience of the students. At the same time, it generates an inadequate practice in the laboratory and insecurity in the functioning of the equipment; since it flooded in the bottoms [2].

The modeling and control of a binary distillation column for the separation of the ethanol-water mixture [3]. Its contribution is due to the dynamic model, and the assumptions and considerations adopted, with the nonlinear differential equations of each of the models are solved using fourth order Runge Kutta. The behavior of each of them is analyzed before variations in the parameters in order to determine the most appropriate model for the development of the control algorithm [3]. The thesis presents the dynamic analysis of the system of the binary distillation column by linearization around a point of equilibrium and

Content from this work may be used under the terms of the Creative Commons Attribution 3.0 licence. Any further distribution of this work must maintain attribution to the author(s) and the title of the work, journal citation and DOI.
Published under licence by IOP Publishing Ltd

the study of bifurcations by varying the parameters of the system independently [3].

A thesis of Design and construction of a level control interface, temperature and water flow in a tank for use in laboratory practices is taken as reference. The contribution is to design and build the physical control interface, to model the system, to elaborate a computer program for the acquisition of data and control, was used the program Simulink for the numerical simulation [4].

The emulation of a continuous binary distillation column for purposes of training in control with educational purposes allows experimentation without having the real system. This article worked the emulation, the response of the composition of the product under disturbances in the feed is obtained using Matlab and Simulink [5].

For this reason, this work arises under the premise of giving continuity to the process previously performed in the distillation tower where was included the temperature control loop of the top of the second column of the continuous distillation tower [2].

This work consisted of designing and implementing a level control loop with the purpose of stabilizing the tower decreased from three hours to one hour to concerning the manual level control, keeping the controlled variable at a level of 0.18 meters, preventing the distillation tower from flooding due to excessive liquid inlet or on the contrary if the flow of liquid is low, the tower can be emptied. First, was done the identification of tower parameters is presented, following by the simulation of the system in Aspen Hysys to obtain the steady state of the system. The mathematical model of the system is presented to analyze the disturbances of the controller implemented with Simulink, the data acquisition consists of sensors, electro valve, data acquisition card and a PC with programmable software in LabVIEW.

2. Identification of tower parameters

The operation of the plant starts with the feed stream of an ethanol-water mixture, passes through a pump, and before entering the process passes directly through an exchanger to preheat, thus modifying the temperature a 55°C, to preheat the mixture is used as service fluid the thermal energy of the bottom of the first distillation column.

The tower has a control board with the signals of temperature indicators through thermocouples that allow the operator to read the temperature in each section of the tower. On the same board are the power buttons of the pumps. It also has pressure indicators type bourdon that allows the pressure to be measured and verify that the boiler supplies the proper steam to the tower. Because it contemplates regulation with a set of manual valves, it generates greater demands on the workforce of students. In this way, its operation must be carried out with a minimum of three people.

As shown in Figure 1, it is observed that the distillation tower consists of two columns:

Figure 1. Distillation tower of CIPI.

Column 1 is distillation (column on the left), consisting of three sections of plates [6] each with four plates per section and two sections of packaging in the upper part of the tower; this column has six temperature sensors. In the case of this plant, the input currents are the feed (A) that is given in the plate 13 (See Figure 2), and the injection of steam that comes into direct contact through a flute at the bottom of the column.

Column 2 is of Rectification (Column on the right) has four sections of the packed bed [6], has five temperature sensors, and an auxiliary steam supply that passes through a coil at the bottom of the column.

The tower was put into operation preparing approximately 200 L of a 15% by volume ethanol-water solution. The 200 L allow the operation of the column continuously, including the reflux system approximately six hours without implementing any control system.

Three tests were carried out (See Table 1). For the preparation, 30 L of 86% ethanol was used. Test one was performed without reflux, test two with low reflux and high distillation speed and test three with high reflux and low distillate speed, obtaining greater recovery of ethanol with high reflux at low distillation speed.

At the time of its manual operation, it was evidenced that the tower distills continuously when the temperature of the top of the first tower is maintained at 86 °C and the top of the second tower at 72 - 74 °C (see Figure 2), where the best distillate is obtained. In the Figure 2 was shows the inlet and outlet streams of the tower.

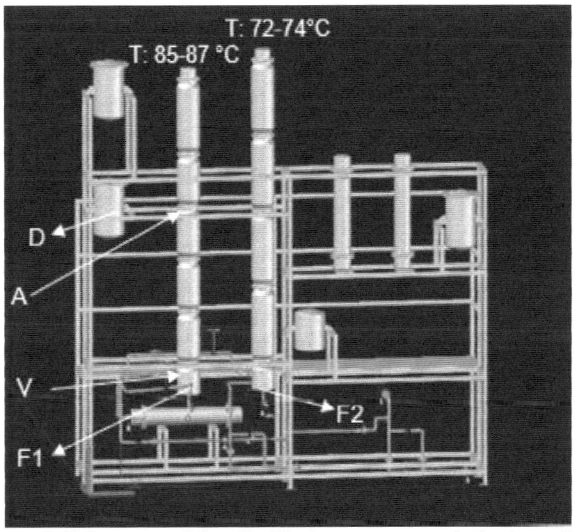

Figure 2. Diagram of the distillation tower made in AutoCAD.

Figure 2 shows the input and output currents of the tower. Each stream was named as follows:

A: Food of the tower
D: Distillation of the tower
V: Steam coming from the boiler
F: Bottoms of the tower
F1: Bottoms of the first column
F2: Bottoms of the second tower

Table 1. Results of open loop tests.

Parameter	Test one	Test two	Test three
Temperature at the top of the second column (°C)	74.0	74.0	72.0
Reflow (No. of turns)	0.0	2.0	4.0
Distillate obtained in practice (L/h)	4.0	1.8	2.0
Ethanol recovery (%)	77.0	84.0	85.0
Practice time (h)	5.0	6.0	6.0
Prepared food (L)	200.0	200.0	200.0

3. Simulation binary distillation tower in the steady state

The distillation tower was simulated in the steady state using the Aspen Hysys software [7], that is to say, the variables that define its behavior, with respect to time remain invariant, and corresponds to the initial conditions for the dynamic state [8].

For the estimation of thermodynamic properties, the model selected in the Aspen Hysys software was NTRL (Non-Random-Two-Liquid) that works non-ideal binary mixtures, based on the Gibbs energy equation with two or more binary parameters, that give reliable results, the model is not mathematically complex, it is of empirical and semi-analytic type [9].

The equipment used in the Aspen Hysys simulator was the distillation column as shown in Figure 3, simulated without reboiler and with 22 real stages that make up the equipment [2].

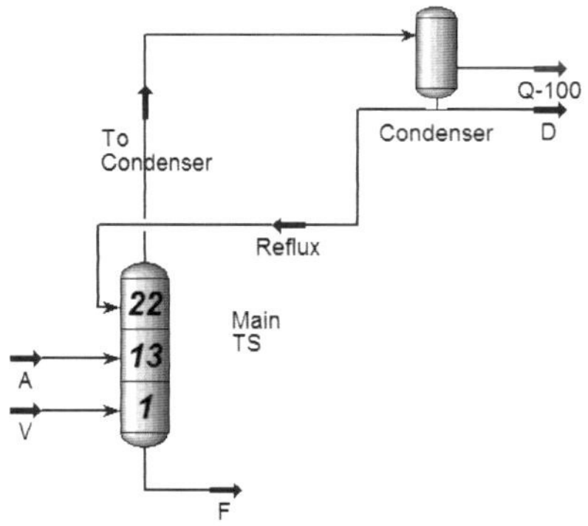

Figure 3. Environment of the simulation in Aspen Hysys.

Table 2 shows the results obtained from the mass balance and in Table 3 shows the distillate composition. The letters correspond to the currents of Figure 2 and Figure 3.

Table 2. Mass balance according to simulation of the distillation tower in Aspen Hysys.

Variable	Reflux	Condenser	A	D	F	V
Vapor Fraction	0.00	1.00	0.00	0.00	0.00	1.00
Temperature [°C]	76.76	76.77	55.00	76.76	92.32	92.32
Pressure [kPa]	73.00	73.00	76.53	73.00	76.53	76.53
Molar Flow [kgmole/h]	0.87	0.88	0.81	0.03	1.71	0.89
Mass Flow [kg/h]	37.24	37.37	16.19	0.13	32.23	16.17

Table 3. Distillate composition in Aspen Hysys.

Variable	Reflux	Condenser	A	D	F	V
Comp Mole Frac (Ethanol)	0.87	0.87	0.15	0.87	0.03	0.00
Comp Mole Frac (H_2O)	0.13	0.13	0.85	0.13	0.97	1.00

4. Assumptions and considerations

The mathematical model of the binary distillation column must represent the dynamic behavior of the real process. In order to obtain this model, the following assumptions and considerations were made:

- At the bottom of the column, the density, pressure, and temperature of the currents are constant.
- The temperature influences the energy, the problem is simplified assuming that the enthalpy specifies of the liquid can be expressed as a product of the temperature and its heat capacity (H = Cp T) [1].
- The reference temperature is $T_r = 298.15$ K
- The steam enthalpy of the section is constant and is expressed as ($H = Cp\,T + \lambda$), where λ is the heat of vaporization [1].
- The heat loss is considered negligible.
- The bottom of the tower is cylindrical $V=hA_T$, where h: Level in the tank [m] and A_T: Total area [m^2].

5. Balance and transfer equations

To make the model, Figure 4 represents a diagram concerning the balance of mass and the energy balance in the bottoms of the column [10]. Where the variables are defined:

Variable to control (h): Level of the bottoms of the first column [m]
Variable to manipulate(q_{L_b}): Volumetric flow of the liquid that leaves the section [m^3/h]
Disturbance variables: (q_{V_o}) Volumetric flow of steam coming from the boiler [m^3/h] and (q_{L_m}) Volumetric flow of liquid entering the section [m^3/h].

The flows can be calculated through a total mass balance (1), a balance per component Water (2) and a balance by component Ethanol (2) around the contact stage can be written as follows [1].

$$V_o + L_m = V_f + L_b \tag{1}$$

$$V_o Z + L_m X_i = V_f Y_i + L_b X_i \tag{2}$$

Where:
V_o: The mass flow of steam coming from the boiler [Kg/h]
L_m: The mass flow of liquid entering the section [Kg/h]
V_f: The mass flow of steam leaving the section [Kg/h]
L_b: The mass flow of the liquid that leaves the last section of the column by the gravity [Kg/h]
Z: Steam composition coming from the boiler [Kg water/Kg total]
X_i: Liquid composition of component *i* that can be water or ethanol [Kg del component /Kg total]
Y_i: Steam composition of component *i* that can be water or ethanol [Kg del component /Kg total]

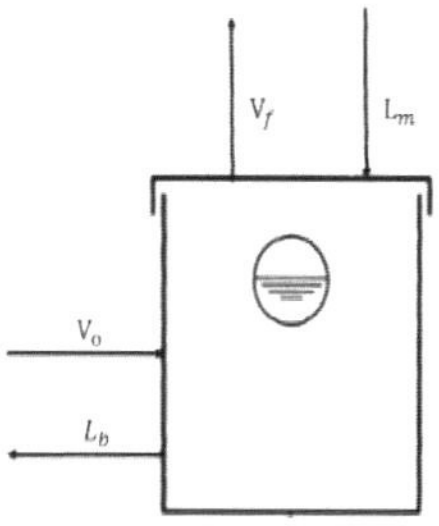

Figure 4.Bottoms of the first column.

The energy balance contemplates the enthalpies of the vapor and liquid streams can be expressed as (3).

$$V_oH_o - V_fH_f + L_mH_m - L_bH_b = \rho A_T CpT\frac{dh}{dt} \quad (3)$$

Where:
H_o: Mass enthalpy of steam coming from the boiler. [J/kg]
H_m: Mass enthalpy of liquid entering the section [J/kg]
H_b : Enthalpy vapor mass leaving the section [J/kg].
H_f: Enthalpy mass of steam the bottoms of the column [J/kg]

Through (1) was eliminated V_f from (3) and was get the (4)

$$V_oH_o - (V_o + L_m - L_b)H_f + L_mH_m - L_bH_b = \rho A_T CpT\frac{dh}{dt} \quad (4)$$

The model of the valve (5) is introduced into the system of equations. The rate of flow through the valve depends on the size, pressure drop, position and fluid properties [1]. The flow of the liquid through a valve is given by [11].

$$q_{L_b} = \alpha C_V(vp)\sqrt{\frac{\Delta P}{G}} \quad (5)$$

Where:
q_{L_b}: Flow of liquid that leaves the last section of the column due to gravity [m^3/h]
α : Coefficient of conversion to international system.
C_V: Valve coefficient [gpm/$psi^{1/2}$]
ΔP: Pressure drop through the valve [kPa]
vp: Opening of the valve; if its value is 0, the valve is closed; if the value is 1, the valve is open.
G: Specific gravity of the liquid flowing through the valve, without dimensions

For this process, the pressure drop (6) through the valve is given by [11].

$$\Delta P(t) = P_1 + \rho gh - P_2 \quad (6)$$

Where:
P_1: Pressure on the liquid [kPa]
ρ: Density of the liquid [kg/m^3]
g: Acceleration due to gravity [9.8 m/s^2]
P_2: Valve outlet pressure [kPa]

The equation (7) is obtained by replacing q_{L_b} in (4) through (5) y (6).

$$\frac{dh}{dt}=\frac{q_{Vo}\rho_{Vo}\left(C_{pVo}(T_{Vo}-Tr)+\lambda_V\right)}{\rho A_T CpT}+\frac{q_{Lm}\rho_{Lm}Cp_{Lm}(T_{Lm}-Tr)}{\rho A_T CpT}-$$

$$\frac{\left(q_{Vo}\rho_{Vo}+q_{Lm}\rho_{Lm}-\left(C_V(vp(t))\sqrt{\frac{P_1+\rho gh-P_2}{G}}\right)\rho_{Lb}\right)\left(Y_{et}(Cp_{et}(T_{Vf}-Tr)+\lambda_{et})+Y_{ag}(Cp_{ag}(T_{Vf}-Tr)+\lambda_{ag})\right)}{\rho A_T CpT}-$$

$$\frac{\left(C_V(vp(t))\sqrt{\frac{P_1+\rho gh-P_2}{G}}\right)\rho_{Lb}Cp_{Lb}(T_{Lb}-Tr)}{\rho A_T CpT} \quad (7)$$

Where:
Cp_{et}: The heat capacity of ethanol in the vapor state as a pure component [J/kg °C]
Cp_{ag}: The heat capacity of water in the vapor state as a pure component [J/kg °C]
T_{Vf}: Temperature of the top of the bottom section of the first column [°C]
λ_{ag}: Enthalpy of vaporization of water as a pure component to conditions of Bogota [J/kg]
λ_{et}: Enthalpy of vaporization of ethanol as a pure component to conditions of Bogota [J/kg]
T_{Lb}: Temperature of the bottom section of the first column [°C]
T: Temperature of the section [°C]

For the linearization of (8) use the expansion of the Taylor series, with the point of operation presented in Table 2.

$$\frac{d\Delta h(t)}{dt}=\left(\frac{\partial f}{\partial V_o}\right)\Big|_{op}\Delta q_{V_o}(t)+\left(\frac{\partial f}{\partial L_m}\right)\Big|_{op}\Delta q_{Lm}(t)-\left(\frac{\partial f}{\partial vp}\right)\Big|_{op}\Delta vp(t)-\left(\frac{\partial f}{\partial h}\right)\Big|_{op}\Delta h(t) \quad (8)$$

The deviation variables are defined:

$$Q_{Vo}(t)=q_{Vo}(t)-q^s_{Vo},\ Q_{Lm}(t)=q_{Lm}(t)-q^s_{Lm},\ VP(t)=vp(t)-vp^s,\ \bar{h}(t)=h(t)-h^s$$

The deviation variables are replaced in the linearized equation (9)

$$\frac{d\bar{h}(t)}{dt}=\alpha 1Q_{Vo}+\alpha 2Q_{Lm}-\alpha 3\bar{h}-\alpha 4\bar{h}-C_1VP \quad (9)$$

Where the values of the constants are:

$C_1=C_V\sqrt{\frac{P_1+\rho gh-P_2}{G}}$	$C_2=C_V\rho gvp^2\left(\sqrt{\frac{P_1+\rho gh-P_2}{G}}\right)^{-\frac{1}{2}}$

$\alpha 1 = \frac{\rho_{\mathrm{Vo}}(C_{\mathrm{pVo}}(T_{\mathrm{Vo}} - \mathrm{Tr}) + \lambda_{Vo})}{\rho A_T CpT}$	$\alpha 2 = \frac{\rho_{\mathrm{Lm}} \mathrm{Cp}_{\mathrm{Lm}}(T_{Lm} - \mathrm{Tr})}{\rho A_T CpT}$
$\alpha 3 = \frac{(q_{\mathrm{Vo}}\rho_{\mathrm{Vo}} + q_{\mathrm{Lm}}\rho_{\mathrm{Lm}} - C_2\rho_{\mathrm{Lb}})(Y_{et}(Cp_{et}(T_{\mathrm{Vf}} - \mathrm{Tr}) + \lambda_{\mathrm{et}}) + Y_{ag}(Cp_{ag}(T_{\mathrm{Vf}} - \mathrm{Tr}) + \lambda_{\mathrm{ag}}))}{\rho A_T CpT}$	
$\alpha 4 = \frac{C_2\rho_{\mathrm{Lb}} \mathrm{Cp}_{Lb}(T_{Lb} - \mathrm{Tr})}{\rho A_T CpT}$	

The above allows to obtain the transfer functions (11) through the Laplace transforms of the first order (10), for the analysis of the dynamics of processes and the design of control systems [11].

$$\bar{\mathrm{h}}(\mathrm{S}) = \frac{K_1}{\tau s+1} \mathrm{Q}_{Vo} + \frac{K_2}{\tau s+1} \mathrm{Q}_{Lm} + \frac{K_3}{\tau s+1} \mathrm{VP} \tag{10}$$

Where,

$$\tau = \frac{1}{\alpha 3+\alpha 4}, K_1 = \frac{\alpha 1}{\alpha 3+\alpha 4}, K_2 = \frac{\alpha 2}{\alpha 3+\alpha 4}, K_3 = \frac{C_1}{\alpha 3+\alpha 4} \tag{11}$$

The transfer functions are represented in a block diagram [12] for the control system as shown in Figure 5.

Where the transfer function (Gp) (12) involves the transfer function of the bottom of the distillation column and the solenoid valve.

$$Gp = \frac{K_3}{\tau s+1} = \frac{2.86}{1.61s+1} \tag{12}$$

The sensor transfer function (Gs) (13) defines the gain as the change in the output or response variable between the change in input or forcing function [11], the output is an electronic signal from 0 to 24 Volts, and the Inlet is the level of the liquid from 0 to 0.18 m.

$$G_s = \frac{Voltaje\ salida}{Altura\ del\ liquido} = \frac{24-0}{0.18-0} = 133.33 \tag{13}$$

Figure 5 also identifies the transfer functions of the disturbances for the level control loop in the bottoms, which for this system are those presented in (14).

$$\frac{\mathrm{X(S)}}{\mathrm{Q}_{Vo}(s)} = \frac{23.08}{1.61s+1}, \frac{\mathrm{X(S)}}{\mathrm{Q}_{Lm}(s)} = \frac{5.43}{1.61s+1} \tag{14}$$

IOP Conf. Series: Materials Science and Engineering **519** (2019) 012001 doi:10.1088/1757-899X/519/1/012001

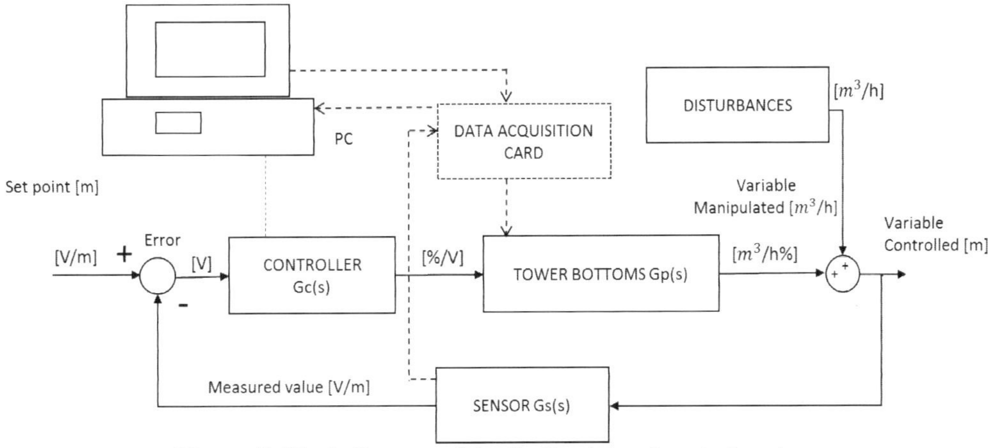

Figure 5. Block diagram of the feedback level control system.

6. Disturbance analysis

To analyze the robustness and stability of the control system we performed simulations of the behavior of the system in different scenarios with the Simulink program. To do that the controller (Gc) was modeled with a Relay block, as shown in Figure 6, creating a hysteresis controller which allows the output to change between two specified values as implemented (see next section).

The first scenario was to make the stationary error zero in the operating conditions. Where the difference of the measurements of the reference point and the desired output is known as the error, which is converted into a value and conditioned to be used in the opening or closing of the final control element [13]. For the simulation to be close to reality a measurement noise was included; which is caused by the bubbles that form due to the phase change, modeled with a Band-Limited White Noise show in the Figure 7a. In its steady state, it is maintained around the desired level of 0.18 m with a noise of 0.02 m above and below the desired point as shown in Figure 7a. Figure 7b shows the control signal, in which it can be seen that although there is noise in the reading signal, the on and off frequency of the valve is not very high with a percentage of 54%.

The second scenario analyzed were the change in the operation of the variable considered as disturbance. In this case, the steam coming from the boiler has influence in the system to generate the formation of bubbles, and the increase of the temperature in the top of the column. As shown in Figure 7c, the set point is moved and does not maintain the desired level of 0.18 m. For other way, it was found that the controller is robust before changes of +/- 0.001.

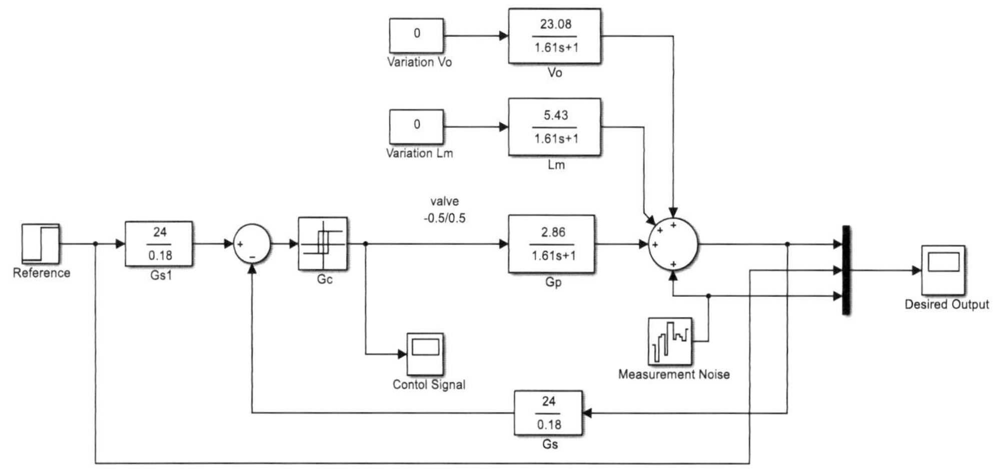

Figure 6. Block diagram in Simulink.

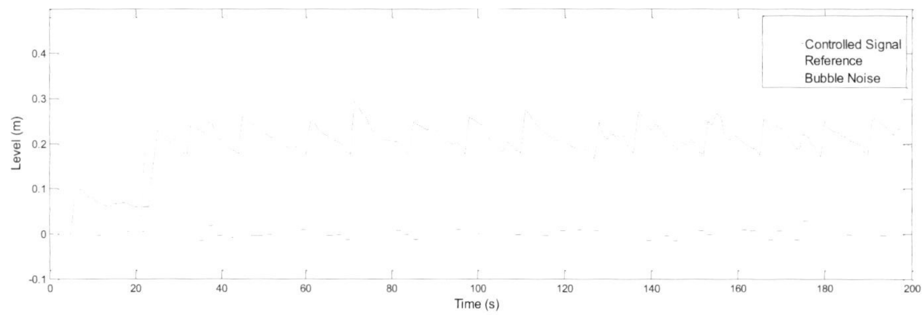

Figure 7a. Level (m) vs Time (s) point of operation with the Simulink program.

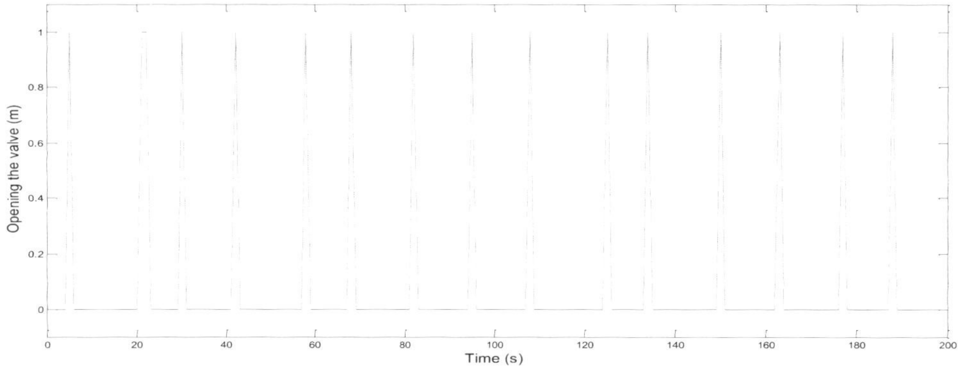

Figure 7b. Control signal over point of operation with the Simulink program.

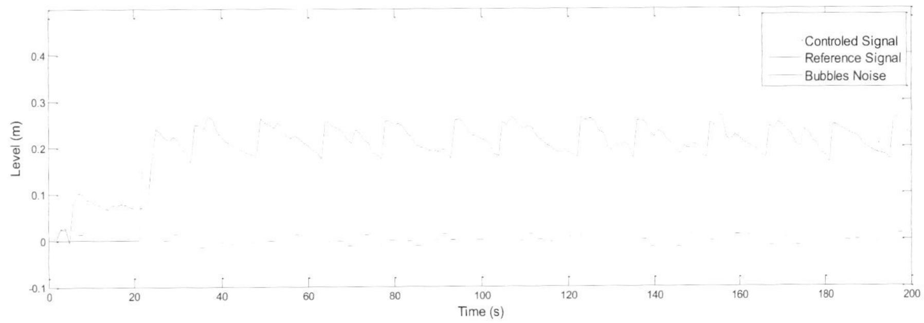

Figure 7c. Level (m) vs Time (s) of the vapor phase with the Simulink program.

7. Level control strategy

Before implementing the assemblies, the instrumentation selection was carried out according to process conditions, instrumentation options that the university has in its robotic laboratory and permitting facility (Appendix A).

Two assemblies were carried out. The first was an ON-OFF control, as shown in Figure 8a under ISA-5.1-2009, where the capacitive proximity sensor, and the information generated by this measurement element is treated using a signal conditioner and converted to a digital format through NI USB-6009 [14] (Appendix A), where LabVIEW graphic programming language process the information. The controlled variable was a normally close ON-OFF valve activated electrically by LabVIEW [15]. Figure 8b show the implementation.

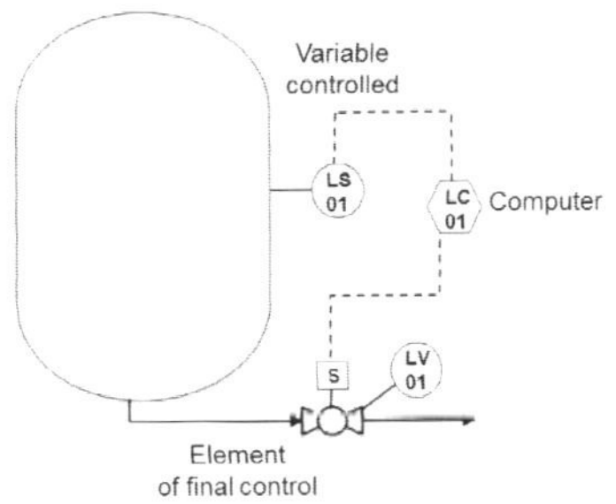

Figure 8a. ON-OFF control diagram.

Figure 8b. ON-OFF control implemented.

The control strategy consists of generating a delay in the opening of the solenoid valve according to a parameter established by the interface in seconds, through a time function. This did it because it was not feasible to install an analogous sensor due to the turbulence and vapors generated.

For assembly one, the control system has two fixed states, with no intermediate values. As shown in Figure 9 the valve remained 97% of the time open, which generated a little correction to the system, temperatures will increase, the tower will be destabilized, and the distillate will not be constant. Note, that this control signal is inverted with respect to the simulation because the valve is normally closed.

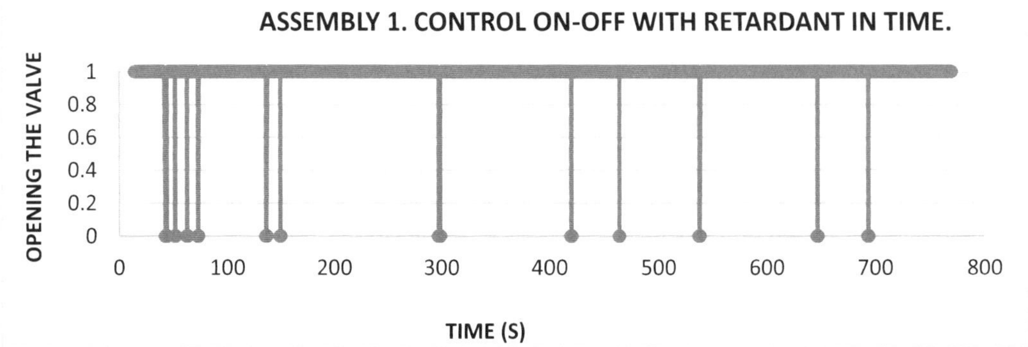

Figure 9. Data collection assembly one.

An improvement was made to the first assembly with an on-off control with hysteresis. This time two sensors were implemented as shown in Figure 10a y Figure 10b.

For its implementation (Appendix A), a source was needed to energize the solenoid valve and sensors, as the solenoid valve has a voltage of 24 VDC it is necessary to use an LM2596 converter that reduces the signal to 5 VDC to power the card, likewise for capacitive type sensors, relays were used that carry the signal to the card.

Figure 10a. Control assembly with hysteresis.

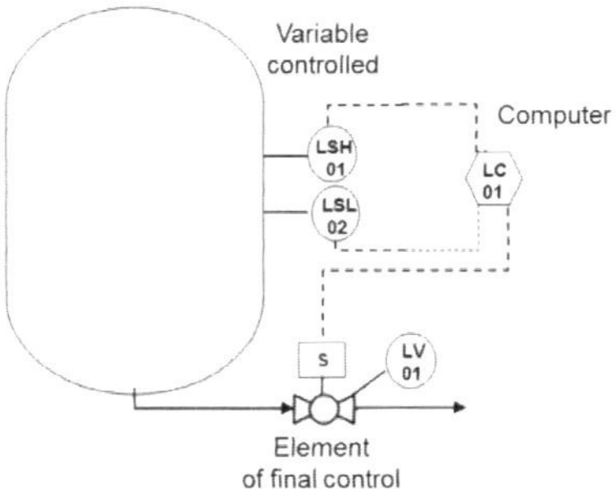

Figure 10b. Diagram for hysteresis.

Within the LabVIEW environment (Figure 11a), the low sensor was called an LSL when it did not register anything, it sent the signal to the valve that had to close preventing the tower from emptying, at this point, it begins to fill until the sensor High LSH was lit. Only at the highest point did the valve open, allowing the tower to be emptied and not flooded.

As shown in Figure 11b, the program receives a low and a high signal, only until it receives the high signal, it decides to activate the valve so that it opens and does not allow the tower to flood this is done using a subroutine called RS flip-flop. The valve only closes if the low signal registers a value of zero. The waveform chart function was used to represent the behavior of the signal graphically.

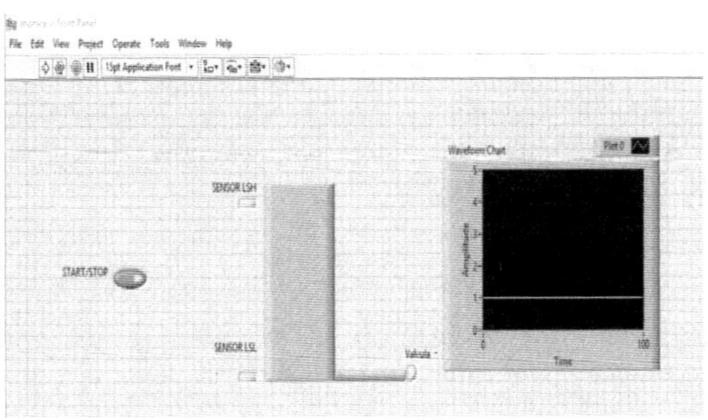

Figure 11a. Front panel in LabVIEW.

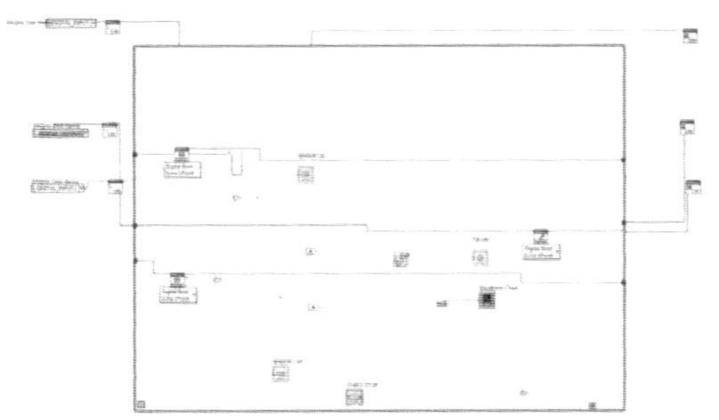

Figure 11b. Block diagram ON-OFF control in LabVIEW.

However, the turbulence that appeared showed a constant record of one, with the valve remaining open at 64% during the 3 hours of operation as shown in Figure 12; unlike the first assembly, a better control to the system is generated.

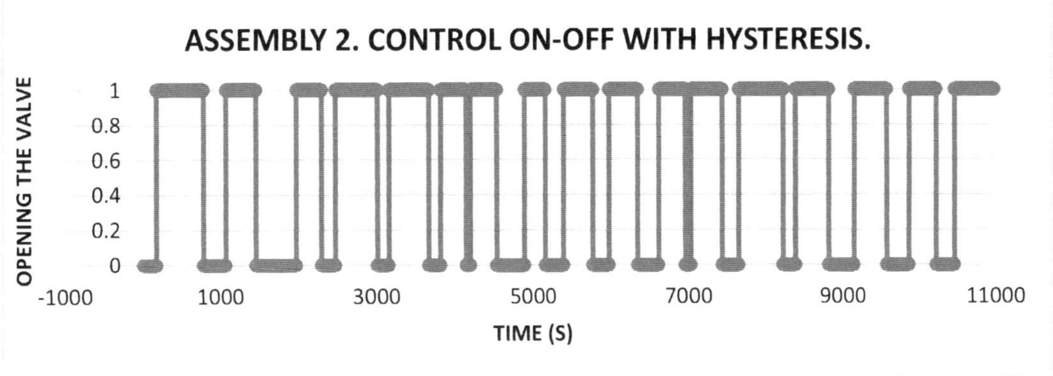

Figure 12. Data collection assembly two.

Finally, to validate the behavior of the two assemblies, the 200 L solution of an ethanol-water solution at 15% by volume was again prepared. For the preparation, 30 L of commercial grade 74% ethanol are used; these results are recorded in table 4; the input of food is one liter per minute that according to the practices is where the meter takes a better record.

Table 4. On-off control (assembly 1) vs on-off control with hysteresis (assembly 2).

Parameter	Assembly 1	Assembly 2
Average temperature at the top of the second tower (°C)	74	72
Reflow (No. of turns)	4	4
Quantity of distillate in practice (L/h)	5	7
Ethanol recovery (%)	60	77
Practical time (h)	4	3
Prepared food (L)	200	200

When comparing the results of Table 4, assembly 1 took more time in practice and did not reach the recovery of ethanol from assembly 2, this was due to the fact that the first assembly was destabilized increasing the liquid intake, which meant that the temperature of the top would increase, the tower would be destabilized and the distillate will not be constant. For the second assembly another sensor was implemented in order to operate with hysteresis, having a high and low control, obtaining a stable recovery of ethanol and maintaining the temperatures at the top of each column.

8. Conclusions and recommendations

The design and implementation of a level control with hysteresis in the bottom of the first column was achieved, which reduces the labor of the students who manipulate the tower. Also, students reduce high-risk conditions by avoiding the manual valve management of the bottom that reaches temperatures above 90 °C with the replacement of the solenoid valve implemented.

With the control loop implemented, a chain reaction is generated, and indirectly it controls the temperature change, once it stabilizes, it is possible to decrease the stabilization time of the tower, going from 3 hours with practices without a control system to 1 hour with the implementation of the control system.

The capacitive sensors are low cost and easy to adapt as they work by proximity; however, the readings were influenced by the turbulence generating that the solenoid valve remained in 64% of the open time, which in the future generates that the valve coil can be reheated. In this way, it is proposed to improve the strategy of control with a mechanical adaptation at the tower, an external tube is connected to put a differential pressure transmitter and measure the pressure difference between of level lower and upper [16]. For money and time, its implementation was not possible.

References

[1] Luyben, W L 2014 Process Modeling, Simulation and Control for Chemical Engineers (Vol. Second Edition) McGraw Hill Education India Private Limited

[2] Cardenas Tovar, L N, & Rodríguez Rivera, N D 2016 Automatización De Una Torre De Destilación Para Operación Continua A Escala Laboratorio Caso De Estudio: Centro De Investigación En Procesos De Ingeniería (C.I.P.I.) (Bogotá, Colombia: Universidad Jorge Tadeo Lozano)

[3] Alzate Ibañez, A M 2010 Modelado y control de una columna de destilación (Manizales, Colombia: Universidad Nacional de Colombia - Sede Manizales)

[4] Richmond Salazar, E 2009 Diseño y construcción de una interfaz de control de nivel, temperatura y flujo de agua en un tanque para uso en prácticas de laboratorio (San Jose, Costa Rica: Universidad de Costa Rica)

[5] Franco Ocampo, D F, & Franco Mejía, E 2015 *Emulación de una columna de destilación binaria continúa para propósitos de formación en control* UIS Ingenierías, 14(2), p 7-17

[6] Navas Herrera, S. J 2014 Control de Columnas de Destilación. Universidad de Sevilla. Retrieved from http://bibing.us.es/proyectos/abreproy/20374/fichero/Proyecto+completo.pdf

[7] AspenTech Inc 2005 Aspen Hysys: simulation basis Retrieved from http://www.aspentech.com

[8] Acedo Sánchez, J 2003 Control avanzado de procesos (Teoría y práctica) (España: Ediciones Díaz de Santos, S. A.)

[9] Poling, B. E, Prausnitz, J M, & O'Connell, J P 2004 The Properties of Gases and Liquids (Fifth Edition ed.) McGraw-Hill Companies

[10] Ogunnaike, B A, & Harmon Ray, W 1994 Process dynamics, modeling, and control (New York: Oxford University Press)

[11] A Smith, C, & B Corripio, A 2000 Control Automático de procesos. Teoría y práctica (México D.F: Editorial Limusa, S.A de C.V Grupo Noriega Editores)

[12] Peter Harriot, & Peter Harriott 1964 Process Control (United States of America: McGraw-Hill, Inc)
[13] Lázaro Castillo, I 2008 Ingeniería de Sistemas de Control Continuo (México: Universidad Michoacana De San Nicolás De Hidalgo) Retrieved from http://isidrolazaro.com/wpcontent/uploads/2013/01/Preview_Control_Continuo_1ed.pdf
[14] National Instruments 2014 Adquisición de datos Retrieved from http://www.ni.com/data-acquisition/what-is/esa/
[15] Villacruz Guijarro, G E 2011 Automatización y puesta en marcha de una columna de destilación continua de platos perforado tipo experimental para el sistema etanol-agua mediante PID y plataforma LabView denominada (UDCC) (Guayaquil, Ecuador: Universidad de Guayaquil)
[16] Creus Solé, A 2011 Instrumentación Industrial (Vol VIII) (México: Marcombo)

Acknowledgment
To Carlos Arturo Céspedes Zambrano in charge of laboratory of the CIPI for his support in the runs made with the tower, Andrés Felipe Ángulo Sánchez assistant of robotics laboratory for the loan of elements to assemble the control loop, this integration of careers allowed to carry out this work, obtaining the implementation of level control.

Appendix A

1. Solenoid valve.

The element final of control selected was a solenoid valve [16], are economical and only require an adequate energy source, where its purpose is to regulate the flow manipulated in the control system [11].

The flow capacity of a control valve is determined by its capacity factor or flow coefficient (Cv), introduced in 1944 by valve manufacturers Masoneilan International, Inc. [11], clearing from equation 5, where the bottoms in the first column are obtained at a rate of 0,269 gpm, considering that the valve is open vp is equal to 1, its get that the value of Cv is 0.31 gpm/psi$^{0.5}$ with this capacity factor and the characteristics shown in table A1, the solenoid valve was selected.

Table A1. Characteristics of the solenoid valve.

Parameter	Value
Medio	Aggressive, Steam
Internal material	Stainless steel
Body	Bronze brass
Seals	EPDM
Actuator action	NC
Connection	1/2”
Temperature	140 ºC
Voltage	24 V

2. Capacitive Sensor.

Sensors of capacitive type [16] were used (See Table A2), responsible for manipulating the high, and low-level variable with the support of the dielectric constant of water. When feeling the presence of the fluid, the light diode (LED) lights up.

Table A2. Characteristics of the sensor

Parameter	Value
Permissible operating voltage	12 – 48 V DC
Switching output	PNP, normally open
Material Front cover	PTFE
Body material	Nickel-plated brass
LED for the supply voltage	Green
Switching element «ON» LED	Yellow

3. Materials and connection.

Figure A1 shows the materials that integrated the final assembly of the control system implemented by hysteresis. The materials were obtained by the robotics area of the University Jorge Tadeo Lozano.

IOP Conf. Series: Materials Science and Engineering **519** (2019) 012001 doi:10.1088/1757-899X/519/1/012001

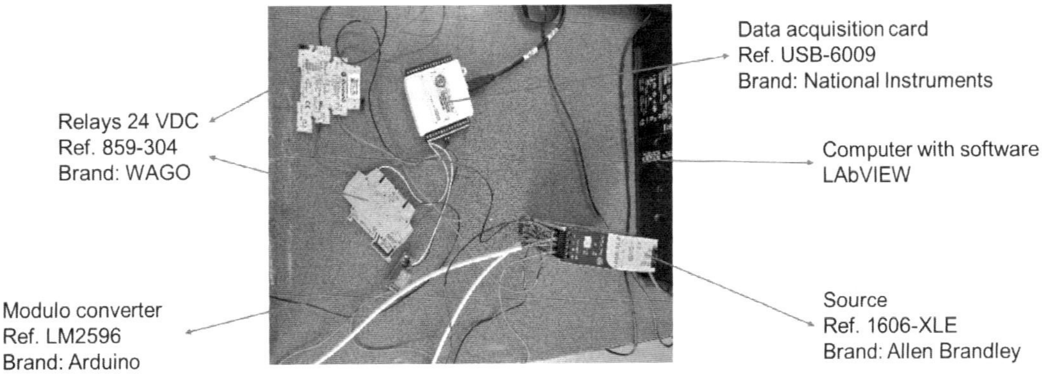

Figure A1. Assembly of the electrical connection of the control system.

To give clarity, Figure A2 shows the electrical connection of the materials for assembly.

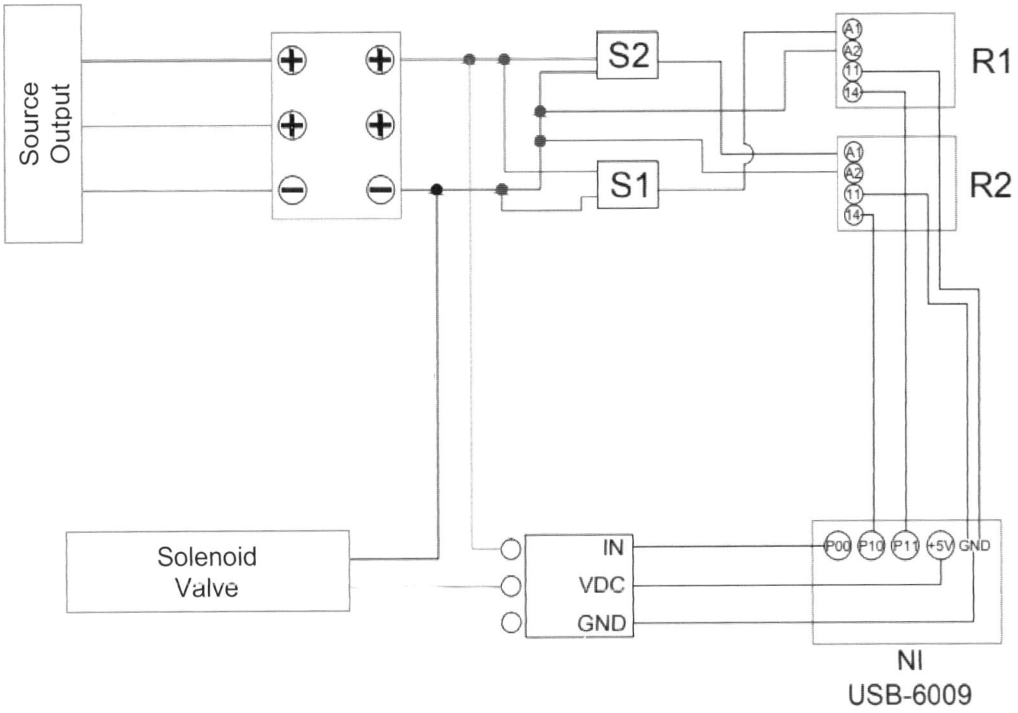

Figure A2. Schematic of electrical connection of the control system.

Preparation and characterization of composite materials polylactic acid (PLA) and Oxide Reduced Graphitic (RGO), for use in 3D printing small scale

A C Cordero[1], A P De la Cruz[1], J B Restrepo[1] and O A Vargas[2]

[1]Program of Chemical Engineering, Faculty of Engineering, Universidad del Atlantico, Barranquilla, Colombia.
[2]Metallurgical engineer before school and Materials Science, Faculty of Engineering physiochemical, Industrial University of Santander, Bucaramanga, Colombia.

accordero@mail.uniatlantico.edu.co , apdelacruz@mail.uniatlantico.edu.co

Summary. Composite materials polylactic acid/Reduced graphitic oxide (PLA/RGO) were prepared using the method of mixing chloroform solution. the chemical structure of chemically modified RGO obtained after reduction GO by infrared spectroscopy (FT-IR) was studied. PLA composites with different contents of RGO (1-3 wt%) in terms of distribution and morphology were compared, using optical microscopy. RGO better dispersion in the matrix of the PLA using chloroform as a solvent and 3% by weight of RGO was obtained.

1. Introduction

The use of polymers from oil has substantially disrupted the ecosystem due to their slow degradation, causing serious pollution problems in addition to the difficulty of recycling materials obtained as decrease its thermomechanical properties. Recently, it increased attention in biodegradable and biocompatible polymers, due to its environmental advantages because their mechanical properties can compete with conventional plastics[1].

Polylactic acid (PLA) is a highly versatile polymer and, in turn, is environmentally friendly (biodegradable) so is constantly used as the core of research and studies; either in the textile industry, the medical industry, among others.

James Lunt (1997) based his research on the properties and commercial applications owned by the PLA comparing the various ways of obtaining it. Properties such as melting point, flexural modulus, heat sealability and high surface energy were key to the PLA could be used in various applications[2].

Likewise, Lunt emphasized biodegradability of PLA particularly for storing products and in applications requiring food contact, wherein the PLA polymers degrade rapidly in environments of high humidity and temperature (55-70 ° C) and compared the degradation rate according to the temperature at which it is[2].

However, with the development of industry, came the need to seek alternatives that would allow the improvement of raw materials for products with higher quality or with specific properties for specific applications. Oksman et al. (2003) took advantage of the need for improvement and development of the materials commonly used in this case, the PLA. They based their study on strengthening and PLA polymers based on renewable raw materials in general, using natural fibers. Using flax fibers at different concentrations, it is sought to increase the mechanical properties (such as tensile strength) to check if

Content from this work may be used under the terms of the Creative Commons Attribution 3.0 licence. Any further distribution of this work must maintain attribution to the author(s) and the title of the work, journal citation and DOI.
Published under licence by IOP Publishing Ltd

there was an improvement in such properties[3]. Oksman et al. (2003) compared the maximum stress, elongation at break and modulus of PLA and polypropylene (PP) when were added different compositions flax and comparing the modulus and the tensile strength of PLA and PP with different compositions of fiber flax used. With these results, achieved check the properties of the compounds of PLA and flax were promising and resistance was 50% better than other compounds used in industry, in addition, the tensile modulus of PLA is increased from 3.4 to 8.4 GPa with an addition of 30% of flax fibers[3].

It was also investigated in other materials that could strengthen the properties of PLA. By the properties possessed by the carbonaceous particles to further research that had occurred over the decade. Fu et al. (2014) proposed the use of functionalized graphene polymeric hardener, as new interface modifier for polylactic acid nanocomposites. Improving viscoelasticity and thermal stability they gained interfacial interactions and the distribution and dispersion of functionalized graphene with a rubber phase containing epoxy[4]. With the manipulation of the functional groups were attached to surfaces of graphene, Fu et al. (2014) could tailor interactions and miscibility with graphene PLA matrix and achieve this a stable dispersion and uniform distribution of graphene in the polymer matrix, providing a future design of multifunctional and high-performance[4].

Ho et al (2015) used carbon produced from bamboo (BC) to increase the properties of PLA. The tensile strength, flexural and impact of compounds of PLA reinforced with particles of BC were the main analyzed properties of which a maximum tensile and bending and a significant increase in the rate of ductility compound was obtained, demonstrating that the characterizations with carbon particles were effective for improving the properties[5].

The PLA often mixed with flexible polymers to increase and improve its impact resistance, has shown that the mixture of this polymer with carboxyl groups is more effective [6]. The graphenes modified with different functional groups can also play the role of crosslinking and develop as a curing of the polymer.

All these studies that have demonstrated improved properties PLA involve an improvement in the products obtained from this, thus having a positive impact on the daily or industrial use of these polymers. In turn, these improvements would provide, to small-scale 3D printing, products with improved properties and a raw material will continue to have a low cost.

A preliminary study is presented in this work with the aim of developing a composite (PLA/RGO), which has an appropriate to serve as an input in 3D printing small-scale thermomechanical properties.

2. Experimental methodology

2.1. Reagents

- Natural graphite powder, with 96% purity and particle size less than 1 mm, commercial grade used as ink in photocopiers, brand FIXER machines.
- PLA 2992D was supplied by Grethsell Products as pellets

Table 1 is made an understanding of the reagents used, all provided by the house MERCK ©.

Table 1. Reagents used.

Reagent	*Concentration (%)*	*chemical structure*
Sulfuric acid	98	H_2SO_4
Nitric acid	65	HNO_3
Hydrogen peroxide	35	H_2O_2
Potassium permanganate	90	$KMnO_4$
Hydrochloric acid	38	HCl
Chloroform	30	$CHCl_3$

2.2. Preparation of Compound PLA/RGO

Synthesis of graphite oxide (GO) by Modified Hummers method.

In a beaker of 1 L, 2 g of natural graphite, 120 ml of H_2SO_4 (98%) and 80 of HNO_3 (65%) were added. The resulting slurry has a pale color black, maintaining stirring for 20 minutes in an ice bath. After this time 10 g of $KMnO_4$ (oxidant), are added in small portions. Once you finish adding, removed from the ice bath and for the next 2 hours it is kept under vigorous agitation[7].

The suspension acquires a dark brown color. 400 ml of distilled water are added, causing the temperature to rise quickly. Then, 8.6 ml of H_2O_2, which cause effervescence and the suspension cause is a change in color from dark brown to yellow mustard are added. Add H_2O_2, it aims to reduce the residual $KMnO_4$ and MnO_2, a soluble $MnSO_4$.

100 ml of aqueous solution (10%) of HCl to the mixture are added when the temperature has dropped, and the suspension allowed to stand for 12 hours.

Proceeds to make ultra-pure water washes, by continuous centrifugation cycles to remove dissolved species until the residue has a neutral pH. The material settling in the centrifuge tubes happens to have an acid pH and yellowish appearance, to be a shiny brown gel. Finally placed to dry the gel for 12 hours at 80 ° C in a vacuum oven, obtaining the graphitic oxide (about 3.5 g).

Graphitic oxide reduction

Graphitic oxide reduction was carried out via thermal exfoliation, in a tube furnace under N_2 flow. The heating rate from room temperature to peeling temperature (300 °C) was 5 °C/min, for 2 hours.

Composite Formation PLA/GRO

All samples of PLA and RGO compound was prepared by the method of intercalation solution. Synthesis processes for the preparation of the compounds have different contents of RGO (1-3 wt%).

In a beaker of 100 ml the appropriate amount of PLA is placed (*e.g.* 2g to prepare material 1% RGO) and 30 ml of chloroform; the mixture is stirred for 24 hours at room temperature. In another beaker, he sonicated appropriate amount of RGO (*e.g.* for material 0.02g 1% RGO) in chloroform for 8 hours; this mixture to the PLA solution was added with vigorous stirring for 12 hours at room temperature, to achieve a homogeneously dispersed system[8].

The resulting solution was cast on circular and rectangular plates glass at 45 °C for 24 hours, likewise, 13x75 mm cylinders were molded. The resulting, drying was completed at 80 °C for 24 hours in a drying oven.

2.3 Characterization

technique Infrared Spectroscopy Fourier transform (FTIR) was used to characterize the functional groups of the carbonaceous materials used (graphite, GO, RGO), for this a Cary 630 Agilent equipment was used, with ATR (attenuated total reflectance) between 4000 and 650 cm^{-1}, with eight scans per sample and absorbance mode.

Optical microscopy was performed to study both morphology RGO dispersion in the polymer, in an optical polarized light microscope at 10x BA210E Motic. The samples correspond to the thin cuts made by molding cores obtained above without additional treatments.

3. Results and Discussion

3.1. Infrared Spectroscopy Fourier transform (FT-IR)

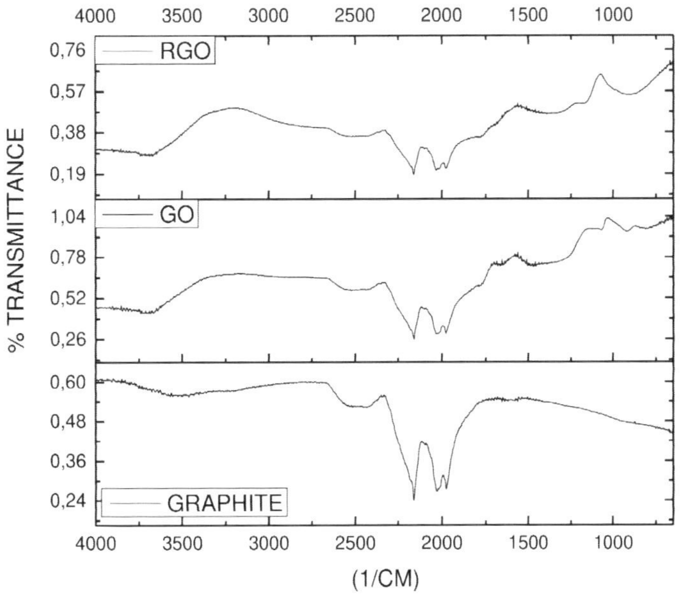

Figure 1. FTIR spectra of graphite, graphite oxide (GO) and reduced graphitic oxide (RGO).

Figure 1 shows the FTIR spectra of pure graphite, graphite oxide and the reduced graphitic oxide, where structural changes of the samples was observed. Pure graphite does not show many peaks characteristic of functional groups, only in the region between 2500-1800 cm^{-1} band according to the vibration of C=O bond, which likewise is present in samples of GO and RGO. The GO spectrum presents characteristic bands of its structure, a band around 1911 cm^{-1} which corresponds to the stretching vibration of the C=C vibration bands associated stretch link between 3300-2500 cm^{-1} O-H and C-O-H bending link in the plane to 1458 cm^{-1}, respectively; another band corresponding to the vibration characteristics of ether (1100 cm^{-1}) groups. The band is also observed at 1240 cm^{-1} attributed to stretching vibrations of the C-O bond, and the band between 1765-1540 cm^{-1} corresponding to the characteristic stretching of the carbonyl group [9].

In the spectrum of the reduced graphitic oxide (RGO), shows a small decrease in the intensity of the bands associated with oxygen containing functional groups such as O-H (3300-2500 cm^{-1}), the group of ethers (1100 cm^{-1}) and the link C-H (2900-2700 cm^{-1}). Moreover, the band of the epoxy groups (1250-800 cm^{-1}) also show a decrease in relation to the GO[9].

3.2. Graphene nanoparticle dispersion (RGO) in organic solvents

Figure 2. Photography of dispersed solutions: (a) RGO in Toluene and (b) RGO in chloroform.

When RGO dispersed in toluene and chloroform, respectively, it shows that the graphite is not dispersed homogeneously in each solvent.

Samulski et. the provided information on GO dispersion in water at pH ranging from 3-10 at concentrations of 2 mg/ml[10]. Moreover, Ruoff et. al, reported a homogeneous dispersion of an aqueous solution of GO could be obtained with concentrations up to 7 mg/ml[11].

Figure 2(a) shows the dispersed solution RGO with toluene, the synthesized RGO is a black powder, while the toluene solution appears to decrease the color as the amount of RGO is increased in the solution and causes increased sedimentation 10 mg/ml; above this content, gelation occurs. Conversely, the synthesized GRO can be easily dispersed in chloroform at concentrations ranging from 2 mg/ml to 10 mg/ml without sedimentation; as shown in Figure 2(b).

3.3. Optical microscopy of composite PLA/RGO

Light microscopy was performed at the different compounds PLA/RGO, which were prepared at different ratios of RGO making the dispersion of both materials in chloroform, in order to analyze the effect of homogeneity within the polymer matrix.

The compounds had no further treatment to the above, were used as obtained after molding and subsequent drying.

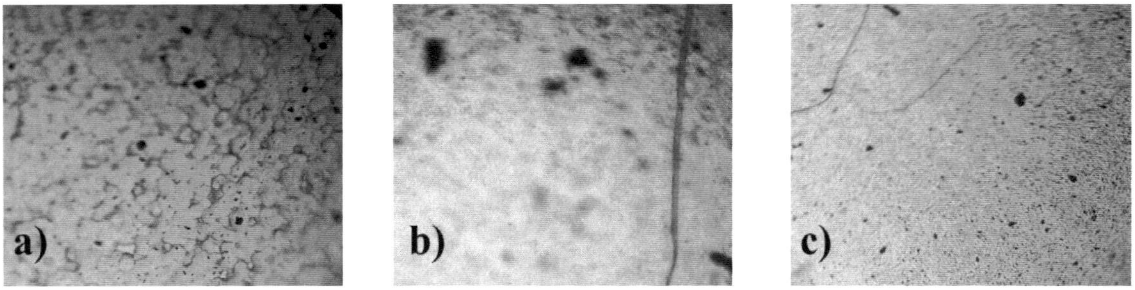

Figure 3. Optical microscopy. (A) PLA + GRO 1% by weight to 10x. (B) PLA + GRO 2% by weight to 10x. (C) PLA + GRO 3% by weight to 10x.

In Figure 3(a), the polymer compound is shown with 1% by weight of RGO, shows that no homogeneous dispersion, there is a diversity of particle sizes is RGO where it is seen that the polymer does not fit the composite matrix.

Figure 3(b) where there was an increase in the content of RGO, the slightly more uniform display is achieved, although a lack is also observed in the distribution uniform in particle size.

By contrast, although not a defined particle morphology and a homogeneous size thereof shown, Figure 3(c) shows better dispersion of RGO in PLA. The dispersion appears to be improved with increasing concentration of GRO in chloroform, which at first seems a contradiction; however, it is possible that as many RGO has caused greater interaction between the particles and the solvent, via oxygenated functional groups, and therefore greater stability of the suspension is generated; less concentrated suspensions, less stable in consequence, tend more easily to settle and form lumps. Such that when mixed with the PLA these lumps are interacting with the polymer, resulting in a worse dispersion. Obviously, the suspension of the RGO in chloroform will have a stability limit, a value of RGO above which is no longer stable and also forms lumps, but that study is beyond the scope of this investigation.

4. Conclusion

According to the results obtained by infrared spectroscopy (FT-IR), it is suggested that the GO prepared has abundant oxygen based functional groups, consistent with its high dispersibility in aqueous medium.

The degree of dispersion in organic solvents RGO is optimal at certain concentrations, demonstrating that chloroform is one of the best solvents to large amounts of solute without presenting gelation or sedimentation.
Optical microscopy it failed to identify a defined morphology of the particles within the polymer matrix, but the results suggest good dispersion RGO the PLA, by higher concentrations of RGO.

References

[1] W. Chavez Montes 2016 *Study of the thermomechanical properties of the nanocomposite PLA/C30B with different molecular weights after being subjected to accelerated weathering conditions* Chihuahua, Chih. p 26.
[2] J. Lunt 1998 Large-scale production, properties and applications of polylactic acid commercial polymers *Polym. Degrad. Stab.* **59** 145-152.
[3] Oksman K., M. Skrifvars, and JF Selin 2003 Natural fibers as reinforcement in polylactic acid (PLA) Composites *Compos. Sci. Technol.* **63** 1317-1324.
[4] Y. Fu, L. Liu, J. Zhang, and WC Hiscox 2014 Functionalized With graphenes novel polymer toughener as interface modifier for polylactic acid-tailored property / graphene nanocomposites *Polym.* **55** 6381-6389.
[5] MP Ho, KT Lau, H. Wang, and D. Hui 2015 Improvement on the polylactic acid properties of (PLA) using bamboo charcoal particles *Compos. Part B Eng.* **81** 14-25.
[6] H. Liu and J. Zhang 2011 Research progress in toughening modification of poly (lactic acid) *J. Polym. Sci. Part B Polym. Phys.* **49** 1051-1083.
[7] OA Vargas C., Á. Knight and J. Morales 2012 Can the performance of graphene nanosheets for lithium-ion storage batteries in Li be Predicted? *nanoscale* **4, 6** 2083-2092.
[8] J. Heo Cheol-HC, Hyun-Gon Moon, Choon Sup Yoon 2011 Nanocomposite Films ABS Based on Functionalized Graphene-Sheets ***6*** 8.
[9] F. Red 2000 Tables Infrared Spectroscopy *Chemical Fac. - Audiovisual Programs Dep.* 14.
[10] Yongchao If and Edward T. 2008 Graphene Synthesis of Water Soluble *Nano Lett.* **8** 1679-1682.
[11] S. Park *et al.* 2008 Aqueous suspension of chemically modified and characterization graphene sheets Chem. Mater. **20** 6592-6594.

Computer aided environmental analysis of a large-scale production of chitosan micro-beads modified with thiourea and magnetite

F Bertel[1], G Cogollo[1], S Meramo-Hurtado[1,2] and A González-Delgado[1]

[1]Nanomaterials and Computer Aided Process Engineering Research Group (NIPAC), Chemical Engineering Department, Faculty of Engineering, University of Cartagena, Av. Del Consulado Calle 30 No. 48-152, Cartagena, Colombia.
[2]Engineering PhD. Program, University of Cartagena, Av. Del Consulado Calle 30 No. 48-152, Cartagena, Colombia.

smeramoh@unicartagena.edu.co

Abstract. In this work, environmental assessment of a large-scale production of chitosan micro-beads modified with thiourea and magnetite was performed using Waste Reduction Algorithm (WAR). This methodology evaluates eight impact categories, four toxicological and four atmospherics. Initially, the simulation of production process of micro-beads was realized. Then, WAR GUI software, which is based on WAR algorithm, was employed to quantify potential environmental impacts (PEI) generation and output rate for this process under 4 different cases. The synthesis of micro-beads was analysed using computer-aided process engineering, this is related to the fact that there is no bibliography reported for the industrial scale-up, simulation and environmental evaluation of this type of process Results shows that photochemical oxidation potential is the evaluated category which emits the biggest output rate of PEI, contributing in all cases with more than 49% of the total rate. In general terms, the process has a negative generated PEI which means that the process has a convenient environmental performance. This leads to the conclusion that the process is environmentally beneficial because it consumes environmental impacts.

1. Introduction

In recent years, water pollution by heavy metals, dyes, solids and hydrocarbons represents a serious threat to the environment and humans [1]. As a solution to this problem, several technologies have been proposed, where adsorption process becomes in an important alternative because its practical, economic and environmentally friendly methods [2]. Recently, the use of biomass for the production of bioadsorbents has attracted a lot of attention due to its abundance, non-toxicity and low cost [3]. In this sense, chitosan-based adsorbents have shown a relatively high removal yield and kinetics for a lot of metals due to their high nitrogen content and porosity [4]. In addition, it has been found that a modification with thiourea and magnetite nanoparticles gives these bioadsorbents a greater adsorption capacity, selectivity towards several heavy metals and high efficiency in recovery processes, this at a laboratory scale [4, 5].

The environmental evaluation is an important tool in the design and analysis of processes because its development allows to determine the degree of pollution generated in a chemical process, since it

Content from this work may be used under the terms of the Creative Commons Attribution 3.0 licence. Any further distribution of this work must maintain attribution to the author(s) and the title of the work, journal citation and DOI.
Published under licence by IOP Publishing Ltd

identifies the potentials toxicological and atmospheric environmental impacts [6]. Waste Reduction Algorithm (WAR) is one of the most used tool for environmental assessment, which is useful because it allows quantifying the generation of potential environmental impacts based on 8 impact categories (Bicer et al. al., 2016). In this study, the simulation of a large-scale production of chitosan micro-beads modified with thiourea and magnetite was performed using computer aided process engineering. Then, from the extended mass and energy balance obtained from simulation, the environmental assessment of the process was performed using WAR GUI software which is based on the WAR algorithm [6]. First, total impacts were calculated for different cases; then, toxicological and atmospheric impacts and, finally, the effect of three energy sources (coal, oil and gas). All these in order to find the viability of the large-scale process regarding to environmental perspective.

2. Methods

2.1. Process description

As a first stage of the process, the synthesis of the magnetite nanoparticles is carried out, and in a second stage, the production of the microbeads is developed by the co-precipitation method under alkaline conditions. Figure 1 represents the production process of the modified microbeads.

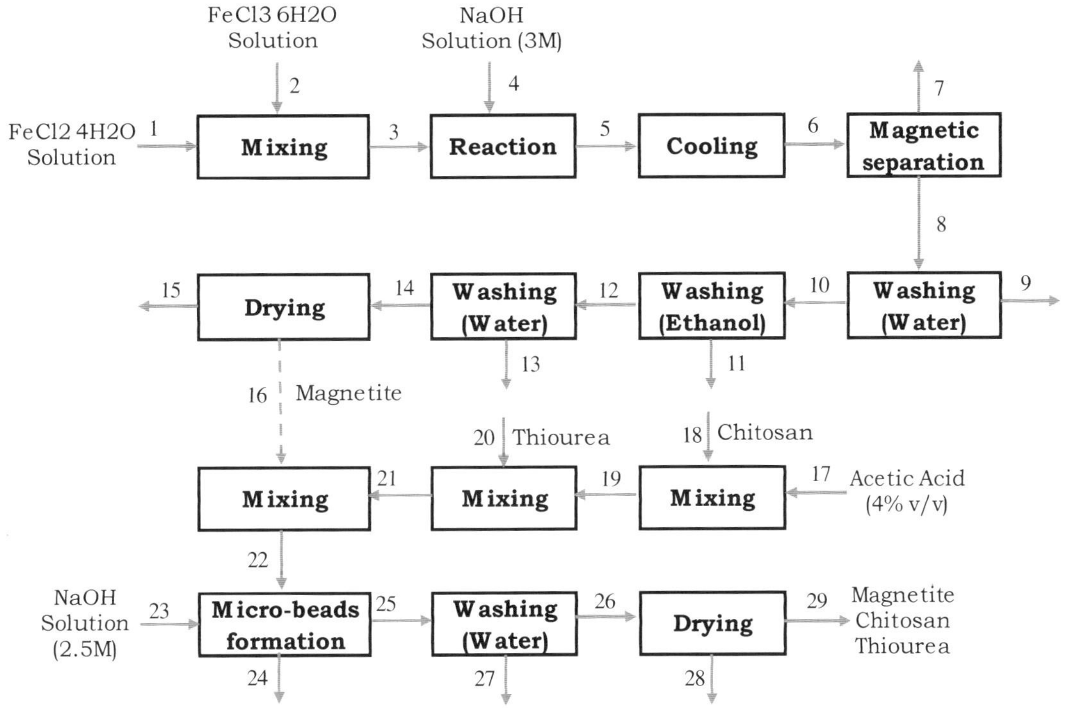

Figure 1. Block diagram of production of chitosan micro-beads modified with thiourea and magnetite.

2.1.1. Production of magnetite nanoparticles. Initially, the solutions of iron trichloride hexahydrate and iron dichloride tetrahydrate are mixed. The output stream (stream 3) is sent to the reaction stage where sodium hydroxide is added (stream4). The current leaving the reactor is cooled and then it is sent to a magnetic separator in order to remove impurities from the magnetite nanoparticles current. Finally, the nanoparticles (stream 8) are sent to a washing train with water, ethanol and water in order to stabilize the pH of the medium, and are sent to a dryer (stream 15).

IOP Conf. Series: Materials Science and Engineering **519** (2019) 012003 doi:10.1088/1757-899X/519/1/012003

2.1.2. Preparation and modification of chitosan microbeads with thiourea and magnetite nanoparticles. Initially, the chitosan (flow 18) is mixed with a solution of acetic acid (flow 17) and subsequently with thiourea (flow 20). In a next mixing step, the magnetic nanoparticles are added (stream 16) and the resulting stream (stream 21) is injected into a sodium hydroxide solution which is constantly stirred until the microbeads are formed. Finally, the residual sodium hydroxide is removed (stream 24) and the microbeads formed are washed with water and dried (stream 29).

2.2. Environmental assessment using WAR algorithm
For the environmental evaluation of the large-scale production of chitosan micro-beads modified with thiourea and magnetite, the waste algorithm reduction (WAR) methodology was selected. The WAR GUI Software was chosen due to the open availability of the software and its ability to quantify the rate of potential environmental impacts (PEI) generation (or consumption) [9], for processes considering the flow of an environmental impact across the boundaries of the system as a consequence of the mass and energy that crosses these limits [8]. This index is considered from two points of view: PEI output rate and PEI generated. The first measures the PEI impact emitted by the process to its surroundings and it is useful to solve questions about the external efficiency of the process; regarding to the second, it measures the generation of PEI within the limits of the process and it is related to the internal environmental efficiency of the process. Lower values for these indices implies that the process has a better environmental performance. The WAR algorithm considers eight impact categories, which are classified into two major groups: local toxicological impacts on humans (HTPI, HTPE) and ecological (ATP, TTP), and atmospheric which divides into global (GWP, ODP) and regional (AP, PCOP) impacts [10].

In this work, four different cases were studied: a base case without taking into account the contribution of energy resources and products flows (Case 1), and 3 cases where was considered the product stream (Case 2), the energy process (Case 3), and the amount of energy-product stream (Case 4). First, total impacts were calculated for the four cases; then, toxicological and atmospheric impacts and, finally, the effect of three energy sources (coal, oil and gas). All these in order to obtain an accurate diagnosis of the environmental viability of the large-scale process.

3. Results and Discussion

3.1. Total Potential Environmental Impacts of the process: Generated and output
From Figure 2, the total generation rate of PEI in four cases studied were negative ($-4.98x10^2$, $-1.44x10^2$, $-4,8x10^2$, $-1.26x10^2$ PEI/h for cases 1, 2, 3 and 4 respectively) which indicates that the process is environmentally friendly since the process itself does not generate environmental impacts. Regarding the total output rate of PEI, it is show that the base case and the case including energy are pretty similar (1.54x103 and 1.56x103 PEI/h) which suggest that the energy inclusion does not generate significant impacts. Otherwise, including products in the analysis entails an identical increase of impacts with respect to analysis without taking account products; this is, totals generation and output rate of PEI for cases 2 (including products) and 4 (including energy and products) are higher than cases 1 (base case) and 3 (including energy).

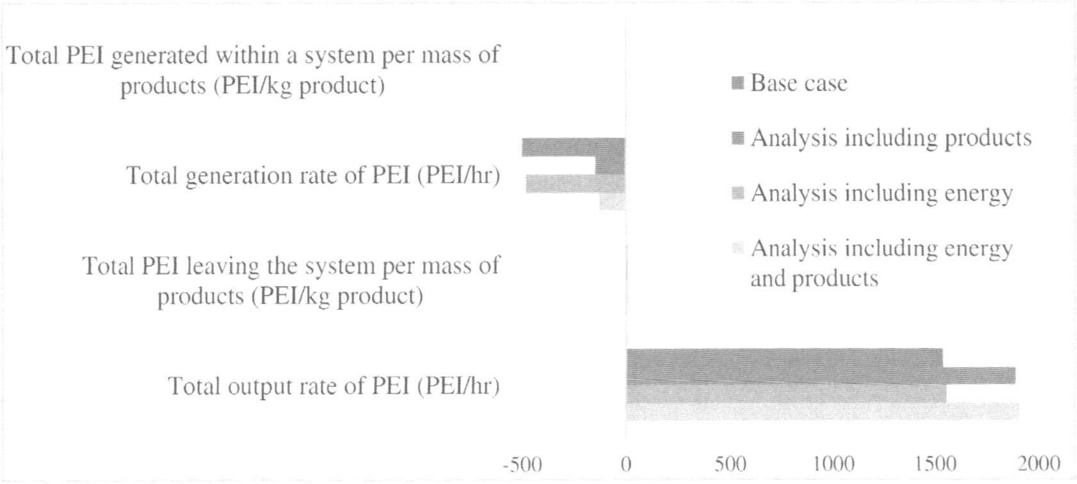

Figure 2. Total PEI generated and output of the large-scale production of chitosan micro-beads modified with thiourea and magnetite.

3.2. Local toxicological impacts of the process

Figure 3 shows the local output and generated toxicological impacts of the process, which include human toxicity potential by ingestion (HTPI), Human toxicity potential by inhalation or dermal exposure (HTPE), aquatic Toxicity Potential (ATP) and Terrestrial toxicity potential (TTP) impact categories. It can be evidenced that the ecological output impacts are more significant compared to the impacts directed on humans for all cases studied. Figure 3 shows that for ATP and HTPE (17.6 and 8.45 PEI/h for case 1) the total output rate of PEI were significantly lower than TTP and HTPI (in both cases equal to 285 PEI/h for case 1) categories. As regard the PEI generated, all values were negatives suggesting that the process have in the product streams, less toxic chemicals with tolerance values limits (TVL) lower than those fed to the system. Furthermore, it was observed that energy inclusion in environmental analysis had little or no effect on results obtained, having identical or very similar values for all toxicological impact categories for cases 1 and 3, and for the cases 2 and 4. It was also found that products inclusion significantly increased generated and output PEI for the process, except for HTPE which has an almost negligible increase.

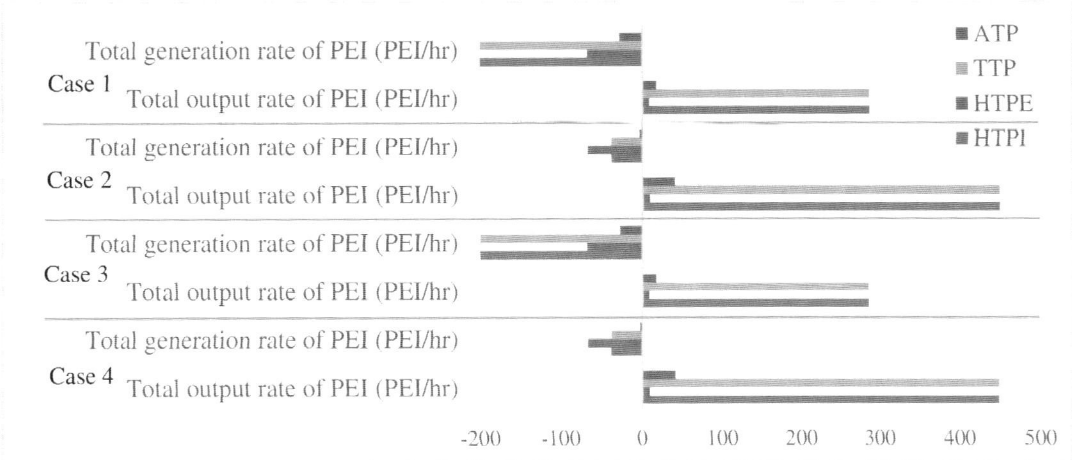

Figure 3. Local output and generated toxicological impacts of the large-scale production of chitosan micro-beads modified with thiourea and magnetite.

3.3. Atmospheric impacts of the process

Figure 4 shows output and generated atmospheric impacts of the process, which include Global Warming Potential (GWP), Ozone depletion potential (ODP), Photochemical Oxidation Potential (PCOP) and Acidification potential (AP) impact categories. For this particular process, it is observed that all the values for AP, ODP and GWP, for cases 1 and 2 are equal to zero and for cases 3 and 4 are considerably low; which leads to the conclusion that this process is environmentally neutral under these categories, so the greatest contribution to PEI out for atmospheric categories comes from the use of fuels in the process as energy sources, as in cases 3 and 4.

PEI output for GWP (1.55 PEI / h) and AP (16.1 PEI / h) impact categories, in cases 3 and 4, indicates that this process emits chemical substances that persist longer in the environment due to their low oxidation and also can contribute to the generation of acid rain.

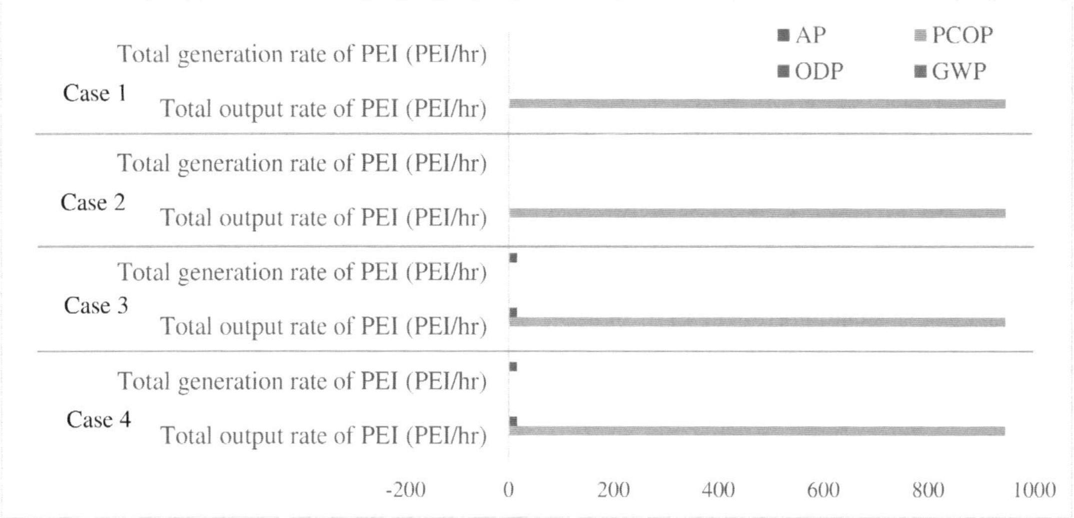

Figure 4. Output and generated atmospheric impacts of the large-scale production of chitosan micro-beads modified with thiourea and magnetite.

3.4. Effect of energy source

Figure 5 shows the effect of energy source on the eight impact categories total output rate of PEI, for three types of fuel (gas, oil and coal), including energy and excluding product stream. It was found that most of the impact categories show very little or no variation due to the fuel used, except for AP impacts where coal usage increases the impact (96.9 PEI/h). Compared to the global output rate of PEI for the process, the PEI from energy source do not represent many concerns because its contribution is lower considering other effects previously described, so this process has a good energetic performance from an environmental point of view. It can be observed that gas had a better performance compared to the others energy sources because it has less or equal impact on all categories taken into account.

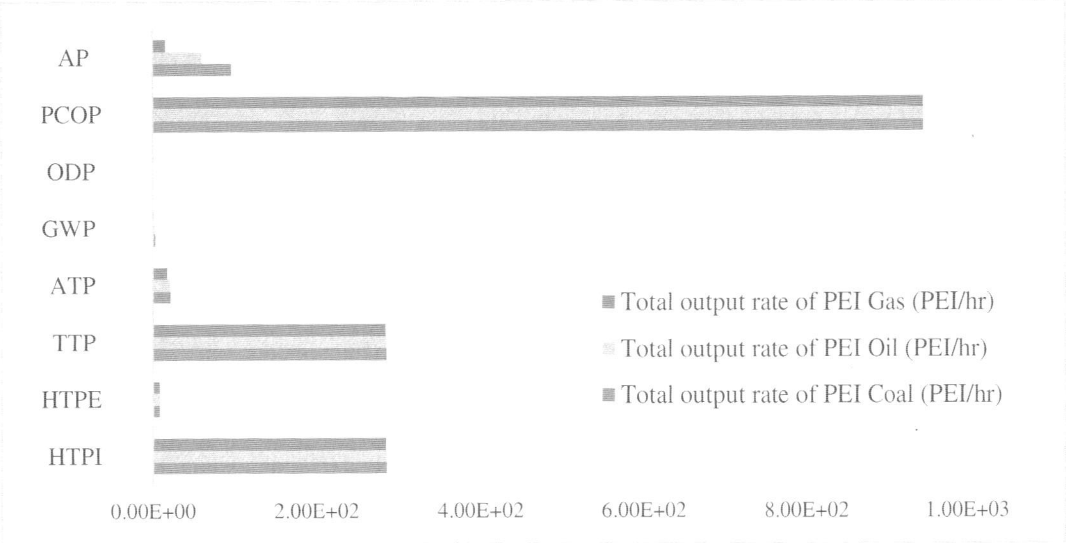

Figure 5. Effect of energy source on the output rate from energy usage for the large-scale production of chitosan micro-beads modified with thiourea and magnetite.

4. Conclusions

Waste Reduction Algorithm (WAR) was implemented for environmental assessment of a large-scale production of chitosan micro-beads modified with thiourea and magnetite. From results obtained, it can be said that the process is beneficial in environmental terms because it transforms the feed streams of high PEI into final products of lower PEI, which is manifested by the negative values of total generation rate of PEI in all the cases. It was also found that energy does not generate significant impacts on total PEI.

Regarding to toxicological impacts, it can be evidenced that the ecological output impacts are more significant compared to the impacts directed on humans for the four cases studied. For these impact categories, all PEI generated values were negatives implying that product streams are less toxic chemicals than those fed to the process. With respect to atmospheric impacts it can be affirmed that the process is environmentally neutral under ODP and GWP categories, so the highest contribution to PEI output for atmospheric categories occurs by the usage of fuels in the process as energy sources. Results shows that PCOP is the evaluated category which emits the most output PEI, contributing in all cases between 49.4 and 61.43% of the total output PEI, which is attributed to the presence of ethanol in the washing stages. Therefore, a way to reduce these environmental impacts involves replacing the ethanol by another solvent which has less environmental impact. Furthermore, this process emits chemical that persist longer in the environment due to their low oxidation and can also contribute to the generation of acid rain.

Acknowledgments

The authors thank to the Colombian Administrative Department of Science, Technology and Innovation COLCIENCIAS for the financial support of the project "Removal of polycyclic aromatic hydrocarbons (PAHs), present in coastal waters Cartagena bay by using shrimp exoskeleton as a source of nanoparticle-modified bioadsorbents", code 1107748593351 CT069/17. Thanks to the University of Cartagena for the knowledge and the support provided.

References

[1] Burgos-Núñez S, Navarro-Frómeta A, Marrugo-Negrete J, Enamorado-Montes G, Urango-Cárdenas I 2017 Polycyclic aromatic hydrocarbons and heavy metals in the Cispata Bay, Colombia: A marine tropical ecosystem *Marine Pollution Bulletin* **120** 379–386

[2] He J, Li Y, Wang C, Zhang K, Lin d, Kong L, Liu J 2017. Rapid adsorption of Pb, Cu and Cd from aqueous solutions by β-cyclodextrin polymers *Applied Surface Science* **426** 29–39

[3] Sun X, Li Q, Yang L, Liu H 2016 Chemically modified magnetic chitosan microspheres for Cr(VI) removal from acidic aqueous solution *Particuology* **26** 79–86

[4] Zhou L, Liu J, Liu Z 2009 Adsorption of platinum (IV) and palladium (II) from aqueous solution by thiourea-modified chitosan microspheres *Journal of Hazardous Materials* **172** 439–446

[5] Karimi MH, Mahdavinia GR, Massoumi B, Baghban A, Saraei M 2018 Ionically crosslinked magnetic chitosan/κ-carrageenan bioadsorbents for removal of anionic eriochrome black-T *International Journal of Biological Macromolecules* **113** 361–375

[6] Herrera-Aristizábal R, Salgado-Dueñas J, Peralta-Ruiz Y, González-Delgado A 2017 Environmental Evaluation of a Palm-based biorefinery under North-Colombian Conditions *Chemical Engineering Transactions* **57**

[7] Bicer Y, Dincer I, Zamfirescu C, Vezina G, Raso F 2016 Comparative life cycle assessment of various ammonia production methods *Journal of Cleaner Production* **13** 1379-1395.

[8] Arteaga S, Sanjuan M and Gozález-Delgado A 2018 Computer-Aided Environmental Evaluation of Bioethanol Production from Empty Palm Fruit Bunches using Oxalic Acid Pretreatment and Molecular Sieves *Chemical Engineering Transactions* **70**

[9] Meramo, S., Bonfante, H., De Avila-Montiel, G., Herrera-Barros, A., Gonzalez-Delgado, A 2018 Environmental Assessment of a Large-Scale Production of TiO2 Nanoparticles via Green Chemistry, Chemical Engineering Transactions **70** 1063-1068

[10] González-Delgado A and Peralta-Ruiz Y 2016 Environmental Assessment of a Crude Palm Oil Production Process under North-Colombian conditions Using WAR Algorithm *International Journal of ChemTech Research* **9** 833-843

IOP Conf. Series: Materials Science and Engineering **519** (2019) 012004 doi:10.1088/1757-899X/519/1/012004

Element Free Galerkin (EFG) sensitivity study in structural analysis

J A Martinez T[1,2] and E W Arrieta O[2]

[1]Faculty of design and engineering, Fundación Tecnológica Antonio Arévalo, Col.
[2]Department of Mechanical and Mechatronic Engineering,Universidad Tecnológica de Bolívar, Col.

E-mail: jmartinezt@tecnar.edu.co

Abstract. The present study shows a parametric analysis of the meshfree method, Element Free Galerkin (EFG), on the elastic analysis of a cantilever beam. This study allows us to determine the best convergence conditions of the solutions varying characteristic. EFG is based on the construction of Moving Least Squares (MLS) approximations using the weighted residual method on the weak formulation, with MLS form functions as the same weighting functions. We consider the parameters of the method such as the order of the basic functions of MLS functions, the size of the support domain of the local MLS functions and the density of Gauss points against errors calculated according to the L_2 norm and processing time. It is shown that by increasing the order of basic functions it is possible to obtain more precise results, however, a larger support diameter and Gauss points higher density are required in order to stabilize the solution, considerably increasing processing times. Therefore, it is only advisable to use high-order base functions when the precision in the results is the priority and a high computational resource is available.

1. Introduction

As FEM (The Finite Element Method) has proven its efficiency and capacity in all areas of engineering, from a practical point of view, its use on real models is always associated with numerous man-hours dedicated to the generation of meshes. Geometric irregularities in mechanical engineering parts usually represent great difficulties to obtain proper geometric discretizations as required by FEM. Therefore, meshfree methods have arisen, which try not to base the approximation on meshes, but on independent points that do not require a pre-defined interconnection [1].

Meshfree methods would allow us to solve problems with large deformations, and complex geometries, of topological optimization without the great difficulties that afflict finite element method. All this, thanks to the fact that the approach is based on nodes that have associated support domains, see Fig. 1. These nodes do not require a pre-defined connectivity within of the general domain [2], which implies that the approach functions do not require a mesh.

On the other hand, not requiring a mesh to achieve a good approximation results, in a greater difficulty when imposing border conditions [3]. This aspect has been overcome by applying more elaborate methods to guarantee the imposition of values at the borders. Methods such as penalization and lagrange multipliers are widely used with meshfree methods achieving good results [4][5][6].

Content from this work may be used under the terms of the Creative Commons Attribution 3.0 licence. Any further distribution of this work must maintain attribution to the author(s) and the title of the work, journal citation and DOI.
Published under licence by IOP Publishing Ltd

The approximation functions by Moving Least Squares (MLS) to allow local discretization, with arbitrarily located points, has become one of the most outstanding options for use in the meshfree methods. For that reason there are currently a good number of researchers developing and demonstrating the advantages of it's use. Examples of this can be observed in the works developed by E. Shivanian [7],where you test a methodology based on local approaches with MLS functions (*"Meshless Local Petrov-Galerkin"*, MLPG) to solve wave problems in three dimensions, by Hongping R. [8] and F.X. Sun [9] demonstrating the convergence of MLS functions in the study of problems in two and n dimensions, respectively, by Mehdi D. and Vahind M. [10] that use the generalized formulation of MLS (GMLS) to simulate problems of scalar fields in the theory of quantum physics and by E. Dabboura [11] who demonstrates the versatility of MLS functions by solving the nonlinear equation of *"Kuramoto-Sivashinsky"*, among other works that are found in the literature ([12], [13], [14], [15], [16]).

One of the most used and robust mesh free methods is the Element Free Galerking (EFG) [17]. Which seeks to solve the weak formulation of a partial differential equation, using the same MLS functions as approximation functions and as weighting functions. EFG, although is one of the most used methods, uses of several parameters that must be adjusted to ensure good convergence. The present work contains a sensitivity study of these parameters of the method.

2. Methodology

2.1. Support domain

The general domain will be discretized using randomly located points, which will be associated with a support domain with the surrounding nodes. These domains will be constructed with simple geometric forms (Circles or rectangles, see Fig. 1) and from their contributions it will be possible to construct the approximation on the general domain.

At the moment of generating the support domains, we have to think how big it should be. If we can locate as many nodes as terms are in the base function for the approximation ($p(x) = [1 \;\; x \;\; y]$), we will achieve a better approximation that satisfies the $Kronecker$ delta condition. However, the idea of MLS approach is to use randomly distributed nodes so that it can not always be predictes how many nodes will there be in support domains. For this reason the $Kronecker$ Delta property is not always met. It is necessary to emphasize that very small support domains can cause the momentum matrix $M(x)$ (See Eq. 5) to be singular; and very large domains require a great computational resources.

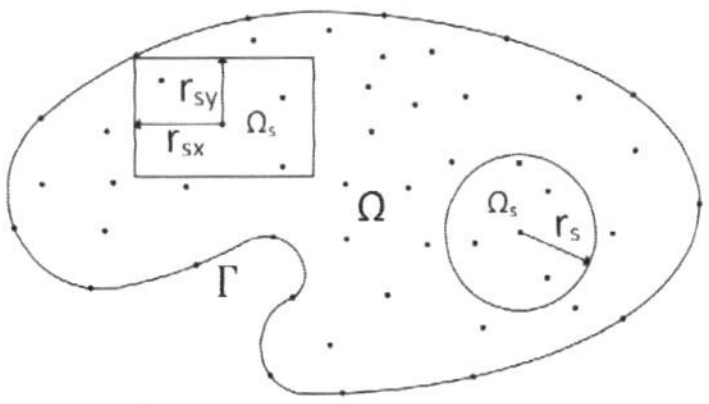

Figure 1. Support domain circular and rectangular mean in a general domain limited by boundary Γ.

2.2. MLS approximation

We want to approximate a function $u(x)$ in a domain Ω to use MLS approximation functions, as defined in Eq. (1).

$$u(x) = p^t(x)a(x) \tag{1}$$

where,

- $u(x)$ is the function to approximate.
- $p^t(x) = [p_1(x) \quad p_2(x) \quad p_3(x) \ldots p_m x)]$: are functions created from polynomial bases, defined for a point $x^t = [x, y]$, with m equal to the number of terms that will depend on the degree of the polynomial constructed with the pascal triangle.
- $a(x)$ is the vector of constant coefficients of the base functions.

2.3. ϕ function.

To obtain the value of the coefficients $a(x)$ it is necessary to define the functional in Eq. (2) [18].

$$J = W(x - x_i)[P^T(x)a(x) - u^*]^2 \tag{2}$$

The function $W(x - x_i)$ is a diagonal matrix of weight functions(See Eq. (3)).

$$W_{kj}(x - x_i) = \delta_{kj} W_k(x - x_i) \tag{3}$$

By minimizing the functional $J(x)$, we get:

$$P^T(x)W(x - x_i)P(x)a(x) = W(x - x_i)P(x)u^* \tag{4}$$

Which can be represented as the linear system in Eq. (5):

$$M(x)a(x) = B(x)u^* \tag{5}$$

Where,

- $M(x) = P^T(x)W(x - x_i)P(x)$: is defined as the moment matrix and has dimensions [m x m].
- $B(x) = P^T(x)W(x - x_i)$: is a rectangular matrix.

Finally it is possible to rewrite Eq. (1) as shown in Eq. (6).

$$u(x) = \Phi(x)^T u^* \tag{6}$$

2.4. EFG formulation

The equation in partial derivatives describing the elastic behavior of a 2D body and its boundary conditions are posed as follows [19]:

$$L^T \sigma + b = 0 \tag{7}$$

$$u_i = \bar{u_i} \qquad en \quad \Gamma_u \tag{8}$$

$$\sigma_{ij} n_j = \bar{t_i} \qquad en \quad \Gamma_t \tag{9}$$

Where, L is a differential matrix operator [18] like the form in Eq.(10).

$$\mathbf{L} = \begin{bmatrix} \frac{\partial}{\partial x} & 0 \\ 0 & \frac{\partial}{\partial y} \\ \frac{\partial}{\partial y} & \frac{\partial}{\partial x} \end{bmatrix} \tag{10}$$

Using Galerkin's method with weighted residues [20] on Eq. (7) and imposing boundary conditions through the penalty method [21], we obtain :

Table 1. Properties of the study case.

Propiedad	Valor
Young's modulus (E)	$30e6$ Pa
Poisson's ratio (υ)	0.3
Load (P)	-1000 N
Length (L)	48 m
Heigh (D)	12 m
Thickness (t)	12 m

$$\int_{\Phi} (L\Phi)^t D(L\Phi)u \cdot d\Omega + \int_{\Gamma_u} (\Phi^t \alpha \Phi)u \cdot d\Gamma = \int_{\Omega} \Phi b \cdot d\Omega - \int_{\Gamma_t} \Phi T \cdot d\Gamma + \int_{\Gamma_u} \Phi \alpha \bar{u} \cdot d\Gamma \tag{11}$$

Solving the integrals of the Eq. (11) we obtain a system of linear equations:

$$(K + K^{\alpha})u = F + F^{\alpha} \tag{12}$$

Where,

$$K = \int_{\Phi} (L\Phi)^t D(L\Phi)u \cdot d\Omega \tag{13}$$

$$K^{\alpha} = \int_{\Gamma_u} (\Phi^t \alpha \Phi)u \cdot d\Gamma \tag{14}$$

$$F = \int_{\Omega} \Phi b \cdot d\Omega - \int_{\Gamma_t} \Phi T \cdot d\Gamma \tag{15}$$

$$F^{\alpha} = \int_{\Gamma_u} \Phi \alpha \bar{u} \cdot d\Gamma \tag{16}$$

3. Numerical results and discussions

The case of a "Cantilever beam" [20, 22, 23], which refers to a structural element embedded in one of its ends with a load perpendicular to the axis of the element (See Fig. 2), will be studied. The mechanical properties of the model and the load conditions are defined in Table 1.

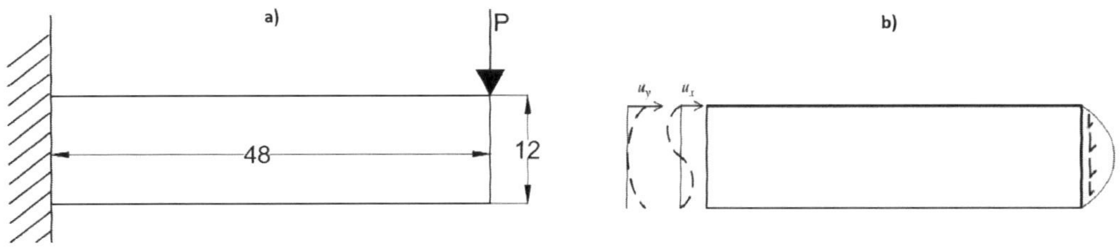

Figure 2. a)Cantilever beam model. b) Scheme of natural and essential boundary conditions.

3.1. Boundary Condition.
For Cantilever beams the exact solution is the Timoshenko and Goodier beam, as found in the literature [22], we will take specified displacements at x = 0, which can be written as Eqs. (17)-(18).

$$u(x,y) = -\frac{Py}{6EI}\left[(6L-3x)x + (2+v)\left(y^2 - \frac{D^2}{4}\right)\right] \tag{17}$$

$$v(x,y) = -\frac{P}{6EI}\left[3vy^2(L-x) + (4+5v)\frac{D^2x}{4} + (3L-x)x^2\right] \tag{18}$$

For the natural boundary condition the load P located in the free ending (x=48) will generate a shear stress distributed in a parabolic form along the beam, according to the Eq. (19).

$$\tau_{xy} = -\frac{P}{2I}\left(\frac{D^2}{4} - y^2\right) \tag{19}$$

Finally, both boundary conditions are shown in the Fig. 2.

3.2. Analytical solution
The solution proposed by Timoshenko and Goodier [23] is given by Eqs. (17)-(21) for displacements, shear stress and normal stress.

$$\sigma_x = -\frac{P(L-x)y}{I} \tag{20}$$

$$\sigma_y = 0 \tag{21}$$

The analytical results given by Timoshenko and Goodier can be plotted as in Fig. 3.

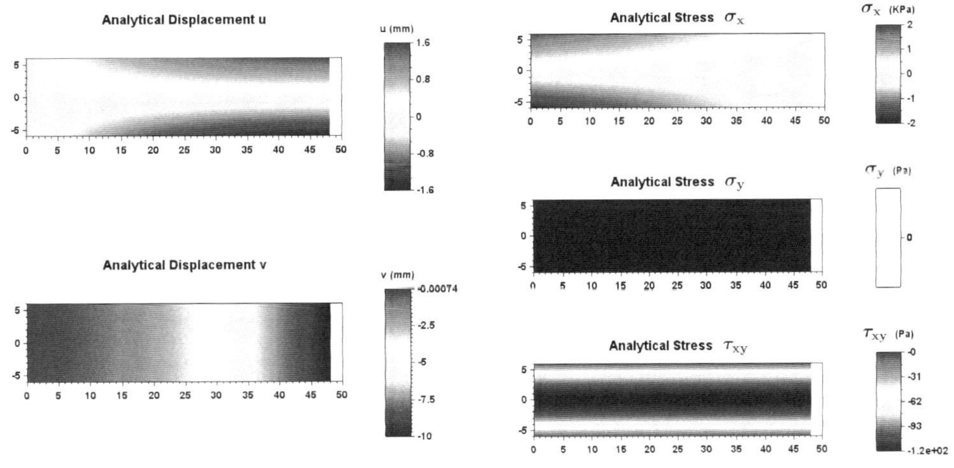

Figure 3. The Cantilever beam analytical solution.

IOP Conf. Series: Materials Science and Engineering **519** (2019) 012004 doi:10.1088/1757-899X/519/1/012004

3.3. Convergence analysis.
Having defined load conditions for the model and an given an analytical solution, we procceed to perform a sensitivity analysis on the parameters of EFG method, to define configurations that will give good solutions without requiring excesive computational resources (measured as processing time). In this sensitivity analysis the parameters to be studied are:

- **Order of base functions** $p(x)$**:** Increasing the order of polynomial in the base functions of EFG method should increases the speed of convergence and the accuracy of EFG [24], which is why it is proposed to study their effect on the quality of the approximation.
- **Size of Support domains**: Rectangular regions whose size is measured according to the percentage of nodes that are covered when taken as support domains. For example, in a domain with 200 nodes, a support domain of 10% means that on average the support domains cover 20 nodes.
- **Number of Gauss points:** The number of Gauss points will be an important factor in solving the integrals of the EFG formulation. It will be shown how many Gauss points are required to achieve a good convergence and how to adjust it depending on the support domain size and the order of polynomial base functions. Experiments were performed using up to 64 Gauss points within in each cell, and varying this density.
- **Processing time and error:** Processing time and the relative and absolute error will be taken as control parameters to test solutions in order to find solutions not so expensive in processor time but also ones that generate acceptable errors. Error will be measured in comparison with the analytical response of the model, and calculated according to the energy standard, as being one of the most used norms in elasticity [25], see equation 22.

$$\epsilon_{\%} = \frac{\left(\int (\sigma_{exac} - \sigma_{efg}) D^{-1} (\sigma_{exac} - \sigma_{efg}) dA\right)^{1/2}}{\int \sigma_{exac} dA} \tag{22}$$

The graphs 4, 5, 6 and 7 are the results obtained by varying each of the mentioned parameters in a sample space of 128 iterations. The vertical axis of the left graphs details the error percentage, the vertical axis of the right graphs sets the processing time, the horizontal axis alludes to the number of Gauss points, the colors of the lines represent the size average percentage of media domains and each graph shows the results for different order polynomials base function. The studies were performed with uniform's distributed nodes using a 10x10 rectangular grid through the beam.

It is possible to appreciate how, as the polynomial basis functions order increases, the response improves, however, a good density of Gaussian points and large support domains are required to stabilize it, on the other hand, this causes a considerable increase in the processing time. Given this, it is only recommended to use higher order functions when higher precision is required and we have great computational resources.

From the graphs, it is possible to recover the support domains and the number of Gauss points that produced the least processing times with admissible errors, these best configurations are found in the table 2.

Table 2. Best results for the second order polynomial base funciont

Order	S. Domain	# Gauss Point	Time (s)	Error
1	54%	2304	22,94	0.485%
2	36%	2304	18,86	0.392%
3	54%	2304	32,58	0.215%
4	107%	3106	107,79	0.099%

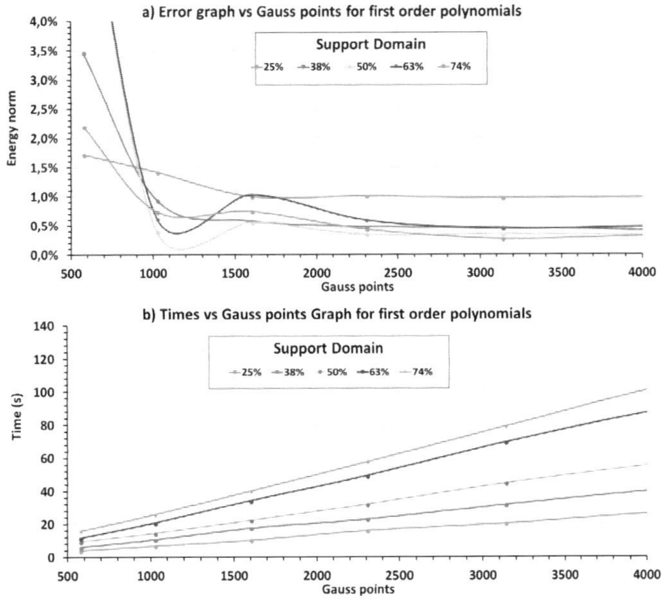

Figure 4. Results of elastic solution to the Cantilever beam problem using EFG with first order polynomials basis functions.

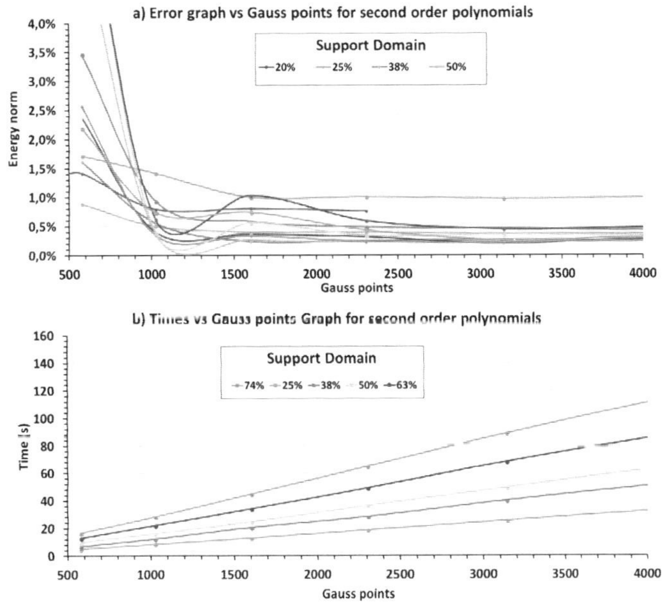

Figure 5. Results of elastic solution to the Cantilever beam problem using EFG with second order polynomials basis functions.

3.4. EFG Solution

From the convergence analysis of the previous section it can be seen that the best configuration, in table 2, was for second order polynomials basis function with rectangular support domains. The results of this configuration can be seen in Fig. 8. It is worthwhile to clarify that the σ_y

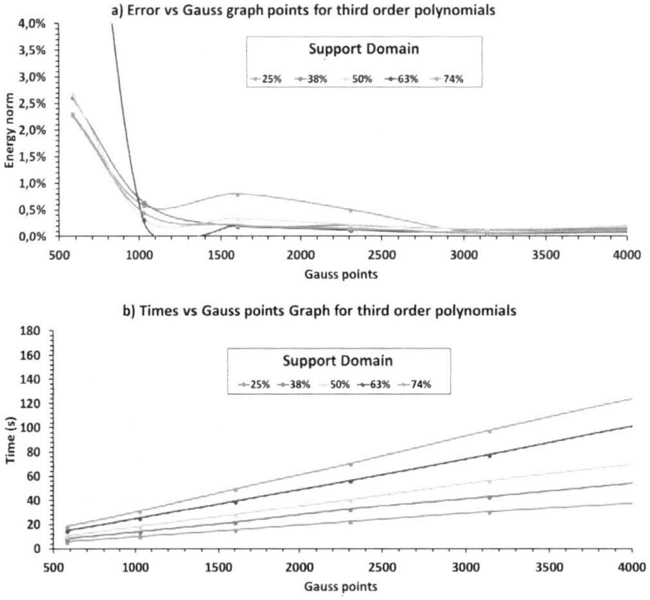

Figure 6. Results of elastic solution to the Cantilever beam problem using EFG with third order polynomials basis functions.

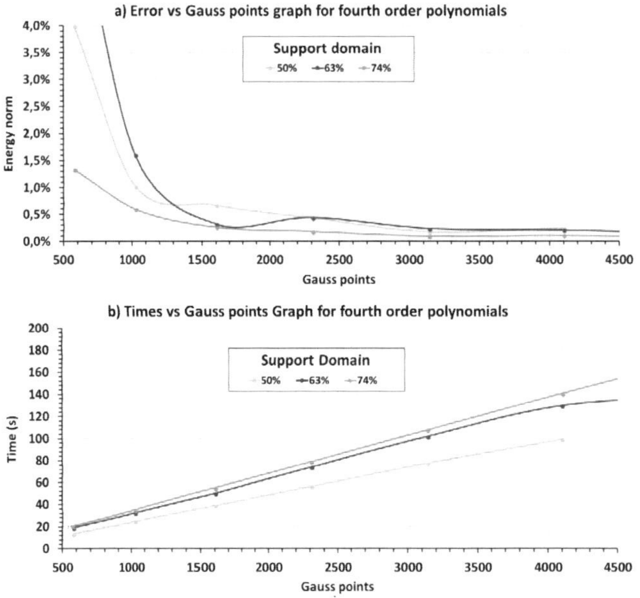

Figure 7. Results of elastic solution to the Cantilever beam problem using EFG with fourth order polynomials basis functions.

absolute error obtained is below 0.0005 when put in the scale of loads applied.

As a last test it is possible to compare the results obtained by EFG in Fig. 3 with the results of the analytical solution of the Fig. 8 and we can appreciate the great similarity between them, as well as the continuity in the solutions. So finally EFG achieved a valid solution to the

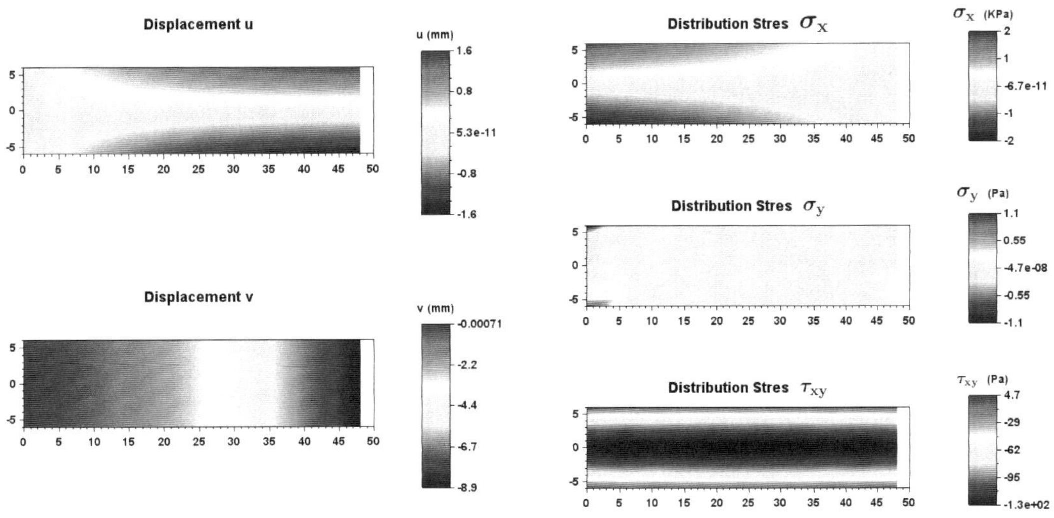

Figure 8. Elastic solution for Cantilever beam using EFG.

Cantilever beam.

4. Conclusions

The EFG method can improve its convergence by varying parameters such as the Gauss points density, the size of support domains, and the polynomial basis functions order $(P(x))$ of the functions Φ. The higher the order of polynomial basis functions, the more Gaussian points and larger support domains will be required to stabilize the method, resulting in longer processing time.

It is advisable to work with initial support diameters of at least 38% of the total nodes on average and with first order polynomials basis functions, in order to achieve a rapid convergence of the problem. For example if we use a 1000 node domain we could to stard, in average, with support domain of 380 nodes.

For the number of Gauss points, it is recommended to use at least 8 times the number of variables (In mechanical structural cases u and v) to solve when using first and second order polynomials basis functions and for third and fourth order polynomials basis function at least 15 times the number of variables.

5. Acknowledgement

Author is grateful to the Sergio Rivera and Edgardo Arrieta engineer for depositing their trust and unconditional support.

References

[1] Metsis P, Lantzounis N and Papadrakakis M 2015 *Computer Methods in Applied Mechanics and Engineering* **283** 782–805

[2] Mhamed M and Idir B 2016 *World Academy of Science, Engineering and Technology, International Journal of Chemical, Molecular, Nuclear, Materials and Metallurgical Engineering* **10** 646–654

[3] Belytschko T, Lu Y Y and Gu L 1994 *International journal for numerical methods in engineering* **37** 229–256

[4] Hosseini S, MALEKAN M, PITANGUEIRA R L D S and SILVA R P 2017 *Latin American Journal of Solids and Structures* **14** 1017–1039

[5] Ramirez S J F, Mesa M E, Branch B J W and Boulanger P 2011 *Revista Avances en Sistemas e Informática* **8**
[6] Wu C K C and Plesha M E 2002 *International Journal for Numerical methods in engineering* **53** 499–514
[7] Shivanian E 2015 *Engineering Analysis with Boundary Elements* **50** 249–257
[8] Ren H, Pei K and Wang L 2014 *Applied Mathematics and Computation* **238** 527–546
[9] Sun F, Wang J, Cheng Y and Huang A 2015 *Applied Numerical Mathematics* **98** 79–105
[10] Dehghan M and Mohammadi V 2016 *Computers & Mathematics with Applications* **71** 892–921
[11] Dabboura E, Sadat H and Prax C 2016 *Alexandria Engineering Journal* **55** 2783–2787
[12] Cheng Y, Bai F, Liu C and Peng M 2016 *International Journal of Computational Materials Science and Engineering* **5** 1650023
[13] Chowdhury H A, Wittek A, Miller K and Joldes G R 2016 *Journal of Scientific Computing* 1–15
[14] Arzani H, Kaveh A and Taromsari M T 2017 *Scientia Iranica. Transaction A, Civil Engineering* **24** 143
[15] Ghasemi M and Amrollahi R 2016 *Engineering Analysis with Boundary Elements* **67** 126–137
[16] Trask N, Maxey M and Hu X 2016 *arXiv preprint arXiv:1611.03911*
[17] Li H and Mulay S S 2013 *Meshless methods and their numerical properties* (CRC press)
[18] Liu G R and Gu Y T 2005 *An introduction to meshfree methods and their programming* (Springer Science & Business Media)
[19] Mase G 1970 *Schaum's outline of continuum mechanics* (McGraw Hill Professional)
[20] Dolbow J and Belytschko T 1998 *Archives of computational methods in engineering* **5** 207–241
[21] Zhu T and Atluri S 1998 *Computational Mechanics* **21** 211–222
[22] Augarde C E and Deeks A J 2008 *Finite elements in analysis and design* **44** 595–601
[23] Timoshenko S and Goodier J 1970 *Theory of elasticity* (McGraw-Hill New York)
[24] Yao Y Bhat R B P M 2005 *NaCoMM* 249–254
[25] Luis J and Ochoa R 1998 *Revista Universidad EAFIT* **34** 107–118

IOP Conf. Series: Materials Science and Engineering **519** (2019) 012005 doi:10.1088/1757-899X/519/1/012005

BPM analysis and security case study: dining rooms Cartagena military, Colombia.

E Polo-Carrillo[1], Y Cabarcas-Ariza[1], C A Marrugo-Arnedo[2,3]
[1]Fundación Tecnológica Antonio de Arévalo.
[2]Grupo de Investigación Fundovida. Fundovida IPS.
[3]Grupo de Investigación en Economía de la Salud (GIES) – Universidad de Cartagena.

emilia.polo@tecnar.edu.co

Abstract. Foodborne diseases constitute an important public health and productivity problem for companies, since it increases the level of employee absenteeism and decreases productivity. Objective, analyse the conditions of good practices of manufacturing and security in the dining rooms of a military complex located in the City of Cartagena, Colombia. Descriptive study based on the application of a standardized instrument in six dining rooms of a military complex located in Cartagena. Analysing 164 aspects contained in Resolution 2674 of 2013, it was possible to observe total compliance in 58 aspects, for its part, 36 partially comply and finally, 70 aspects do not comply with the provisions of the legislation. It is necessary to consider that the implementation of good manufacturing practices is constituted as the primary instrument for the adequate handling of food. The study is the basis for the implementation of internal policies aimed at improving good practices in manufacturing and security in the dining halls of the military complex.

1. Introduction

Foodborne diseases (ETA) are an important public health problem due to the increase in their occurrence, the emergence of new forms of transmission, the emergence of vulnerable population groups, the increase in the resistance of pathogens to antimicrobial compounds, Inter alia. The incidence of these diseases is a direct indicator of the hygienic-sanitary quality of food, and it has been shown that contamination of these can occur during processing or by the use of contaminated raw materials, since some bacteria pathogenic for man form part of the normal flora of birds, pigs and cattle [1-5].

According to Yunesky (2012) [2], the safety is one of the elements that together with the nutritional, organoleptic and commercial characteristics make up the quality of the food; this has been defined by the Codex Alimentarius as the guarantee that the food will not cause harm to the consumer when it is prepared and / or when consumed, according to the use to which it is destined. In that sense, the health and life of people depend in large part on the nutritional quality of the food they consume daily, which in turn depends on the hygienic and sanitary quality to which they are subjected throughout the production chain, from the field to the consumer's table [6-9].

Although the lack of hygiene and sanitation in the processing and preparation of food is a problem that can occur anywhere in the world [10-12], the incidence of diseases caused by poorly processed or poorly prepared foods is a critical, severe problem that is found more frequently in developing countries, according to global disease burden statistics, in In 2015, one in ten people get sick every year when they

Content from this work may be used under the terms of the Creative Commons Attribution 3.0 licence. Any further distribution of this work must maintain attribution to the author(s) and the title of the work, journal citation and DOI.
Published under licence by IOP Publishing Ltd

eat contaminated food and 420,000 die as a result of these diseases. Children under 5 are particularly at risk and 125,000 children die every year from foodborne diseases [13].

The regions of Africa and South-East Asia of WHO have the highest burden of foodborne diseases Almost one third (30%) of all deaths from foodborne diseases occur in children under 5 years old, despite the fact that children of that age represent only 9% of the world population [13].

In Colombia, for the year 2010 there were approximately 11,589 cases of foodborne diseases, in the following years the behavior in the notification was the increase, 2011 (13,961 cases), 2012 (11,836 cases), 2013 (9,326 cases) and in in 2014 (11,425 cases). As of 2015, 10,243 cases of ETA were reported in 162 outbreaks.

Up to the thirteenth epidemiological period of 2016, Sivigila was notified of 9,781 cases of ETA, involved in 679 outbreaks. Of the total of cases notified to epidemiological week 52 of 2016, 51.3% of the cases of foodborne diseases were registered in the male sex; 21.2% of the cases were registered in the group of 10 to 14 years [13].

Empirical studies affirm that outbreaks transmitted by food have important economic repercussions for companies [3, 14]. It is important to mention that, of every ten medical disabilities reported by the military personnel on board the complex and civilians on the payroll, six correspond to infections caused by poor food handling, which produces high rates of absenteeism and decrease in the labor productivity of the workers.

In this order of ideas, the objective of the present investigation is to analyze the conditions of good practices of manufacturing and security in the dining rooms of a military complex located in the City of Cartagena de Indias, Colombia.

2. Materials and methods

A descriptive investigation was developed, with the objective of characterizing the conditions of Good Manufacturing Practices (GMP) and safety within the dining rooms, taking as reference Resolution 2674 of 2013.

2.1. Collection of information

For data collection, the direct observation technique was used, where the conditions of the dining rooms and Good Manufacturing Practices (GMP) were rigorously inspected for 8 months (July 2017 - February 2018).

A standardized form (sanitary inspection certificate to food factories) of the National Institute for Drug Monitoring (INVIMA) was used as a measuring instrument. Based on Resolution 2674 of 2013, it provided the parameters of compliance with the regulations for the food industry. food which are (conditions of buildings and facilities, equipment and utensils, personal manipulator, hygiene and manufacturing requirements, quality control and safety, training plans, among others).

The product qualification of the application of the standardized form on compliance scale takes values from one to five, where one (1) is equivalent to non-compliance, three (3) is equivalent to partial compliance and five (5) to total compliance.

2.2. Sample

6 dining rooms of the Cartagena military complex D.T.H Colombia were inspected.

3. Results and Discussion

In Table 1, it is important to highlight that the hygiene and safety conditions of the buildings and facilities are well constructed given that they have enough space to carry out the work and for the circulation of personnel, however a negative aspect is the non-compliance of the standards of good practices: There is no physical separation between areas nor a logical sequence between them which shows that the facilities are designed in such a way that they do not facilitate cleaning and disinfection work since a large percentage of the processing areas do not They have good hygiene and sanitation conditions

Table 1. Conditions of hygiene and security

Manufacturing practices	Sanitary requirements	Resolution 2674 of 2013	Score
BUILDING AND INSTALLATIONS	Localization and access, supply of water, health services, disposition of waste, methods of cleaning and disinfection.	Act. 6	3
	Floors and drains: pipes.	Act. 7	1
	Sanitation: programs and procedures.	Act. 26	3
	Protection control.	Act. 26, Act. 28	3
	Conditions of processes and manufacturing: design, construction, adaptation, cleaning and lighting and ventilation.	Act. 7	3
EQUIPMENT AND UTENSILS	Design, construction, adaptation, cleaning and disinfection. Prevention of cross contamination.	Act. 9	3
HYGIENIC MANUFACTURIN G REQUIREMENTS	Raw materials / supplies: cleaning, storage conditions, conservation. Prevention and cross contamination.	Act. 16, Act. 20, Act. 28	3
MANUFACTURIN G OPERATIONS	Safety assurance: duly established procedures, PCC, Ph, Aw, among others. Packaging and packaging: characteristics-labeled	Act. 19.	3
ASSURANCE AND QUALITY CONTROL	Control systems, technical sheets.	Act. 22, Act. 16	1

The canteens have enough drinking water reserve, however, one of them does not have a structure that facilitates washing and cleaning, in addition to this, the lid of this facilitates the entry of insects, pests and animals. Due to the aging of these storage tanks they present surfaces that generate toxic pollutants, the covers are not hermetic and none of them have identification as required by the norm.

The management of liquid waste in the dining rooms is not carried out under quality parameters that prevent contamination of the food, the sanitary facilities of the kitchens are not suitable for use, given that the bathrooms are not in optimal operating conditions, they remain clean, they do not have the minimum resources required for personal hygiene, there is no signage. The chemical products for disinfection are found in places that promote cross-contamination, there are no hand soaps and towels for cleaning and drying. The utensils are cleaned and disinfected properly but it is not enough because most of them have cracks and glue stuck to their surface, with bumps and bruises that make proper disinfection impossible.

Similarly, solid waste that there is no order and cleanliness inside the kitchens and that in practice there is no collection system within the kitchen in an orderly manner of solid waste, there is no separation at source and waste in Most of them are deposited in a single tank / tank, and are not removed frequently to avoid contamination. It should be noted that there are cold rooms of organic waste easily decomposed, however, are not used for that purpose.

The control of pests (arthropods, rodents, birds) in each building have official contracting for pest control in full compliance with the legislation applied in this BPM and safety analysis; However, in total contrary to the standards allowed in said legislation, it was found that the kitchens do not have barriers, screens and other devices allowing the free passage of pests, insects and rodents, which makes it common to see the entry of animals into the area. Kitchen is worth mentioning that good practices go hand in hand with cleaning and disinfection, the analysis showed that they do not have ecological points

IOP Conf. Series: Materials Science and Engineering **519** (2019) 012005 doi:10.1088/1757-899X/519/1/012005

and inside the kitchens there is no separation of waste, even basically: liquids and solids. The cleaning and disinfection procedures do not satisfy the particular needs of the process and the final product.

Consequently, the cleaning, cleaning and disinfection of areas, the conditions of process and manufacturing have great importance to prevent incidents and accidents of work, in this aspect the characteristics of floors walls and drainages are a fundamental aspect, product of this analysis could be establish that the floors are mostly worn and greased with cracks allowing pollutants to be generated, the walls specifically at the junction with the roof, most do not notice the original color they are painted because they look dark accumulated dirt, between walls and floors are holes where dirt accumulates. The ceilings of kitchens and dining rooms are not built with materials that facilitate cleaning and disinfection. In one of the audited kitchens, roof failures were observed.

The windows are dirty and neglected there are no barriers between insects, pests, with the kitchen, wooden doors are mostly worn and do not have automatic closing devices and tight fitting.

A positive aspect that could be determined at a glance are the conditions of the stairs, elevators and complementary structures are located and built strategically, allowing the mobility of personnel and not generate, or cause pollution. Electrical, mechanical and fire prevention installations must be properly designed and marked; in the same way it happens with the lighting conditions because it is sufficient for the tasks.

The ventilation of the buildings has openings for air circulation, however these are not protected with anti-insect nets of non-corrosive material which easily allows the passage and housing of animals. The extractor hood of one of the dining rooms that is used for the evacuation of air, is dirty and rusty.

On the other hand, the equipment and utensils that predominate in the kitchens are: kettle, stoves, extractor hood, blender, oven, blender, spoons, knives, ladles, pots, pans, plates, chopping boards, plates, glasses, scales, drawers, silversmiths, bain marie, trays, coffee makers, auxiliary cart, quadrants, drawers, industrial blender, juicers, industrial juicer, freezers, fridges, fryers, coolers, pans, strainers, centrifuges, skimmers, skimmers, spatulas, corkscrews, forks , peelers, mortars, scissors, tweezers, can opener and funnel. Most are not designed for easy cleaning and disinfection, their surfaces do not allow it. In general terms the kitchen area mostly surfaces have cracks, porosities and defects for deposit of microorganisms.

As regards hygienic manufacturing requirements, raw materials and supplies are included, as well as manufacturing, preparation, processing, packaging and storage activities that comply with the requirements to guarantee food safety. As there is a site for the storage of raw materials and finished products, however, there is no adequate technique of conservation and control of frozen products, the storage of raw materials for food in relation to food preservation that is of great importance in Regarding compliance with the current regulations that guarantee food safety are conservation techniques, in this type of establishments is adequately met, the existence of cold rooms is highlighted, but these do not have temperature measurement instruments and do not have the conditions of order and cleanliness for its use.

An equivalent compliance with good manufacturing practices "BPM" is the assurance and control of quality and safety in terms of control systems and laboratory; the controls carried out on the different products and receipt of the raw materials are carried out in an appropriate manner. However, there are no manuals, guides, procedures or any type of documented indications for equipment or rooms. As far as the Laboratory is concerned, there are no manuals and procedures, nor is there a verifiable system for analysis of critical points or critical stages (conservation-cold room). The cold room and other equipment do not have a maintenance plan, they are repaired only when they are damaged, and they do not have qualified personnel in charge of the training program.

Table 2. Conditions of hygiene and personal safety manipulator

Personal manufacturing practices manipulator	Sanitary requirements	Resolution 2674 of 2013	Score
Health status	physical	Act. 11	**3**
Education	Training plan of BPM	Act. 1, Act. 13	**1**
Protection elements	Protection elements	Act. 14	**3**
Habits	Uniform, Hand washing	Act. 14	**3**
Visiting staff	Safe hygiene practices	Act. 13	**1**

Discrimination of the variables analyzed in the inspections shown in Table 2, hygienic conditions and personal safety manipulating hygienic practices and protection measures, in the analysis was determining this aspect being a critical factor in terms of good manufacturing practices that can alter the quality of the food handled and that is a generator of pathogenic microorganisms that cause diseases and toxi-alimentary infections; it could be established that there is hiring to perform medical examinations, however, after an absence due to infection there are no follow-up examinations, there is not in all cases a certificate that shows the suitability for food handling, there are no adequate points for disinfection , protocols for conservation are not always implemented. It was observed that not all have mesh or cap, as is also notorious the use of disposable face masks covering nose and mouth while handling food, the use of different accessories, it is common to observe people consuming food within the handling area.

It is observed that the visiting staff has easy access to the food preparation area, that this area is not restricted and is very easy to contaminate since the personnel that enter does not have elements of protection, before, on the contrary, some They arrive with dirty coveralls and go directly into the kitchen.

In the same sense, it could be established that in terms of education and training, the military complex does not have a training plan for the kitchen and dining room staff, therefore, there is no timetable that evidences the execution of these trainings. Not all food court officials are required to certify the fitness for food handling, likewise there are no good handling habits by the manipulative staff and it is very easy to see the staff come out and enter again with the robe with which they manipulate in the kitchen, non-existent controls on the entry of unauthorized personnel into food handling areas, with coveralls working in other activities and without protection elements. Good habits do not predominate, some due to lack of training or because the conditions of adequacy do not facilitate them, such as not having easy access, such as washing hands when leaving the bathroom since they do not have special sinks for this and must be done in the bathrooms. kitchen dishwashers provided for other work, the same happens with hand drying towels that are absent in this process.

In general, it was possible to detail several security flaws in all the facilities of the dining rooms, they are the relevant aspects that are part of the culture of every organization, ensuring the welfare and integrity of the officials and visitors. In Table 3, 24 aspects of general security of the establishments are audited and it can be detailed that only 10 of these are in total compliance with an estimate of 5.

To conclude, the present study demonstrates that in an analysis of 164 aspects contained in Resolution 2674 of 2013, it was possible to observe total compliance in 58 aspects, meanwhile, 36 partially comply and finally, 70 aspects do not comply with that established by the legislation in the dining rooms of a military complex located in the city of Cartagena of Indias.

The canteens in a generalized way have the same characteristics since all are characterized by their work under the same conditions, it is important to bear in mind that these analyzes were carried out based on the Colombian legislation, law 9 of 1979, better known as the sanitary code, the decree 3075 of 1997 whose complement is resolution 2674 of 2013.

Table 3. Synthesis of security aspects

General aspects of establishment	Score
Conditions of permanent order and cleanliness and clearly defined procedures	1
Floor free of objects at all times, constructed in such a way as to avoid the stagnation of liquids and other substances	1
Order with objects, materials and elements.	1
Work surfaces are free of unevenness	1
Classification of materials present in the work area (necessary and unnecessary)	1
There is an easy-to-use medicine cabinet in the restaurant area	5
Ecological points for adequate separation of waste	1
Areas of work and storage are not properly demarcated or marked	1
Means of notice to restrict access by personnel outside certain restricted work areas	1
Warning signs indicating the risks present in the work areas	1
Internal dimensions surface extension and capacity, are suitable for equipment placement	5
Preventive maintenance of equipment and other tools	1
Appropriate tools in such a way that they do not represent a danger to workers	5
Visitors are prohibited from freely accessing the food preparation areas	1
Visiting staff entering the preparation area do so with protection elements	1
It has security regulations displayed in a visible place for the food area	1
Electrical installations visibly in good condition and all electrical circuits signalized	5
The exposed electrical installations are protected against impacts	5
It has electrical insulation (pole to ground)	5
For fire prevention there is water, extinguishing devices	5
For fire prevention there are trained personnel	5
There are enough exit routes	5
Extinguishers with valid recharge date	5

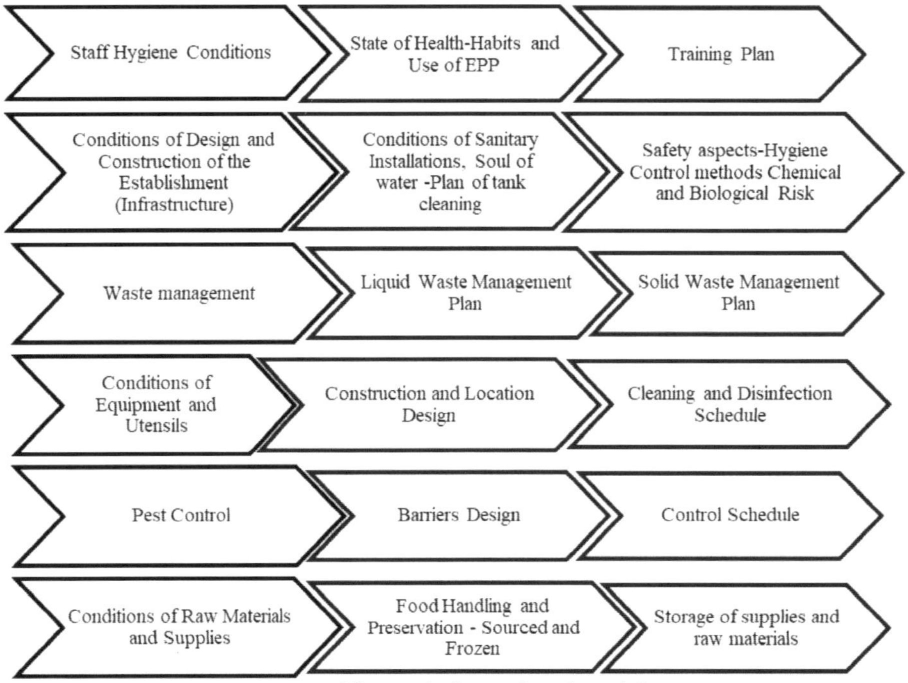

Figure 1. Operational model

IOP Conf. Series: Materials Science and Engineering **519** (2019) 012005 doi:10.1088/1757-899X/519/1/012005

4. Conclusions

The results of this study contextualize a reality little studied and indirectly shows the risks that could increase the existing relationship between good manufacturing practices and labor productivity by the workers of this complex, in this sense, evidence of the increase in disabilities due to poor sanitary control.

This study presents strengths, since it takes as a basis the national legislation to determine the safety conditions and good manufacturing practices inside the canteens of the military complex because its results are the starting point for the implementation of internal policies that improve the practices and positively impact the quality of life of both internal staff and external personnel, since empirical evidence argues that these practices have a direct relationship with risks associated with public health. The reduction of these risks protects the life and health of the people who are concentrated there and coexist most of the time for working conditions as this is a "special regime".

Similarly, analysis was a fundamental contribution, this being the main input that will support the quality management systems and the occupational safety and health management system of the military complex, from which actions included within both systems emerge. Compliance with minimum requirements for management systems, this being the beginning for the establishment of new policies and improvement actions that guarantee the reduction of absenteeism at work due to multiple gastrointestinal infections.

Bearing in mind that safety is synonymous with impeccability and this is necessary in terms of nutritional quality, since the only guarantee to prevent foodborne diseases is to ensure that handling, preservation, cleaning, disinfection and other operations committed to the process of preparation and serving of food are carried out respecting quality parameters.

The analysis was based on a business need, finding that the negative aspects in the dining halls cause absenteeism in the workforce and then decrease productivity resulting from absence at work, it is necessary to affirm that there is a direct relationship between transmitted diseases for food and medical work disabilities, evidencing that it is urgent for the company to establish a tool that guarantees safe products under sanitary conditions: Implementation of Good Manufacturing Practices -BPM duly documented based on the requirements of current legislation regarding the minimum requirements that a company dedicated to the preparation and / or distribution of food must meet.

Said plan must of BPM must draw a clear line in sanitary matter adapted and personalized to the nature of the military complex and its activities as special regime, a scale of name is proposed "Scale Security-BPM", strategically designed based on three levels the first level of blue color that refers to the variables themselves aimed at improving safety conditions and good manufacturing practices, the second and third level we find the criteria that make up each variable. n the same sense, the company in question must design within the documentation of BPM and Security a sanitation plan where it is captured and subsequently executed the conditions of operability regarding proper waste management demonstrating efficient collection of waste from the raw material, fats and oils assuring their final disposal, in this way criteria for cleaning and disinfection with duly established procedures taking into account agents, products, substances and concentrations used for these tasks; demonstrating pest management and control of pests and rodents, indicate and document within this plan that there are controls to ensure compliance with the physicochemical and microbiological requirements of drinking water used and stored in tanks that have a cleaning and disinfection schedule.

The implementation plan of good manufacturing practices must go hand in hand with the Occupational Health and Safety Management System, this cannot be carried out in isolation, that is, that the BPM plan must have a focus on risks and be part of the Occupational Health and Safety Management System, where both safety and hygiene aspects (lighting, temperature, ventilation, noise, etc.) must be taken into account so that this becomes a strategy to reduce work-related disabilities for foodborne diseases ETA and therefore the implementation of BPM becomes a fundamental tool for increasing productivity and decreasing indices of absenteeism.

Conflict of Interest.
The authors report there are no conflicts of interest.

References
[1] Gonzalez T 2005 Foodborne diseases and PCR prevention and diagnosis. *Salud publica de Mexico* 388.
[2] Yuniesky G 2012 Actions for the management of sanitary quality and safety. *Journal of Health Management and Policies* 140.
[3] Martinez, & Ceballos 2007 Outbreaks of foodborne diseases. *Epidemiological Bulletin* 241-252.
[4] World Health Organization (WHO) 2001 Food safety. *Secretary's report Executive Council.*
[5] Ministry of Health of Chile 2010 Sanitary regulation of Chilean foods. Lexnova editions.
[6] Bastías J. M., Cuadra, M., Muñoz, O., & Quevedo, R 2013 Correlation between good manufacturing practices and compliance with microbiological criteria in the manufacture of ice cream in Chile. *Chilean nutrition magazine.*
[7] Kopper G 2009 Diseases transmitted by food and its socioeconomic impact. Rome.
[8] Sansawat and Terry 2011 Review of the standards of food safety management and design and manufacture of food packaging. Sgs.
[9] Vega M., & Guerrero V 2015 Design and proposal of a food safety system based on BPM (Good Manufacturing Practices) for Destiny Hotel in the city of Baños. *Digital Repository - Central University of Ecuador.*
[10] Tafur 2009 Food safety and international trade. *Colombian Journal of Animal Sciences.*
[11] Pérez-Silva Belmonte, & Martínez 1998 Microbiological study of foods prepared in high-risk collective dining rooms. *Spanish magazine of Public Health.*
[12] Alerte V., Diaz J., Vollaire J., & Espinoza 2012 Outbreaks of diseases transmitted by food and water in the Metropolitan Region, Chile (2005-2010). *Chilean Journal of Infectology.*
[13] Ministry of Health of Medellin 2017 Epidemiological Bulletin: Diseases transmitted by food.
[14] Arcos J., & Jeanette A 2013 Risk factors associated with the emergence of foodborne diseases that are sold in school bars in ambato health area. *Digital Repository - University of Ambato.*

IOP Conf. Series: Materials Science and Engineering **519** (2019) 012006 doi:10.1088/1757-899X/519/1/012006

Vibrational analysis of acetaminophen from commercial tablets

J R Castro-Suarez[1,2], M Vásquez-Osorio[2], S P Hernandez-Rivera[1] and A A Pájaro-Payares[3]

[1] ALERT DHS Center of Excellence for Explosives Research, Department of Chemistry, University of Puerto Rico, Mayagüez, PR 00681, USA.
[2] Fundación Tecnológica Antonio de Arévalo-TECNAR, Cartagena, Colombia.
[3] Materials and Fuel, Research Group, Department of Chemistry, University of Cartagena, Cartagena, Colombia.

johncastrosuarez@gmail.com

Abstract. Acetaminophen (AAP) is an active ingredient very used in many pharmaceutical preparations. AAP is a pain reliever and a fever reducer drug. Generally, it is used to treat many conditions such as headache, muscle aches, arthritis, backache, toothaches, colds, and fevers. Standard methods most used for AAP chemical detection are based on chromatography and spectrophotometry techniques, these methods generally imply a sample preparation step, while vibrational spectroscopy based methods do not. Hence, analytical methods based on vibrational spectroscopy are very important for the pharmaceutical industry and law enforcement agency, given that allow obtain a way easy and fast molecular information for its detection. An effort for attaining a reliable identification procedure for the qualitative determination of AAP in different pharmaceuticals product, an additional statistical treatment of ATR-FTIR data is proposed. The proposed method was tested on solid samples containing API. The statistical routine of Hit Quality Index (HQI) values yielded excellent results. The results show that ATR-FTIR, QCL and Raman spectra are useful for detection of AAP in the drugs tested. Using the proposed method allowed to know if the drug is present as a desired product or it is counterfeit drugs.

1. Introduction

Acetaminophen (AAP) or paracetamol is one of the most used active ingredient in pharmaceutical formulations. AAP is generally used as an analgesic and antipyretic to relieve pain and reduce fever, it is used specifically to treat various conditions such as headache, muscle aches, arthritis, back pain, toothaches, colds and fevers. The amount provided in the patient is of the utmost importance, insufficient could prolong the treatment of a certain condition, and even worse, a lot could have adverse consequences for the patient's health.

Due to its easy availability without prescription, your deliberate consumption or overdose can happen. Overdoses of AAP are associated with liver toxicity and renal failure. Liver toxicity begins with plasma levels of AAP in the range of 120 μgmL-1 4 h after ingestion and severe damage occurs with plasma levels of up to 200 μgmL-1 after 4 h ingestion [1]. Many analytical methods have been used and proposed by the scientific community for the determination of AAP, both for pharmaceutical formulations and for biological samples. Among them, chromatographic [2] and optical (UV-VIS) [3] methods are the most used. Other analytical methods such as electrochemicals [4] have recently been reported by the scientific community. These techniques have been shown to be efficient in terms of

Content from this work may be used under the terms of the Creative Commons Attribution 3.0 licence. Any further distribution of this work must maintain attribution to the author(s) and the title of the work, journal citation and DOI.
Published under licence by IOP Publishing Ltd

reproducibility and obtaining low detection limits. However, in their chemical analysis process, sample preparation steps are carried out by lengthening the analysis time and making it difficult to use in the field.

In recent years, analytical methods based on the use of special infrared (IR) and Raman spectroscopy have been shown to be useful for the analysis of different chemical compounds, including drugs in different pharmaceutical products, from the process of its manufacture until validation of the final product [5-7].

M. Mallah, et al. [8], recently reported the development of a transmission FTIR spectroscopic method for the direct, low-cost and rapid quantification of paracetamol content in solid pharmaceutical formulations.

In this study, a simple, rapid and reproducible methodology using (ATR) attenuated total reflectance FTIR spectroscopy, quantum cascade laser spectroscopy and Raman spectroscopy for the direct detection of acetaminophen in different pharmaceutical presentations were carried out. Statistics of chemometrics routines such as hit quality index (HQI) values were applied to the obtained infrared spectra. The results show that ATR-FTIR, QCL and Raman is useful for the detection of AAP in studied drugs. Using the proposed method allows knowing if the drug is present as a desired product or if it is counterfeit medicines.

2. Reagents, samples and sample preparation

The reagents used in this study correspond to standard and samples. As standard, pure analytical grade acetaminophen purchased from Sigma-Aldrich, and was used to obtain the AAP reference ATR-FTIR vibrational spectrum. As samples, some solid formulations in the form of tablets containing AAP as API were used. These samples were obtained from the local markets of Cartagena-Colombia. The samples acquired containing AAP were from different manufacturing laboratories, the laboratories were AG, BEST, Genfar, Lafrancol, La Santé, MK, Sanofi-aventis and GRUNENTHAL.

Unlike other commonly used methods, in this method a simple preparation of the sample (tablets) was used before the acquisition of the FTIR spectra. The only step involved is pulverize of a small portion of the solid pharmaceutical samples (after removing the coating if it had) as a fine powder in a mortar to reduce the particle size of the sample. We record 50 measurements (spectrum) for the eight sample types and 5 independent measurements for the AAP standard, for a total of 55 mid-infrared spectra with molecular information from AAP.

All samples and standards were analyzed using three systems of vibrational detection. Of these, two using infrared spectroscopy and one using Raman spectroscopy.

First Infrared detection system: The spectra taken of AAP from the solid standard and commercial tablets were recorded using the Thermo Nicolet 5700-FTIR spectrometer with ATR accessory and Deuterated Triglycine Sulfate Detector (DTGS). The resolution was 4 cm-making an average of 64 scans (scan) in the Mid IR region of 400-4000 cm^{-1}.

Second Infrared detection system: The infrared vibrational spectra for the detection of AAP, were carried out using a LaserScan ™ from Block Engineering, LLC, Marlborough, MA. These instruments use an IR source of next generation quantum married laser (QCL) widely tunable. This type of sources (laser) increases the sensitivity that are orders of magnitude greater than conventional spectroscopic systems. The spectra Infrared were recorded in the spectral range of 1000-1600 cm^{-1}, all spectra were taken with 2 CoAditions and 4 cm^{-1} resolution.

Raman detection system: the instrument Renishaw InVia Raman Microscope. Objective, 20X; Exposure time (s), 1s; laser, 785nm, 495 mW; Accumulation 1. Figure 1 shows the experimental configuration used in this investigation.

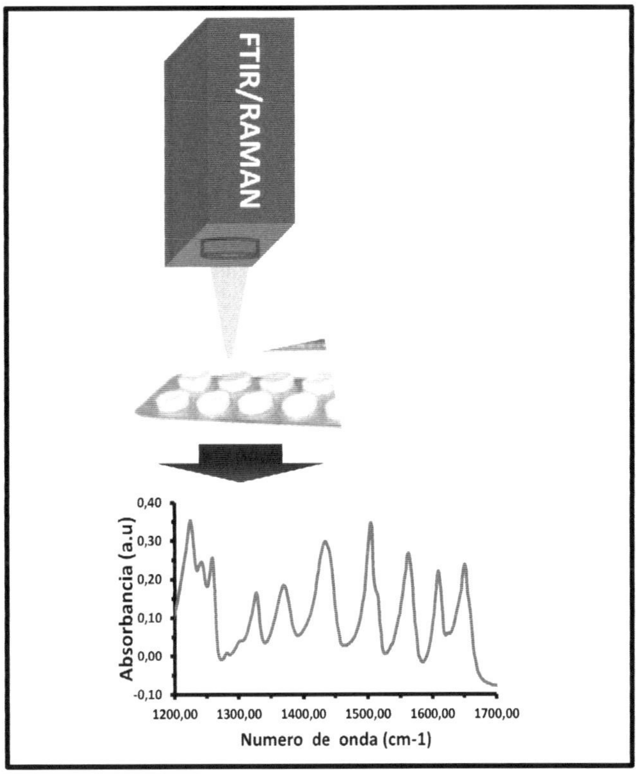

Figure 1. IR spectral acquisition from commercial tablets of APP.

3. Results and Discussion

3.1. Obtaining vibrational spectra from AAP

Vibrational spectroscopy comprises NIR (780-2595 nm), mid-infrared (MIR; 400-4000 cm^{-1}), and Raman spectroscopy (Raman shift; 200-4000 cm^{-1}). The phenomenon of IR spectroscopy (NIR and MIR) is based on the absorption, reflection, and emission of the light from the sample, while Raman spectroscopy is based on the scattering phenomenon. However, bot techniques deliver information about the fundamental vibrational and rotational modes of the molecules [9].

During IR spectroscopy, the sample spectrum is acquired by passing infrared radiation through the sample, which excites the molecular vibrations. The transmitted or reflected incident radiation from the sample is then analyzed by the detector. To obtain a Raman spectrum, the sample is illuminated by a monochromatic laser beam that interacts with the molecular vibrations. The scattered light from the sample is detected by the spectrometer [9].

Spectra of solid acetaminophen (AAP) Standard are shown in Figure 2. Spectra were taken using a attenuated total reflection (ATR) geometry by means of a Nicole 6700 FTIR spectrophotometer, see Figure 2-A; using a back reflection geometry by LaserScan, shown in figure 2-B; using a Raman spectrometer as illustrated in Figure 2-C. To obtain the IR spectra of AAP, the air spectrum was taken as background. We can observe in Figure 2 that the spectra recorded, have the vibrational bands IR and Raman characteristics of AAP when they are compared with the FTIR/Raman spectra of the literature [10]. The intense lines of the benzene ring vibrations at 1606 and 1616 cm^{-1} for AAP were chosen as the characteristic peak in Raman.

IOP Conf. Series: Materials Science and Engineering **519** (2019) 012006 doi:10.1088/1757-899X/519/1/012006

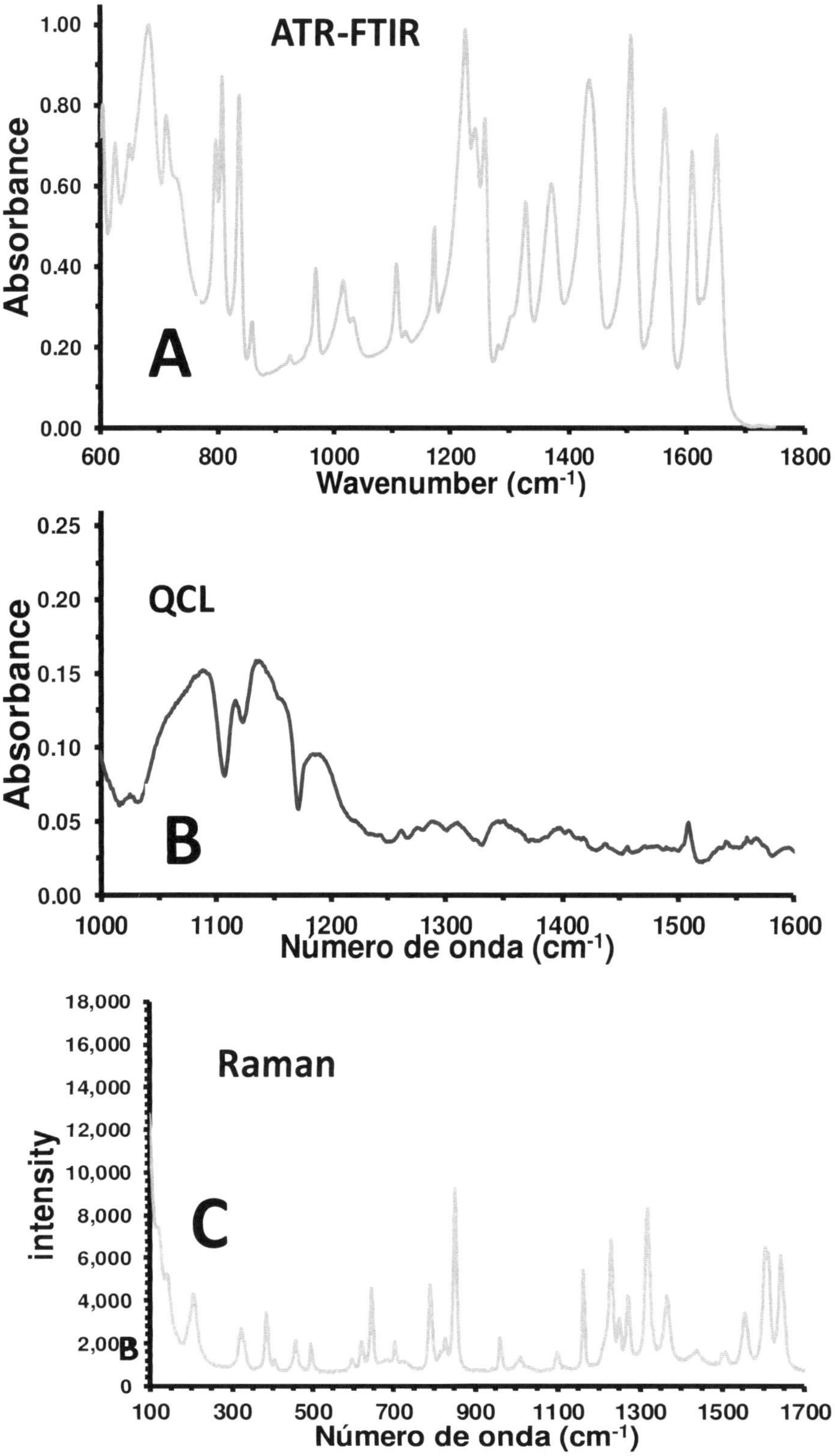

Figure 2. Spectra of standard AAP using different vibrational instrumental techniques. A) ART-FTIR, B) QCL and C) Raman.

IOP Conf. Series: Materials Science and Engineering **519** (2019) 012006 doi:10.1088/1757-899X/519/1/012006

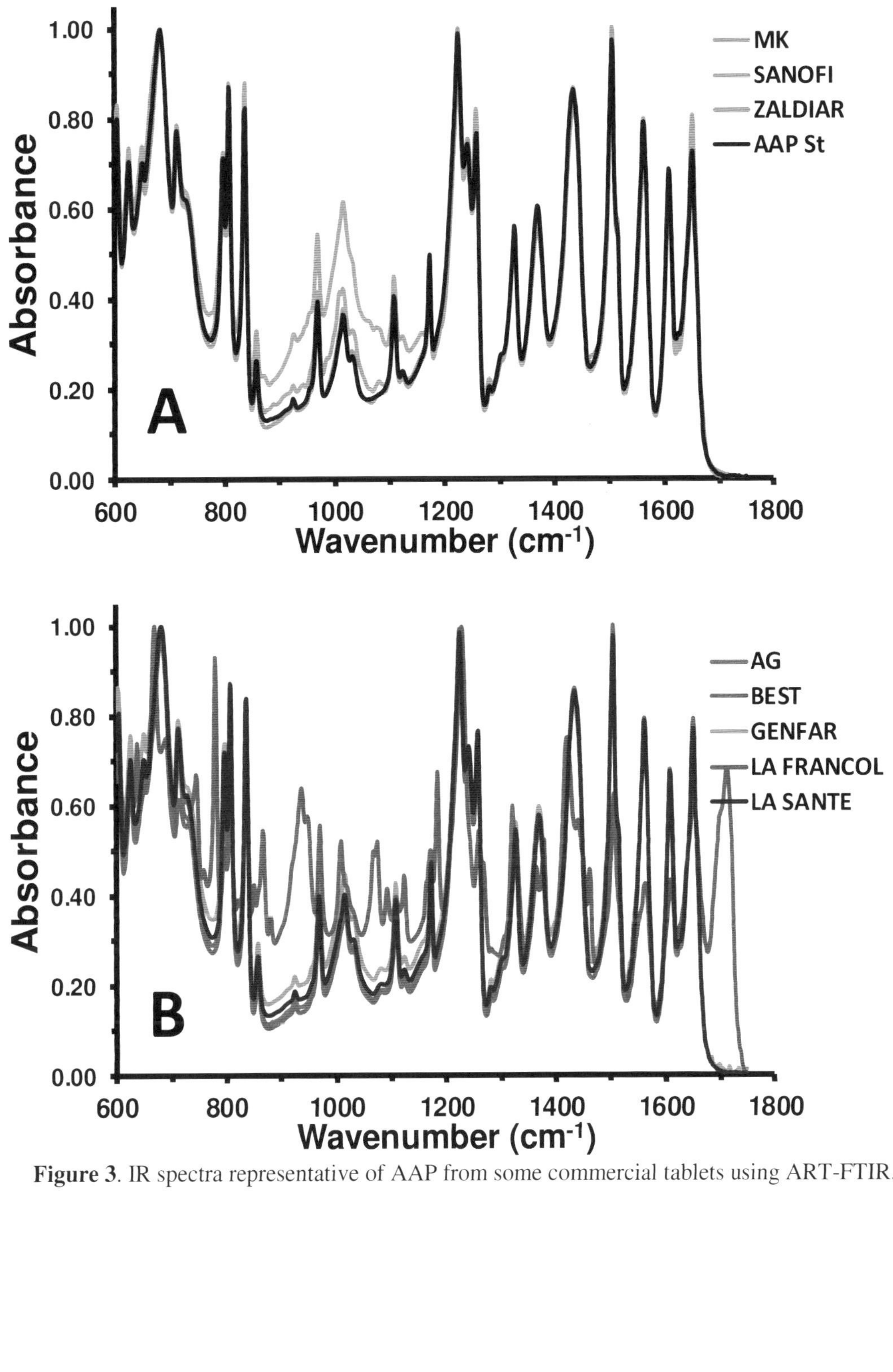

Figure 3. IR spectra representative of AAP from some commercial tablets using ART-FTIR.

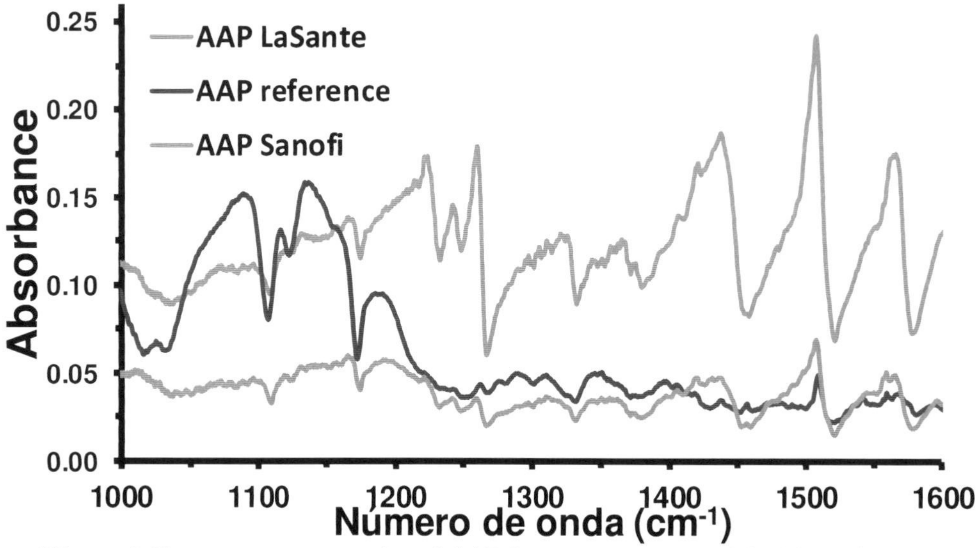

Figure 4. IR spectra representative of AAP from some commercial tablets using QCL spectroscopy.

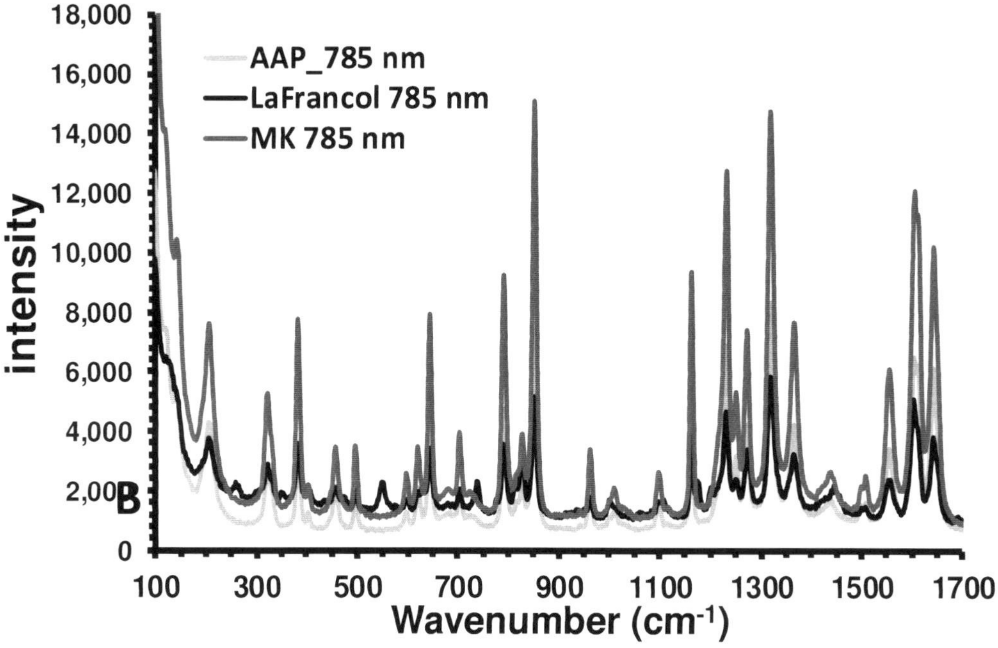

Figure 5. IR spectra representative of AAP from some commercial tablets using Raman spectroscopy.

3.2. Spectral analysis

The simplest spectral search is based on the calculation of the hit quality index (HQI) values [11]. The HQI is a numerical quantity that indicates the correlation between two spectra and has been widely used in spectroscopy to indicate the degree of spectral matching in library searches. HQI values can be calculated using various algorithms, but the two most commonly used are the Euclidean distance and spectral correlation algorithms. In the spectral correlation algorithm utilizes the Pearson product moment correlation coefficient, r_{xy}, which is a measure of the strength and direction of the linear relationship between two variables. This parameter is defined as the covariance of the variables divided by the product of their standard deviations. In this case, the spectral correlation algorithm is applied between two spectra: a reference (spectra form reference AAP) and an unknown spectrum to be identified (spectra form commercial tablets). Table 1 presents the spectral correlation coefficients for spectra form commercial tablets tested containing AAP.

It is generally acceptable to consider a spectrum of an unknown compound to be similar to one of the library when the spectral correlation coefficient is greater than ~ 0.85. Given this restriction, spectra measured form commercial tablets were correctly identified when ATR-FTIR coupled with spectral correlation algorithms were used.

Table 1. Values of spectral correlation coefficients (HQI) for spectra of AAP from various commercial tablets using ATR-FTIR.

	AAP St	AG	BEST	GENFAR	LA FRANCOL	LA SANTE	MK	SANOFI	ZALDIAR
AAP St	1.00	0.998	0.997	0.993	0.645	0.998	0.999	0.952	0.993
AG	0.998	1.00	0.999	0.994	0.649	0.999	1.000	0.960	0.996
BEST	0.997	0.999	1.000	0.992	0.650	0.999	0.999	0.958	0.996
GENFAR	0.993	0.994	0.992	1.000	0.665	0.997	0.994	0.976	0.993
LA FRANCOL	0.645	0.649	0.650	0.665	1.000	0.657	0.648	0.678	0.647
LA SANTE	0.998	0.999	0.999	0.997	0.657	1.000	0.999	0.967	0.997
MK	0.999	1.000	0.999	0.994	0.648	0.999	1.000	0.956	0.995
SANOFI	0.952	0.960	0.958	0.976	0.678	0.967	0.956	1.000	0.970
ZALDIAR	0.993	0.996	0.996	0.993	0.647	0.997	0.995	0.970	1.000

4. Conclusion

A system based on medium infrared spectroscopy using attenuated total reflection, QCL and Raman were used for the detection of acetaminophen as API present in commercial pharmaceutical tablets. Many vibrational characteristics of acetaminophen can be observed in the ATR, QCL and Raman spectra of analyzed tablets, when compared with standard AAP, so the vibrational contributions from the excipients is not a critical factor when deciding whether AAP is present in a certain tablet. In general, we have shown that ATR-FTIR spectroscopy is useful for the development of a rapid methodology for the detection and discrimination of AAP present in tablets when HQI is claculated.

Conflict of Interest.

The authors report there are no conflicts of interest.

References

[1] Bosch M E, Sánchez A R, Rojas F S and Ojeda C B 2006 Determination of paracetamol: Historical evolution *Journal of pharmaceutical and biomedical analysis* **42** 291- 21

[2] Qi, M L, Wang P, Leng Y X, Gu J L and Fu R N 2002 Simple HPLC method for simultaneous determination of acetaminophen, caffeine and chlorpheniramine maleate in tablet formulations *Chromatographia* **56** 295-98

[3] Chafetz L, Daly R E, Schriftman H and Lomner J J 1971 Selective colorimetric determination of acetaminophen *Journal of pharmaceutical sciences* **60** 463-66
[4] Liu B, Ouyang X, Ding Y, Luo L, Xu D and Ning Y 2016 Electrochemical preparation of nickel and copper oxides-decorated graphene composite for simultaneous determination of dopamine, acetaminophen and tryptophan *Talanta* **146** 114-21
[5] Castro-Suarez J R, Pacheco-Londoño L C, Ortiz-Rivera W, Vélez-Reyes M, Diem M and S. P. Hernandez-Rivera 2011 Open path FTIR detection of threat chemicals in air and on surfaces *SPIE Defense, Security, and Sensing. International Society for Optics and Photonics* **8012** 801209
[6] Castro-Suarez J R, Ortiz-Rivera W, Galan-Freyle N, Figueroa-Navedo A, Pacheco- Londoño L C and Hernández-Rivera S P 2013 Multivariate analysis in vibrational spectroscopy of highly energetic materials and chemical warfare agents simulants *Multivariate Analysis in Management, Engineering and the Sciences* InTech
[7] Wartewig S and Neubert R H 2005 Pharmaceutical applications of Mid-IR and Raman spectroscopy *Advanced drug delivery reviews* **57** 1144-70
[8] Mallah M A, Sherazi S T, Bhanger M I, Mahesar S A and Bajeer M A 2015 A rapid Fourier-transform infrared (FTIR) spectroscopic method for direct quantification of paracetamol content in solid pharmaceutical formulations *Spectrochimica Acta Part A: Molecular and Biomolecular Spectroscopy* **141** 64-70
[9] Kandpal L M, Park E, Tewari J and Cho B-K 2015 Spectroscopic Techniques for Nondestructive Quality Inspection of Pharmaceutical Products *Journal of Biosystems Engineering* **40** 394-408.
[10] Moynihan H A, O'Hare I P, Spectroscopic characterisation of the monoclinic and orthorhombic forms of paracetamol *International journal of pharmaceutics* **247** 179-85
[11] ASTM E2310 - 04 2009 Standard Guide for Use of Spectral Searching by Curve Matching Algorithms with Data Recorded Using Mid-Infrared Spectroscopy

IOP Conf. Series: Materials Science and Engineering **519** (2019) 012007 doi:10.1088/1757-899X/519/1/012007

Enhanced RDX Detection Studies on Various Types of Substrates via Tunable Quantum Cascade Laser Spectrometer Coupled with Grazing Angle Probe

J L Ruiz-Caballero[1], L A Blanco-Riveiro[1], I A Ramirez-Marrero[1], L A Perez-Almodovar[1], A M Colon-Mercado[1], J R Castro-Suarez[1,2], L C Pacheco-Londoño[1,3], S P Hernandez-Rivera[1]

[1]ALERT DHS Center of Excellence for Explosives Research, Department of Chemistry, University of Puerto Rico, Mayagüez, PR 00681, USA
[2]Fundación Universitaria Colombo Internacional, Unicolombo, Cartagena, Colombia
[3]Universidad Simón Bolívar, Facultad de Ciencias Básicas y Biomédicas, Barranquilla, Colombia

samuel.hernandez3@upr.edu

Abstract. Owing to scientific advances in the field of materials sciences and engineering, researchers have developed new energy sources used for spectroscopic applications and measurements of properties resulting from the interaction of matter and electromagnetic radiation in the mid-infrared (MIR) region. MIR lasers, such as quantum cascade lasers (QCLs), used for spectroscopy have quickly found numerous applications in a wide cadre of IR techniques. This provides the opportunity to study properties of highly energetic materials (HEM), among many other applications. MIR laser spectroscopy based detection experiments of HEMs were carried out using a QCL optically coupled to compact grazing angle probe mount (QCL-GAP) enabling reflection-absorption infrared spectroscopy (RAIRS) measurements of thin films of HEMs. A saturated solution of RDX in acetone was prepared, and aliquots of subsequent dilutions of the stock solutions were transferred to test surfaces for QCL-GAP back-reflectance measurements. RDX reflectance signals were monitored as function as the decreasing surface concentration until the signal/noise was ~ 3. Stainless steel (SS) plates were used as reflective substrates, and anodized aluminum (AN-Al), cardboard, and Teflon were used as non-reflective (matte) substrates. Using generated calibration curves a low limit of detection (LOD) of 1.7 ng/cm^2 for RDX/SS and 95 μg/cm^2 for RDX/AN-Al were found. Based on the area of laser spot (0.3 cm^2) we conclude the minimum masses detected were 490 pg (RDX/SS) and 28 μg (RDX/AN-Al).

1. Introduction

As part of contingency preparations for awareness and localization of explosives-related threats (ALERT), we have expanded our studies to gain a deeper knowledge about the diverse ways for depositing explosives on substrates for preparing samples and standards used in the remote detection program of highly energetic materials (HEMs). In this regard, it has been a continuous growth of knowledge using various technologies available for preparing test samples. We have explored techniques that include spin coating [1-5], smearing [6], and drop-cast [7] as part as the methodologies used to simulate real-world scenarios of sample preparation and subsequent detection. This part of the

Content from this work may be used under the terms of the Creative Commons Attribution 3.0 licence. Any further distribution of this work must maintain attribution to the author(s) and the title of the work, journal citation and DOI.
Published under licence by IOP Publishing Ltd

IOP Conf. Series: Materials Science and Engineering **519** (2019) 012007 doi:10.1088/1757-899X/519/1/012007

remote detection program is important because the surfaces that come in contact with explosives that are left as residues can evidence potential terrorist acts. [8] Targets include car-bomb surfaces, cardboard substrates evidencing package and letters bombs, clothes and backpacks (or baggage) and other surfaces that may have come in contact with HEM handled in the processes. [9]

In recent years, detection of explosives has become an issue of high importance, particularly in Homeland Security and National Defense applications. This has led to developments in instruments and methods for detection of HEM which includes mass spectrometry [10], vibrational spectroscopy [11], standoff detection [12-13], other sensing schemes [14-15], and trace detection. [16-17] An efficient detection and identification method that provides the low limits of detection (LODs) of HEM on substrates when exposed to different explosives types are always on demand. Developed methodologies would be highly beneficial to airport security, border controls, and military operations, among others.

This work centers on the use of mid-infrared (MIR) quantum cascade lasers (QCLs) which can be tuned to strong absorption bands of HEM such as 1,3,5-trinitro-1,3,5-triazacyclohexane (RDX) deposited on substrates for target (analyte) detection, discrimination, and quantification. QCLs are unipolar semiconductor injection lasers that are based on sub-interband transitions in a multiple quantum-well heterostructure. This type of laser has various advantages over other types of lasers. As a semiconductor laser it can produce varying wavelengths and operate at various temperatures. QCLs can produce from a few tens to hundreds of milliwatts of continuous or pulsed mode power under ambient conditions. [18-21] The beam size can be as small as 2 x 4 mm^2 at the sample, which corresponds to the average beam size at the sample focus of standard laboratory (FT-IR) instruments. [22] The need to develop more powerful MIR sources that enable reflection-absorption infrared spectroscopy (RAIRS) data when a target chemicals or microorganisms are present on a substrate at low surface concentration levels suggests the use of collimated, coherent, and polarized sources [1]. A new setup developed in our laboratory, in which a tunable QCL spectrometer was coupled to a grazing angle probe (QCL-GAP) has provided improvements including the beam size change to 24 x 4 mm^2. This arrangement consists of using a mirror to direct the IR beam to hit the sample at an angle of 82° with respect to the normal vector (Figure 1). [1] This application of QCL spectroscopy presents certain advantages, such as a low limit of detection (LOD) values ranging from 10 to 50 ng/cm^2 of HEM. [9, 23] In another improvement is in acquisition time (10-15 s per scan) whereas typical RAIRS measurements are highly time-consuming (min to h). [18] If a thin film or coating of some sort exits on a smooth surface, the light beam passes through the coating, reflects off the smooth substrate and passes through the coating for the second time. This phenomenon is called double transmission RAIRS. [1, 23] The best experimental conditions for obtaining RAIRS grazing angle (≥ 80°) incidence, with the radiation, polarized parallel to the plane of incidence (p-polarization). It is seen that s-polarized light is almost canceled by reflection at grazing angle (GA). P-polarized light is almost intensity doubled at grazing incidence. Under such conditions, the surface electric field of the resulting stationary wave is normal to the metal plane and represents an enhanced intensity due to constructive interference. [1]

Since the objective of this investigation was to enhance the detection of HEMs, the substrates studied were divided into two classes (a) highly reflective substrates that in token are poor absorbers of the incident electromagnetic radiation (i-EMR) in the MIR region, and (b) non-reflective or matte substrates which are high absorbers of i-EMR. The first class of substrates used are highly reflective metallic substrates such as stainless-steel (SS), gold, highly polished aluminum, and other metals. The second class of substrates used are non-reflective or matte and include Teflon, anodized aluminum (AN-AL), and cardboard. Teflon substrates have an absorption band in the spectroscopic window used in the analyses, but it still allows us to follow other important bands of RDX in the spectral window of 1173 – 1428 cm^{-1}. This research was intended to expand on the results obtained in previous investigations by determining the limits of detection (LOD) for methodologies QCL-based spectrometers for RDX deposited on the mentioned substrates [9] and to test if the QCL based methodology can be used effectively in the field for detecting HEMs on these substrates at low concentration levels.

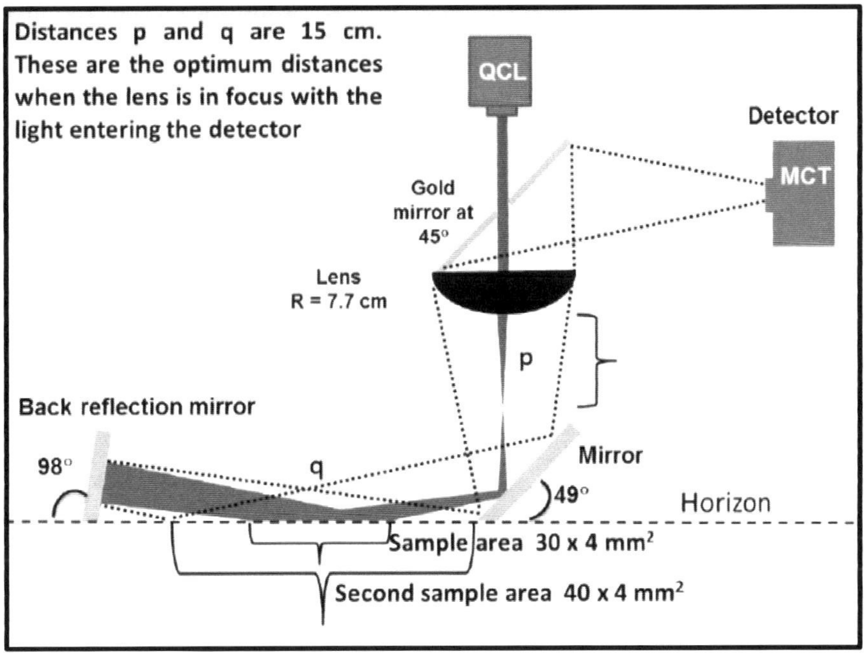

Figure 1. Optical layout for coupling a QCL spectrometer to the GAP.

Ideally, to achieve the purpose, the instrument would have to scan the substrates to detect whether it is contaminated with HEMs or not. If the surface is smooth and reflective, the reflection angle is the same as the incidence angle, and no signal is lost, provided the reflected beams are collimated and directed towards the detector. If the surface is rough and non-reflective, the laser reflection could be lost nearly completely depending on the surface if an integrating sphere is not used to collect all reflection angles. This effect is called diffuse reflectance as shown in Figure 2. The amount of signal loss depends on the deposition method, how homogenous the HEM is and the type of substrate under analysis. The secondary aim of this research is to see how much the LOD is affected by the effect above and if an alternative can be found to optimize the analysis in these conditions. This is particularly important when using AN-AL, Teflon, and cardboard substrates.

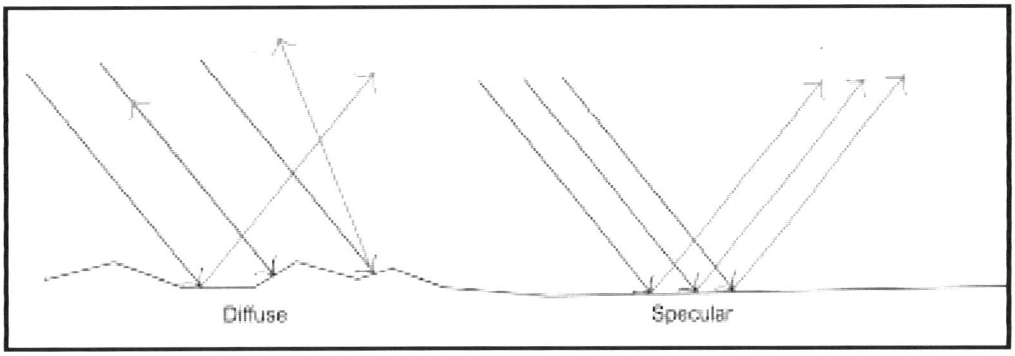

Figure 2. Diffuse vs. specular reflectance.

Deposition methods are a key part of this research. Finding the best way to reproduce real samples such as the case when a terrorist cleaning his hands contaminated with HEMs by wiping them with his clothes, contaminated hair, and fingerprints can prove to be a difficult task. [22] As deposition methods go the following methods have been previously used: spin coating deposition [1-5], smearing, and drop cast. [24] While spin coating was effective in generating a nearly homogenous film on the substrate, the way about the deposition is done a large part of the deposited solution could be lost thus creating a problem in quantifying the amount that was deposited. Sample smearing and drop cast are easy to quantify how much is deposited, but the films obtained are not homogenous throughout the substrate, both leading to errors in the analysis. To solve this situation, an alternative methodology consisting of air spray of solutions of HEM was implemented. The first step of the effort was to guarantee that the right volume of spray is deposited only on the entire substrate. [1]

This report intends to expand the knowledge in the field of preparation of RDX samples and standards based on thin films, for both basic and applied research by depositing traces of target analytes on substrates of high interest using any technique such as smearing, spray deposition and injection methodology to develop the sought HEM/substrates assemblies. The primary goal is to develop methodologies capable of producing solid samples deposited on substrates in controlled size and distribution modes deposited on selected surfaces to generate specimens that would reproduce real contamination samples. Finally, the applications of these well-characterized samples include use in experiments that require fine control of the distribution of loadings of analytes on surfaces.

2. Experimental

2.1. Reagents

Acetone (C_3H_6O, HPLC grade), methanol (CH_4O, HPLC grade), isopropyl alcohol (C_3H_8O, HPLC grade), acetonitrile (C_2H_3N, HPLC grade) were acquired from Sigma-Aldrich (Sigma-Aldrich Corp., Millipore-Sigma, St. Louis, MO, USA) RDX was synthetized according to the Bachmann method with certain modifications [25]. Reagent and methodologies used in the synthesis of the HEMs used are described in detail in previous reports. [1-3, 26-28]

2.2. RDX Samples Prepared by Spin Coating Deposition on Three Substrates Types

RDX was dissolved in acetone to make a saturated solution (stock solution). Dilutions of the stock solution were prepared by dividing in half of the original solution and diluting in acetone, making a solution with half the concentration of the initial one. This process was repeated until the solution showed no appreciable reflectance signals on the QCL-GAP setup. The substrate was placed on the spin coater, and 30 μL of the RDX test solution was transferred. Then assembly was spun at 3000 rpm for 30 s so that the substrate would have enough time to generate a uniform layer of RDX. [1-3] The substrates with RDX crystal films on them were analyzed with the tunable QCL-GAP setup. The process was repeated until the substrate showed no reflectance signal of explosive. The same procedure was performed on the following substrates: SS plates with 26.01 cm^2 areas, An-Al of 1.00 in^2, and Teflon of 1.00 in^2.

2.3. Samples Prepared by Immersion of Cardboard (bottom side) on RDX Solutions

Using the same RDX solutions described in the last section, cardboard substrates (19.74 cm^2) were placed in a 250 mL beaker and immersed in 30 μL of the solution to be transferred and left absorbing until dried. Spectra were acquired as described before.

2.4. Quantum Cascade Laser Setup Overview

This arrangement consists of using a gold-coated mirror to direct the IR-beam to hit the sample at 82° with respect to the surface normal. This enables acquiring spectra under RAIRS conditions. The main objective of the study was to compare the different types of surfaces, their reflectance, and the LOD value for the various types of HEMs. This application of QCL-GAP spectroscopy presents certain advantages, such as a lower LOD of RDX on certain substrate materials. This could be added as another benefit in the optics array and the time-consuming analysis. RDX was used as the target explosive for the experiment since it has been previously studied, its polymorphism is well known, and as it is easy to synthesize. [1-3, 5, 9]

3. Results and Discussion

3.1. RDX Deposited on SS substrates by spin coating

As mentioned before in the experimental section, RDX sample concentrations prepared by spin coating deposition on SS substrates are shown in Table 1. The table also includes the calculations of signal-to-noise relationship which is used to determine the LOD values. These quantities were obtained from spectra similar to the ones shown in Figure 3.

Table 1. Mass deposited at different concentrations on SS and their average

Concentration				Signal to Noise Relationship		
mol/L	**µg/cm²**	**ng/cm²**	**cm²/ng**	**Signal (S)**	**Noise (N)**	**S/N**
1.00E-01	25.62	25620	3.90E-05	1.113	0.006	175.64
5.00E-02	12.80	12810	7.81E-05	0.784	0.006	125.65
2.50E-02	6.405	6405	1.56E-04	1.012	0.009	116.97
1.25E-02	3.202	3202	3.12E-04	1.015	0.011	93.83
6.25E-03	1.601	1601	6.25E-04	0.448	0.006	77.51
3.13E-03	0.801	801	1.25E-03	0.161	0.007	21.57
1.56E-03	0.400	400	2.50E-03	0.155	0.010	15.72
7.81E-04	0.200	200	5.00E-03	0.094	0.006	14.86
3.91E-04	0.100	100	9.99E-03	0.077	0.005	14.97
1.95E-04	0.050	50.0	2.00E-02	0.049	0.005	9.65
9.77E-05	0.025	25.0	4.00E-02	0.047	0.008	5.95

With respect to Figure 3, many depositions of RDX solutions by spin coating stimulated the formation of polymorph β-RDX crystal deposits. RDX polymorphism was something expected since it was documented before by FTIR analysis [1–3]. The spectra represented with red and black dashes lines have been included for purposes of comparison. They correspond to reference spectra of β- and α-RDX, respectively. Spectral signatures from the tunable QCL-GAP setup in reflectance units (R) were obtained in ThermoGalactic Grams/AI "spc" file format. For quantitative analysis, reflectance (R) were changed to log(R) which is proportional to absorbance and can be used for Chemometrics analyses [29]. The rectangular blue frames illustrate the range used to calculate the signal and noise values by integration of peak areas. The β- and α-RDX were acquired using bulk quantities using a 532 nm diode laser as the excitation source (green line) by Raman Microspectroscopy (InVia™, Renishaw PLC, West Dundee, IL, USA). The spectra were adapted to the scales of Figure 3 to illustrate polymorphism of the RDX deposits on SS by this deposition technique. This confirms that RDX deposited on SS at low surface concentration exhibits β-RDX spectral signatures with red-shift of peaks at ~1226 cm^{-1}, ~1277 cm^{-1}, and ~1328 cm^{-1}. The observation of polymorphism confirms previously published findings within the research group [1-2, 22-24, 26-27].

IOP Conf. Series: Materials Science and Engineering **519** (2019) 012007 doi:10.1088/1757-899X/519/1/012007

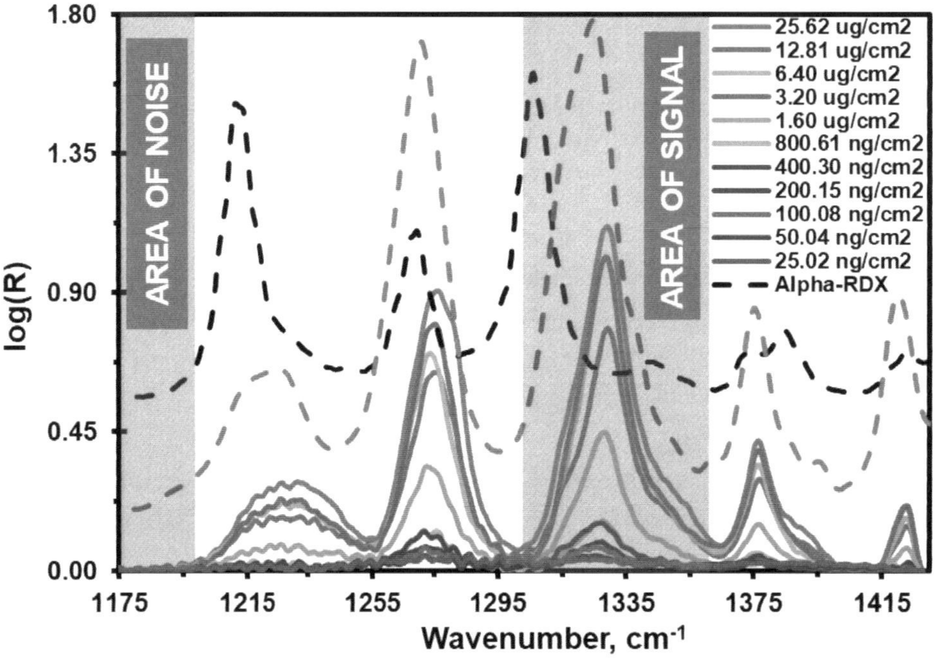

Figure 3. QCL reflectance spectra of RDX on SS at various surface concentrations

Figures 3 and 4 provide the data needed to establish that the LOD value for RDX/SS was 1.64 ng/cm^2. This was obtained by curve fitting the ratio of S/N dependence on the reciprocal of the surface concentration. The lowest quantity of RDX that can be distinguished from the absence of himself comes when the signal raises 3 times the noise. Assuming (S/N) = 3 then the surface concentration (C_S) becomes LOD which it results to be 1.64 ng/cm^2.

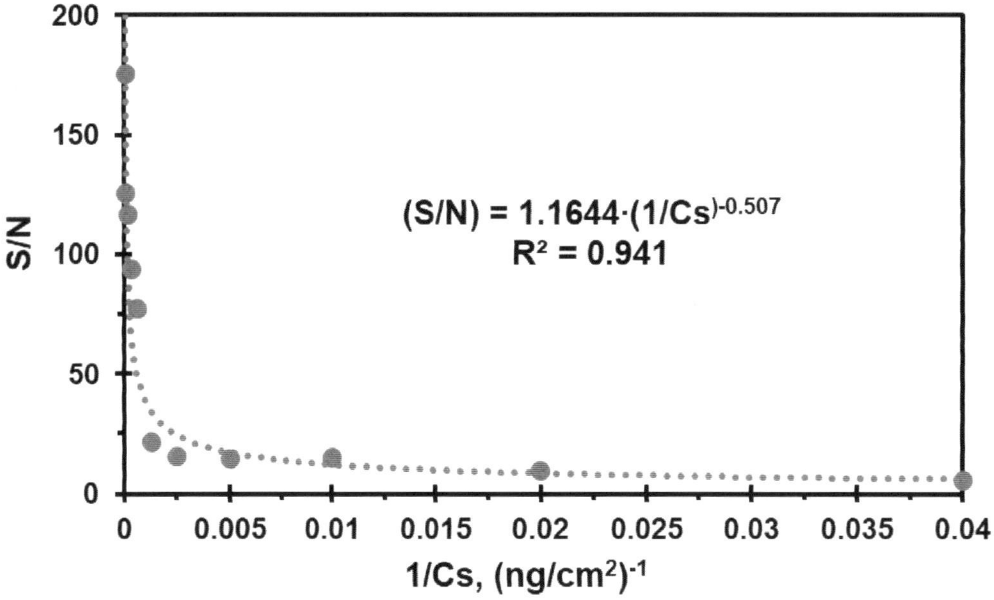

Figure 4. S/N as function of the reciprocal of the surface concentration for RDX/SS samples.

3.2. RDX Deposited on AN-AL Substrates by Spin Coating

Similar experiments were done using the same RDX solutions prepared before but changing the SS substrates to AN-AL substrates of areas 1.00 in^2. Good results were obtained using spin coating technology and acquiring reflectance spectra with QCL–GAP setup. The results are summarized in Figure 5. This spectral window enclosed both the characteristic strong signatures of RDX. Comparing Figure 3 with Figure 5, the reflectance signal was plotted using the unprocessed raw data (no Log(R)) because in this way it was possible to interpret the results directly as they were acquired. LOD values were obtained using the methodology employed before. A blue rectangular frame delimits the area used to find the S/N values. An interesting observation was readily available from Figure 5 by comparing the QCL-GAP reflectance spectra of AN-AL with the FTIR reference spectra of the polymorphic phases of the RDX (in red and black dotted lines). All the acquired spectra at the various surface concentrations studied indicate that the α-RDX polymorph is favored. The important fact is based on the similarity with the peaks at 1330 cm^{-1} and the doublet that forms in the range 1360 – 1395 cm^{-1}. However, our most recent findings published (Figueroa et al., 2016 [2]) confirm that there is a coexistence of both polymorphs on the substrate studied in that project (SS). A key aspect of the spectra taken with the GA arrangement is that these polymorphs which have collinearity with the electric field vector of a polarized source (thermal or laser) will be favored. This feature was confirmed by recent work (Ruiz et al., 2017 [1]) where the optical properties of the polymorph β-RDX follow the selection rules according to the orientation of the electric field vector from the source of excitation. These orientations of the can be directed perpendicular or parallel to the normal surface vector and can be achieved using a field-polarizer (for a non-polarized, thermal source).

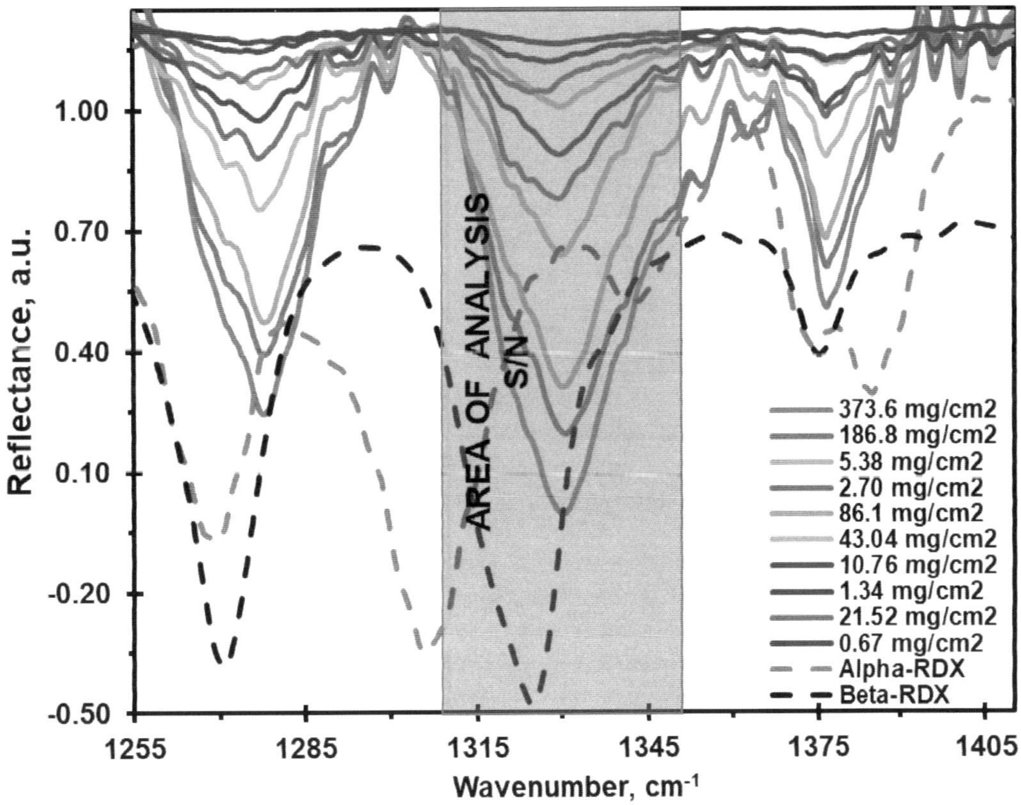

Figure 5. Reflectance spectra of RDX/AN-AL samples at various surface concentrations.

Figures 5 and 6 provide the data needed to establish that the LOD for RDX/AL-AN samples: is 94.7 μg/cm^2. Comparing limit of detection between stainless-steel (SS) and anodized-aluminum (An-Al) substrates, the AN-AL is 215 times larger. In fact, β-RDX is favored by stainless-steel and α-RDX is favored by anodized-aluminum. This could be explained regarding porosity of the anodized aluminum. The cavities of the AN-AL substrates could generate a greater amount of solid deposited (~ bulk concentrations) and in those conditions the α-RDX is the favored conformation. This is an important finding because to be prepare standards against terrorism attack; all databases should include those RDX polymorphs.

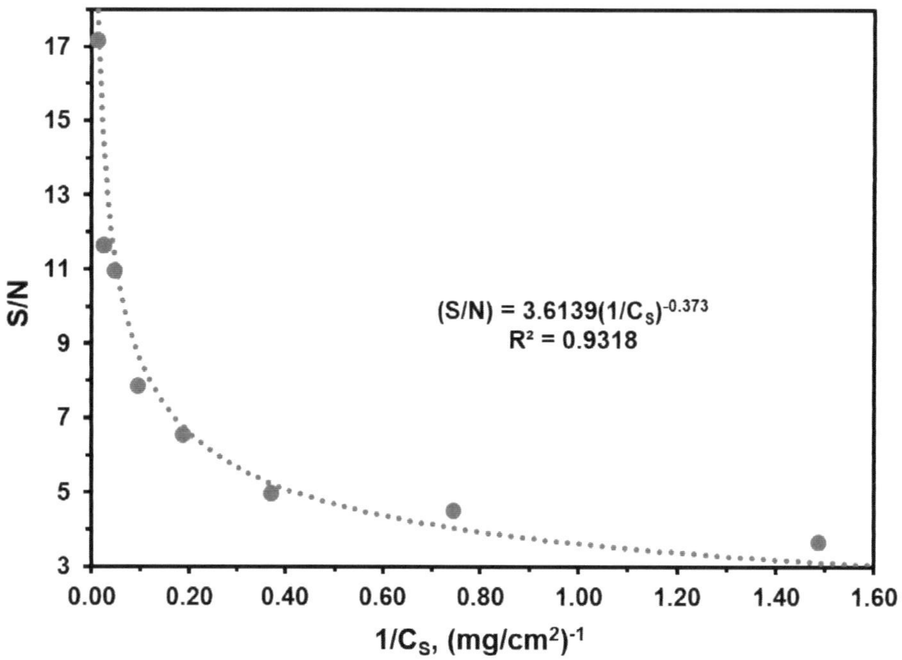

Figure 6. Signal/Noise vs. of reciprocal of RDX surface concentration on AN-AL.

3.3. RDX deposited on Teflon Substrates by Spin Coating
Similar to the previous procedures, experiments were done on Teflon substrates, In this case, RDX deposited on the Teflon substrates using spin coating technology represent largest source of interference to the RDX signal, when compared the other substrates (see Figures 3, 5, and 7). For this reason, the LOD values were not calculated. In agreement with the objective of generating a database that registers substrates with interferences, this information is very important.

3.4. RDX Deposited on Cardboard Substrates by Immersion
From the procedure described in the Materials and Methods section, is important to highlight that the RDX/cardboard samples were prepared by immersing the substrates into a beaker-flask with a fixed volume of a saturated RDX solution. This action was the best way to prepare RDX samples on cardboard. MIR laser reflectance spectra shown in Figure 8 were acquired by the QCL-GAP setup. Spin coating technology was inappropriate because the cardboard absorbed on both layers (front and back) and retained the solvent longer. Also, when using the drop-casting method, a similar situation occurred. However, the cardboard immersed in a beaker allowed controlling that the front layer retained the RDX and assured that the lower layer did not lodge RDX particulates. The problem faced was that it was not possible to quantify the amount accurately absorbed. Only estimated values using the amount transferred

to the beaker-flask by gravimetric analysis were available. Table 2 shows relevant information for the experiments. The injection volume for each sample was of 30 μL.

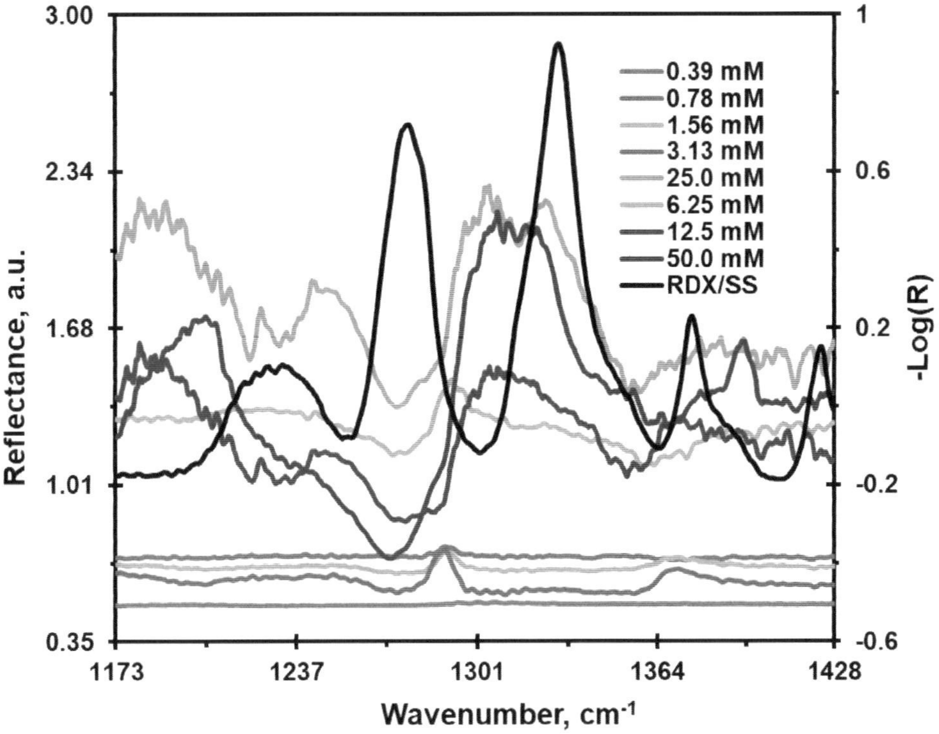

Figure 7. QCL-GAP reflectance spectra for of RDX/Teflon at various surface concentrations.

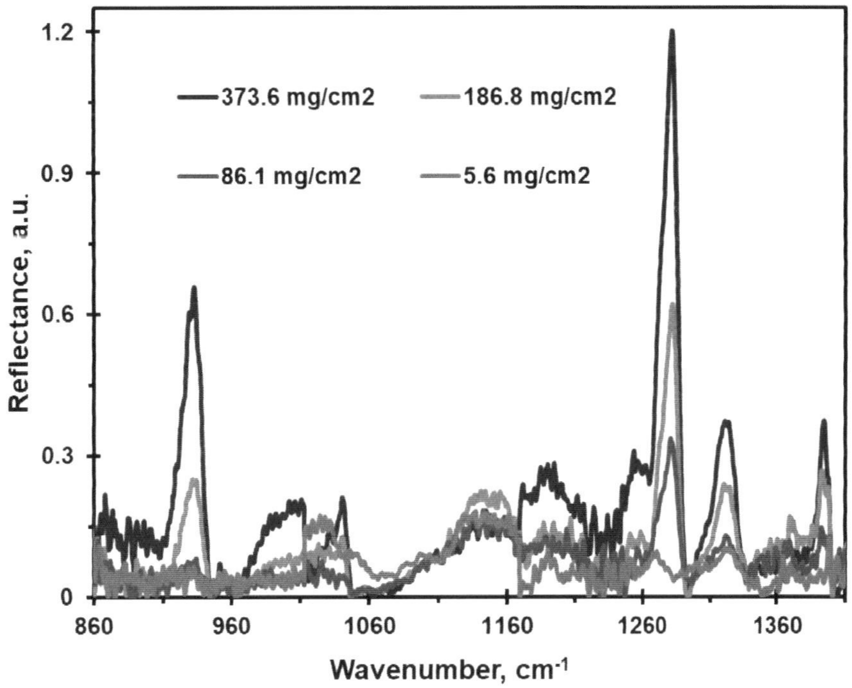

Figure 8. QCL-GAP reflectance spectra of RDX/cardboard at various surface concentrations.

Table 2. Masses of RDX deposited on cardboard substrates

M (mol/L)	mg/cm^2	Area	1/Cs	Signal (S)	Noise (N)	S/N
2.17E-01	3.74E+02	1.97E+01	2.68E-03	1.20E+00	9.85E-03	1.22E+02
1.09E-01	1.87E+02	1.00E+01	5.35E-03	6.21E-01	8.37E-03	7.41E+01
5.00E-02	8.61E+01	5.66E+00	1.16E-02	3.36E-01	1.40E-02	2.41E+01
2.50E-02	5.55E+00	4.52E-01	1.80E-01	7.01E-02	6.52E-03	1.08E+01

LOD values could not be determined because the amount of RDX absorbed on cardboard as a substrate could not be quantified. Similar to Teflon substrates, interferences are present in those signature signals. To solve the problem of quantifying the amount and the homogeneity of RDX deposited over cardboard, we suggest using other methodology named direct-Spray. This improves deposition in a better way. [30-31]

4. Conclusion

The deposition of RDX solutions by spin coating technology predominantly stimulates the formation of β-RDX polymorph crystal deposits. However, results depend on second order on the substrate properties. For example, β-RDX is favored in depositions on SS to 100%; α-RDX is favored in depositions on AN-AL substrates. Roughness and micro cavities of AN-AL seem to generate a greater amount of solid deposits (quasi-bulk concentration), and under these conditions, the formation of α-RDX is favored as expected. LOD values for stainless steel resulted in 1.6 ng/cm^2. LOD for AN-AL substrates resulted in 94.7 µg/cm^2. LOD for the immersion of cardboard in the saturated solution of RDX could not be determined accurately. As future work, we suggest using the spray deposition methodology. This methodology favors solvent evaporation very quickly and the deposition of the analyte on the surface in a controlled manner.

Acknowledgments

This material is based upon work supported by the U.S. Department of Homeland Security, Science and Technology Directorate, Office of University Programs, under Grant Award 2013-ST-061-ED0001. The views and conclusions contained in this document are those of the authors and should not be interpreted as necessarily representing the official policies, either expressed or implied, of the U.S. Department of Homeland Security.

Conflict of Interest.

The authors report there are no conflicts of interest.

References

[1] Ruiz-Caballero J L, Aparicio-Bolaño J A, Figueroa-Navedo A M, Pacheco-Londoño L C, Hernandez-Rivera S P. Optical properties of β-RDX thin films deposited on gold and stainless steel substrates calculated from reflection-absorption infrared spectra 2017 *Appl. Spectrosc.* **71**(8) 1990-2000.

[2] Figueroa-Navedo A M, Ruiz-Caballero J L, Pacheco-Londoño L C, Hernandez-Rivera S P. Characterization of α- and β-RDX polymorphs in crystalline deposits on stainless steel substrates 2016 Cryst. Growth. Des. **16**(7) 3631-3638.

[3] Ruiz-Caballero J L. 2017 Preparation and study of thin RDX films standards on various substrates deposited by spin coating technology *Ph.D. thesis*, The University of Puerto Rico in Mayagüez

Campus, Mayagüez, PR, USA.

[4] Panigrahi S, Waugh S, Rout S K, Hassan A K, Ray A K 2004 Study of spin coated organic thin film under spectrophotometer *Ind. J. Phys.* **78**(8) 823-826.

[5] Primera-Pedrozo O M, Soto-Feliciano Y, Pacheco-Londoño L C, Hernández-Rivera S P 2009 Detection of high explosives using reflection absorption infrared spectroscopy with fiber coupled grazing angle probe/FTIR *Sensing and imaging. An International Journal* **10**(1) 1-13.

[6] Yonkoski R K, Soane D S 1998 Model for spin coating in microelectronic applications *J. Appl. Phys.* **72**(2) 725-740

[7] Liu C H, Yu X. 2011 Silver nanowire-based transparent, flexible, and conductive thin film *Nanoscale Res. Lett.* **6** 75-82

[8] Oxley J C, Smith J L, Kirschenbaum L J, Marimganti S, Vadlamannati S 2008 Detection of explosives in hair using ion mobility spectrometry *J. Forensic. Sci.* **53**(3) 690-693

[9] Castro-Suarez J R, Hidalgo-Santiago M, Hernandez-Rivera S P. 2015 Detection of highly energetic materials on non-reflective substrates using quantum cascade laser spectroscopy *App. Spectrosc.* **69**(5) 1023–1035

[10] Sisco E, Forbes T P 2013 Rapid detection of sugar alcohol precursors and corresponding nitrate ester explosives using direct analysis in real time mass spectrometry *Analyst.* **140** 2785-2796.

[11] Ewinga A V, Kazarian S G 2017 Infrared spectroscopy and spectroscopic imaging in forensic science *Analyst.* **142**(2) 257-272

[12] Pacheco-Londoño L C, Ortiz-Rivera W, Primera-Pedrozo O M, Hernandez-Rivera S P 2009 Vibrational spectroscopy standoff detection of explosives *Anal. Bioanal. Chem.* **395**(2) 323-335

[13] Van Neste C W, Senesac L R, Thundat T 2009 Standoff spectroscopy of surface adsorbed chemicals *Anal. Chem.* **81**(5) 1952-1956

[14] Toal S J, Trogler W C 2006 Polymer sensors for nitroaromatic explosives detection *J. Mater. Chem.* 16 2871-2883

[15] Chen N, Ding P, Shi Y, Jin T, Su Y, Wang H, He Y 2017 Portable and Reliable Surface-Enhanced Raman Scattering Silicon Chip for Signal-On Detection of Trace Trinitrotoluene Explosive in Real Systems *Anal. Chem.* **89**(9) 5072–5078

[16] Moore D S 2004 Instrumentation for trace detection of high explosives *Rev. Sci. Instrum.* **75**(8) 2499-2512

[17] Tripathi A, Emmons E D, Wilcox P G, Guicheteau J A, Emge D K, Christesen S D, Fountain III A W 2011 Semi-automated detection of trace explosives in fingerprints on strongly interfering surfaces with Raman chemical imaging *Appl. Spectrosc.* **65**(6) 611-619

[18] Faist J, Capasso F, Sivco D L, Sirtori C, Hutchinson AL, Cho A Y 1994 Quantum Cascade Laser *Science* **264**(5158) 553-556

[19] Hvozdara L, Pennington N, Kraft M, Karlowatz M, Mizaikoff B 2002 Quantum cascade lasers for mid-infrared spectroscopy *Vib. Spectrosc.* **30**(1) 53-58

[20] Breshike C J, Kendziora C A, Furstenberg R, Nguyen V, McGill R A 2017 Stabilizing infrared quantum cascade laser beams for standoff detection applications *Proc. SPIE* **10111** 101110B-1

[21] Breshike C J, Kendziora C A, Furstenberg R, Nguyen V, McGill R A 2017 Methodology for using active infrared spectroscopy in standoff detection of trace explosives *Proc. SPIE* **10183** 1018302-1

[22] Pacheco-Londoño L C, Aparicio-Bolaño J A, Galán-Freyle N J, Román-Ospino A D, Ruiz-Caballero J L, Hernández-Rivera S P 2018 Classical least squares-assisted mid-infrared (MIR) laser spectroscopy detection of high explosives on fabrics *App. Spectrosc.* doi: 10.1177/0003702818780414

[23] Castro-Suarez J R, Pacheco-Londoño L C, Vélez-Reyes M, Diem M, Tague T J, Hernandez-Rivera S P 2013 FT-IR standoff detection of thermally excited emissions of trinitrotoluene (TNT) deposited on aluminum substrates *Appl. Spectrosc.* **67**(2) 181-186.

[24] Wrable-Rose M, Primera-Pedrozo O M, Pacheco-Londoño L C, Hernandez-Rivera S P 2010 Preparation of TNT, RDX and Ammonium Nitrate Standards on Gold-on-Silicon Surfaces by Thermal Inkjet Technology *Sens. Imaging.* **11**(4) 147-169

[25] Bachmann W E, Sheehan J C 1949 A new method of preparing the high explosive RDX *J. Am. Chem. Soc.* **71**(5) 1842–1845

[26] Infante-Castillo R, Pacheco-Londoño L C, Hernandez-Rivera S P 2010 Vibrational spectra and structure of RDX and its ^{13}C- and ^{15}N-labeled derivatives: A theoretical and experimental study *Spectrochim. Acta, Part A.* **76**(2) 137-141

[27] Infante-Castillo R, Pacheco-Londoño L C, Hernandez-Rivera S P 2010 Monitoring the $\alpha \rightarrow \beta$ solid-solid phase transition of RDX with Raman spectroscopy: A theoretical and experimental study *J. Mol. Struct.* **970**(1) 51-58.

[28] Karpowicz R J, Sergio S T, Brill T B 1983 β-polymorph of hexahydro-1,3,5-trinitro-triazine. a Fourier transform infrared spectroscopy study of an energetic material *Ind. Eng. Chem. Prod. Res. Dev.* **22**(2) 363-365

[29] Shrivastava A, Gupta V B 2011 Methods for the determination of limit of detection and limit of quantitation of the analytical methods *Chron. Young Sci.* **2** 21-25

[30] Barreto-Cabán M A, Pacheco-Londoño L C, Ramirez M L, Hernández-Rivera S P 2006 Novel method for the preparation of explosive nanoparticles *Proc. SPIE.* **6201**(9) 620129-62013

[31] Lu Y C, Chou K S 2010 Tailoring of silver wires and their performance as transparent conductive coatings *Nanotechnology* **21**(21) 215707-215713

IOP Conf. Series: Materials Science and Engineering **519** (2019) 012008 doi:10.1088/1757-899X/519/1/012008

Computational analysis of the thermal behavior on a silica (SiO_2) aerogel coating for applications in the construction industry

A Ángel [1,2], G Suárez [1] and J Niño [2]

[1] Mathematics Research Group, Center of Basic Science
[2] Aerospace Engineering Research Group, Faculty of Aeronautical Engineering
[1,2] Universidad Pontificia Bolivariana, Medellín, Colombia

gustavo.suarez@upb.edu.co

Abstract. A mathematical model was developed to describe the thermal behavior of the aerogel SiO_2 material, which can be used as an architectural surface coating film. These types of materials allow temperature conservation within a panel so that a controlled environment is achieved. Conditions for modeling were considered, such as: porous material properties, microstructure, boundary conditions, etc. From the study of the material's micrometry, an image processing was made, and regular grids of 6192 nodes were built. Algorithms were implemented using Matlab to describe the thermal flow within the porous material. The results obtained indicate the thermal distribution across the pores of the material and showed the zones with greater incidence and thermal conservation in the panel. This research contributes towards the thermal analysis of special insulating materials as alternatives of applications in the construction industry.

1. Introduction

In the construction sector, various types of materials are being implemented to meet specific design and thermal behavior conditions, specifically, there are few materials that gather all the conditions to perform notoriously as a thermal insulator [1]. The current needs for buildings' construction demand a migration from traditional thermal insulators, with their relatively high thermal conductivities being one of the reasons for this change as these thermal conductivities require the insulating films to be thicker in colder seasons [1]. As thermal insulation correlates with energy use of buildings [2], seeking of new insulating materials were thoroughly carried out [3–5], not only to study their viability but also to model and characterize materials according to their thermal performance, more specifically, grasping of silica aerogel heat transfer behavior is an issue of scientific-industrial interest [6,7].

Numerical and experimental results have been presented, showing that aerogel coatings applied to buildings' walls perform better than other insulating materials [6]. Some researches were also developed with regards to the use of aerogels in window glazing taking advantage of both thermal conductivity and transparency properties [8–10]. All of these to achieve the same objective, reducing energy consumption of buildings due to indoor environmental control. It is also worth to be mentioned that not only thermal insulation of buildings has been considered but also aerospace applications including cryogenic fluid containers, thermal barriers, fire retardation, amongst other thermal applications are also a topic of study [11,12].

Content from this work may be used under the terms of the Creative Commons Attribution 3.0 licence. Any further distribution of this work must maintain attribution to the author(s) and the title of the work, journal citation and DOI.
Published under licence by IOP Publishing Ltd

In the present paper a mathematical model is developed to describe the thermal behavior of a silica aerogel wall coating subjected to a certain ambient temperature. Variables such as thermal conductivity and the microstructure of the material were considered on the algorithmic implementation of the model which was built in Matlab to analyze different conditions of the thermal flow within the porous material. The objective of this research is to visualize the heat flow behavior on the inside of the microstructure (SiO_2 aerogel) to determine the efficiency of the material that can be used in construction applications.

2. Mathematical modeling and computational implementation

2.1. Mathematical description

The heat transfer equation in steady-state regime was established to model the thermo-conductive behavior inside of the porous microstructure of the material. The mathematical representation defined for the model was:

$$k_c \nabla^2 T = -\rho h \in \Omega^2 \tag{1}$$

Where T corresponds to the temperature of the material on a specific point, k_c is the thermal conductivity of the aerogel SiO_2 which takes values between 13–14 mW/(mK) and ρh is the rate of heat generation or accumulation of internal energy within the material.

And the heat flow conditions or temperature that is applied over the external surface can be described through:

$$k_a \frac{\partial T}{\partial x} = h_c (T - T_\infty) \in \Omega^2 \tag{2}$$

$$T = \beta \in \partial \Omega \tag{3}$$

Where $\partial T / \partial x$ is the heat flow due to convection through the boundaries, h_c is the convective coefficient of the fluid to which the surface is exposed, k_a is the thermal conductivity of air in W/mK, T is the temperature along the boundary and T_∞ is the ambient temperature.

2.2. Microstructure of the material and implemented boundary conditions

A sample of the material was determined by capturing a micrograph image [12]. An image processing was made on the sample to improve sharpness, shapes and microstructural definitions. Following, it is described by the flow chart the treatment made on the sample, figure 1.

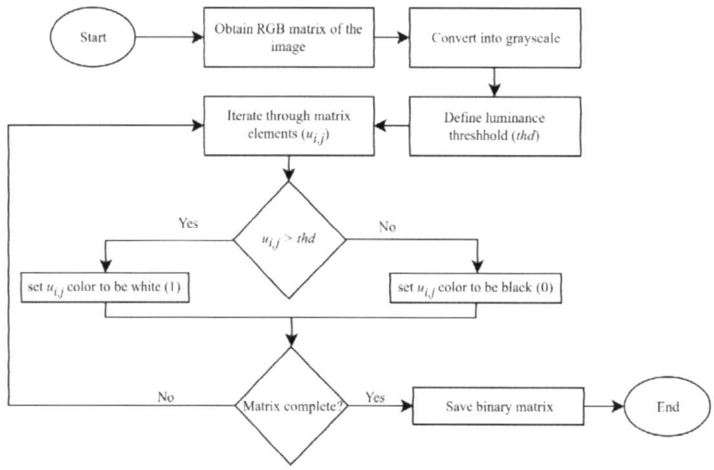

Figure 1. Scheme for the microstructure sample treatment.

A fraction of the microstructure sample was selected according to the degree of porosity and monolithic material content to subsequently numerically fragment the sample through a uniform lattice by squared elements, see figure 2.

(a) **(b)**

Figure 2. SEM micrograph of the porous medium sample [12], where the dashed lines indicate the selected fraction (a), and uniform 6192 node lattice for the numerical development (b).

2.3. *Numerical-computational description*

A computational algorithm was built to develop the differential equations system using Taylor series approximations:

$$T_{i+1,j} + T_{i-1,j} + T_{i,j+1} + T_{i,j-1} = 4T_{i,j} + h_x^2 T''_{i,j}\big|_x + h_y^2 T''_{i,j}\big|_y \tag{4}$$

That after truncation lead to an F.D.M model [13], which was applied to the mathematical model as follows:

$$-2\left(\frac{1}{h_x^2} + \frac{1}{h_x^2}\right)T_{i,j} + \frac{T_{i+1,j}}{h_x^2} + \frac{T_{i-1,j}}{h_x^2} + \frac{T_{i,j+1}}{h_y^2} + \frac{T_{i,j-1}}{h_y^2} = -\frac{\rho h}{k} \tag{5}$$

Where h_x and h_y are the determined partitions to numerically fragment the microstructure.

In the computational implementation, the porous surfaces were defined with lattice nodes to represent energy absorbent voids, and for the monolithic material, lattice nodes were defined to represent thermal energy conductive surfaces.

3. Results and discussion

In figures 3a, 3b and 3c; temperature distribution can be seen through the porous microstructure for three thermal circumstances. The implementation of the material was considered for structures located in cold, tropical and hot regions. At considerably low ambient temperatures the porous material had a thermal conservation of -73.15 °C, at tropical temperatures the material had a thermal conservation-dissipation of 7 °C and at high ambient temperatures the material had a thermal dissipation of 27 °C.

Temperature conservation can be seen in the inside as the thermal flow spreads between the micropores, avoiding energy losses due to constraints in the microsystem. In high-temperature ambient regions, the porous material allows maintaining a conservative state of steady temperature as the pore generates a thermal balance that keeps the system refrigerated.

In general, the simulations performed on the aerogel SiO_2 coating show the ability of the material in retaining heat and therefore the conservation of a certain temperature, generating a comfort ambient inside the architecture in which it was implemented. Also, in the three cases developed by simulations it can be seen that the pores act as a kind of absorbent medium, that is, as the heat flow approaches the pore, the temperature gradient decreases drastically. It is essential to develop the analysis in different locations of the micrograph sample due to high heterogeneity and amorphism of the porous microstructure.

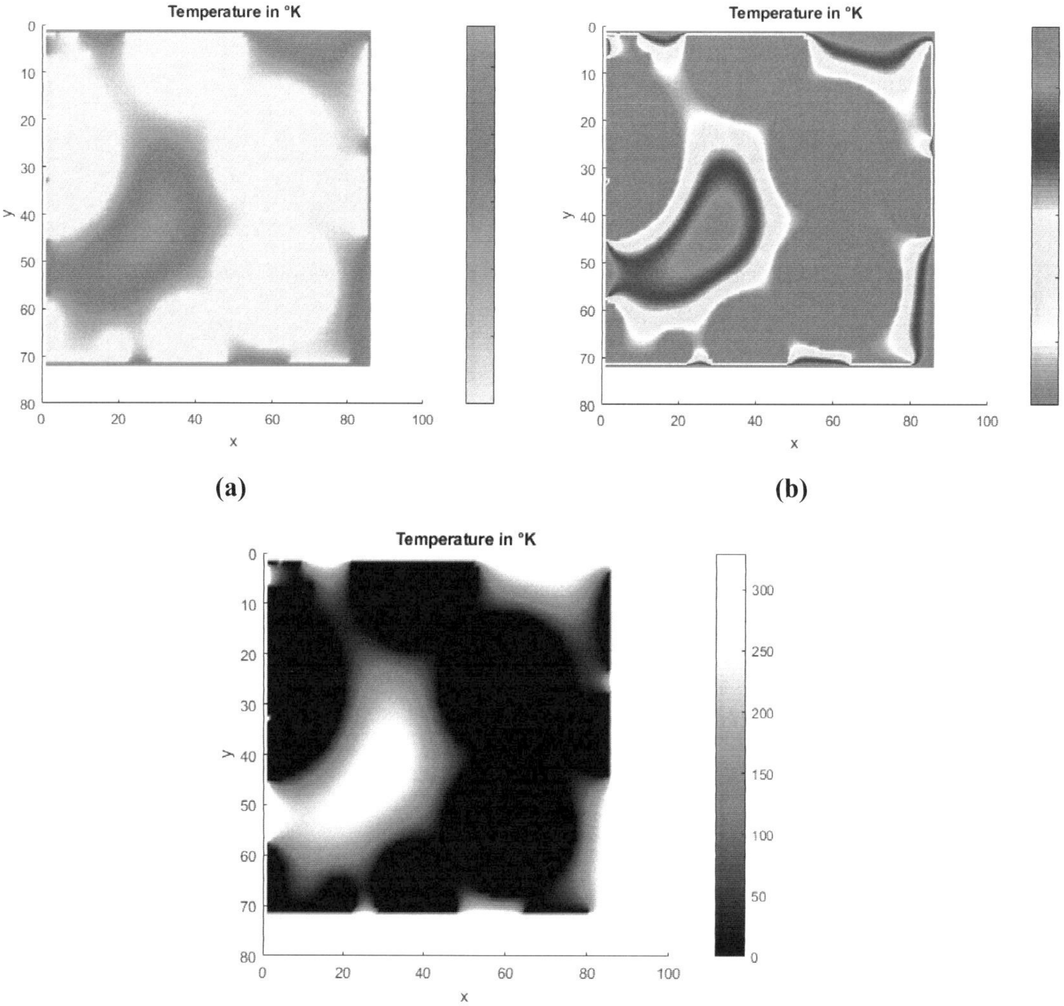

(c)

Figure 3. Numerical solutions at an ambient temperature of -74 °C (a), 25 °C (b) and 56 °C (c).

4. Conclusions

The drastic environmental changes that are happening globally have demanded architectural designs to be built with special materials that guarantee adequate control of temperature in enclosed spaces. Silica aerogel then becomes an appropriate alternative to achieve stable temperature environments. Achieving that the same material meets different performance features is quite complex, nevertheless, the type of material here implemented could become an alternative to commercial use. A versatile computational algorithm was built, which allows evaluating different environmental thermal flows, it is even possible to vary the properties of the material. Porous materials have the benefit of being lighter than conventional monolithic materials. This research offers clarifications about the benefits that the silica aerogel material can provide with the possibility of being implemented in other applications such as aeronautics/aerospace and military industry.

IOP Conf. Series: Materials Science and Engineering **519** (2019) 012008 doi:10.1088/1757-899X/519/1/012008

References

[1] Jelle B P 2011 Traditional, state-of-the-art and future thermal building insulation materials and solutions – Properties, requirements and possibilities *Energy Build.* **43** 2549–63

[2] Ebert H-P 2011 Thermal Properties of Aerogels *Aerogels Handbook* (New York, NY: Springer New York) pp 537–64

[3] Baetens R, Jelle B P and Gustavsen A 2011 Aerogel insulation for building applications: A state-of-the-art review *Energy Build.* **43** 761–9

[4] Hasan M A, Sangashetty R, Esther A C M, Patil S B, Sherikar B N and Dey A 2017 Prospect of Thermal Insulation by Silica Aerogel: A Brief Review *J. Inst. Eng. Ser. D* **98** 297–304

[5] Schiavoni S, D'Alessandro F, Bianchi F and Asdrubali F 2016 Insulation materials for the building sector: A review and comparative analysis *Renew. Sustain. Energy Rev.* **62** 988–1011

[6] Ibrahim M, Biwole P H, Wurtz E and Achard P 2014 A study on the thermal performance of exterior walls covered with a recently patented silica-aerogel-based insulating coating *Build. Environ.* **81** 112–22

[7] Xie T and He Y-L 2016 Heat transfer characteristics of silica aerogel composite materials: Structure reconstruction and numerical modeling *Int. J. Heat Mass Transf.* **95** 621–35

[8] Gao T, Jelle B P, Ihara T and Gustavsen A 2014 Insulating glazing units with silica aerogel granules: The impact of particle size *Appl. Energy* **128** 27–34

[9] Jensen K I, Schultz J M and Kristiansen F H 2004 Development of windows based on highly insulating aerogel glazings *J. Non. Cryst. Solids* **350** 351–7

[10] Buratti C and Moretti E 2012 Glazing systems with silica aerogel for energy savings in buildings *Appl. Energy* **98** 396–403

[11] Bheekhun N, Abu Talib A R and Hassan M R 2013 Aerogels in Aerospace: An Overview *Adv. Mater. Sci. Eng.* **2013** 1–18

[12] Perichaud A A, Iskakov R M, Kurbatov A, Z. T, Prokohdko O Y, V. I, L. S, Apel P Y, Yu. V and Abadie M J M 2012 Auto-Reparation of Polyimide Film Coatings for Aerospace Applications Challenges & Perspectives *High Performance Polymers - Polyimides Based - From Chemistry to Applications* (InTech)

[13] Moitsheki R J and Rowjee A 2011 Steady Heat Transfer through a Two-Dimensional Rectangular Straight Fin *Math. Probl. Eng.* **2011** 1–13

Modeling of dynamic mass coupled system with Runge-Kutta fourth order

L Rojas[1], M Molina-Cardenas[2], E Caballero[2], C Rojas[3] and A Patiño-Vanegas[4]

[1]Fundación Tecnológica Antonio de Arévalo, Cartagena, Colombia
[2]Universidad Libre, Barranquilla, Colombia
[3]Universidad de Sucre, Sincelejo, Colombia
[4]Universidad Tecnológica de Bolívar, Cartagena, Colombia

luis.rojas@tecnar.edu.co

Abstract. This paper shows the mathematical modeling process of a mechanical system of masses coupled by two springs and a shock absorber. The process of capture of movement of the mass system coupled with springs was done with software Tracker video analysis and modeling tool. The motion capture of the coupled masses A and B was made with a duration of 20 seconds. A comparison of the results of the movement of the two masses will be established as a first instance, by adjusting trajectories of curves in time using Matlab software and as a second instance a numerical solution of the Mass-Spring model will be established using the Runge Kutta method of 4 order. With this last method, it is expected to achieve a better precision modeling of the initial value problem. Finally, a comparison of the two mathematical models will be made analyzing.

1. Introduction

Studies in different fields of engineering have shown that mathematical modeling is of great help for the development and control of mechanical and industrial processes that have the basis of production the analysis of movement of objects by its geometric configuration in terms of weight, shape, location space and climatic conditions. Some authors as Qi, C. et al., doing the mathematical modeling to the modeling and identification of nonlinear distributed parameters dynamics of the micro-cantilever [1], Chen, Z. et al., use the mathematical modeling to perform the mathematical model and simulation for motion of underwater vehicle [2], Wang, D. et al., use the modeling in ordinary differential equation (ODE) to modeling, analyzing and controlling long distribution feeder [3], Rath, G. et al., make a direct numerical solution of stiff ODE system in optimal control [4], Lobao, W. J. A. et al., perform a genetic programming and automatic differentiation algorithms applied to the solution of ordinary and partial differential equation [5], Korolova, O. et al., show a performance of Runge-Kutta family numerical solvers for calculation of transient processes in AC machines [6], Krisko, V. et al., use mathematical modeling to nonlinear dynamics of contact interaction of MEMS beam elements accounting the Euler-Bernoulli hypothesis in a temperature field [7], Kern, R. and Gehring N. show a tracking control for a long pneumatic transmission line [8].

The Numerical analyzes and simulations through computer software such as Matlab [9 - 12] are important in the mathematical process modeling since through their calculation tools they allow to optimize time in execution of iterative cycles for solving differential equations problems and in this case in particularly help to solve the initial value problem [13 - 15].

Content from this work may be used under the terms of the Creative Commons Attribution 3.0 licence. Any further distribution of this work must maintain attribution to the author(s) and the title of the work, journal citation and DOI.
Published under licence by IOP Publishing Ltd

IOP Conf. Series: Materials Science and Engineering **519** (2019) 012009 doi:10.1088/1757-899X/519/1/012009

There are many methods to analyze mathematical models and adjust trajectories over time, taking into account the analysis of the initial value problem [13 - 15]. Some of these methods for curve fitting or trajectory adjustment are:

Methods focused on linear regressions. Some authors as Kavitha, S. et al., [16] make a comparative analysis on linear regression to use the correct model for better prediction and accuracy in business consumer interest.

Methods focused on Polynomials and multiple linear regressions to analyze higher-order polynomials. Hirose, H. et al., in [17] use the multiple linear regression to predict the continuous value of target variable using the values of explanation variables in prediction accuracy of the prices for auctions of used cars is drastically improved. Xuang, F. et al., [18] implement a contact temperature prediction of high voltage switchgear based on multiple linear regression model to analyze and process a monitoring point data.

Methods based on Fourier analysis involving fitting periodic functions to data through Fast Fourier transform (FFT). Kong, W. et al., [19] propose a novel grouping scheme of the basic functions within the framework of the fast Fourier transform (FFT) to create a block-sparse structure for the near-matrix on account of the efficient analysis of multiscale problems. Su, T. et al., [20] applied this method to present a power harmonic detection method in renewable power based in Nuttall double-window all-phase FFT algorithm as a basis for the study of harmonics.

Interpolation, estimating intermediate values between precise data points with newton's interpolating polynomial method. Laih, Chi-Sum. Et al., [21] use this method to the design of a single-key-lock mechanism to reduce the average computation time. On the other hand with LaGrange interpolating polynomials method. Van Beeumen, R. et al., [22] write about linearization of Lagrange and Hermite interpolating matrix polynomials as a tool to study the classical approach to investigating the polynomial eigenvalue problem.

Spline interpolation. In this method, Liang, Y. et al., [23] present a class of algebraic trigonometric interpolation splines and applications as an efficient new model for geometric design.

Curve fitting with sinusoidal functions. Pattanadech, N. et al., [24] in this method present a fast curve algorithm for parameter evaluation in lighting impulse test technique for the evaluation of the base curve of lightning impulse voltage and current.

In addition, methods with ordinary differential equations treatment ODEs like Euler's method and Runge-Kutta fourth order (RK4). Wang, X., [25] use Euler's method to find the strong convergence rates of the linear implicit Euler. Simons, T., [26] use RK4 method for the numerical integration of the radial Schrödinger equation. Haut, T. et al., [27] use the method to solve wave propagation problems via the direct construction of an approximate time-evolution operator

In this work, the procedure performed for the validation of a real coupled spring mass model with ODEs is presented. It is expected to show if our model describes the real behavior of the system. For this case it is highlighted that the constants that appear in our solution are related to the initial conditions and the parameters that characterize the system under study.

Additionally, this paper presents the RK4 method as an aid to solve the initial value problem through numerical and computational analysis.

The methodological development of the presented work will be given in the following way: In the second section the mathematical modeling realized to the two coupled masses will be shown with second order differential equations and the modeling through software Matlab, in the third section we present the modeling done with Rung-Kutta of fourth order and its fitting application with Matlab, in the fourth section results are presented. To finalize, the modeling conclusions and references will be presented.

To compare the different series of time obtained in this experiment we will use the MSE (Mean-Squared Error), because it can describe the degree of similarity between two set of time series data.[28].

The MSE between the time series data will be:

$$MSE(x,y) = \frac{1}{N}\sum_{i=1}^{N}(x_i - y_i)^2$$

2. Mathematical model

To start with the modeling process and determine the position of the two coupled masses, laboratory assembly was done see Figure 1, in which two masses are presented coupled to two springs and as a damping element water is used inside a glass container .

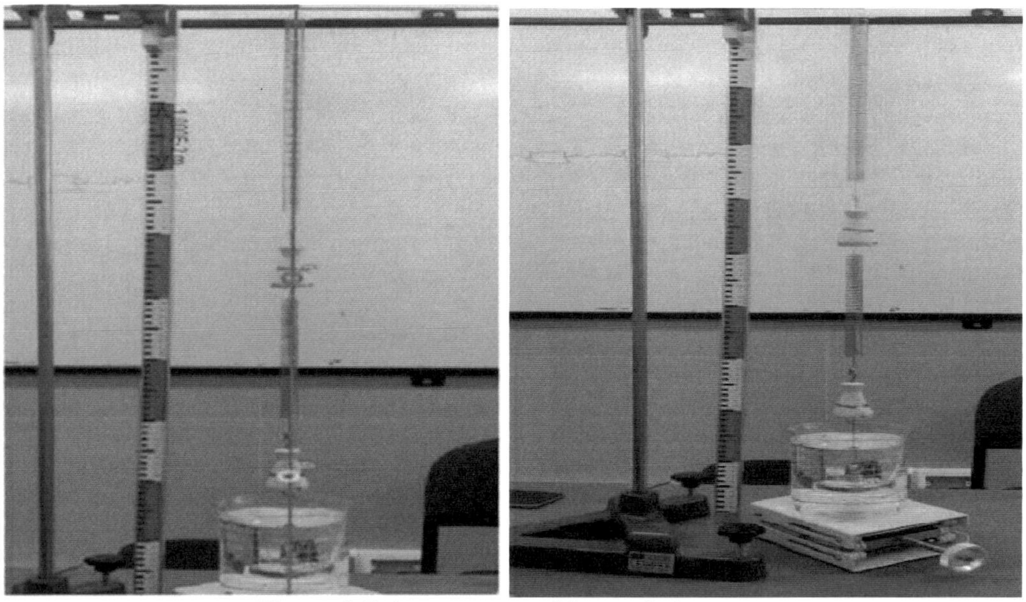

Figure 1. Laboratory assembly coupled masses

The real data of the system are shown in Table 1. The experimental analysis for the capture of the movement of the coupled system was done with the Tracker software (https://physlets.org/tracker/) and the capture of movement was made during 20 seconds in which the two masses were sliding. From the capture process, the displacement graphs of the masses were obtained, see Figure 2, and see Figure 3.

Table 1. Real data of the system.

Physical Data	Values
Mass 1 (m1)	0.2 (kg)
Mass 2 (m2)	0.325 (kg)
Elasticity (k1)	11.136 (N/m)
Elasticity (k2)	18.41 (N/m)
Damping (B)	0.08 (Ns/m)

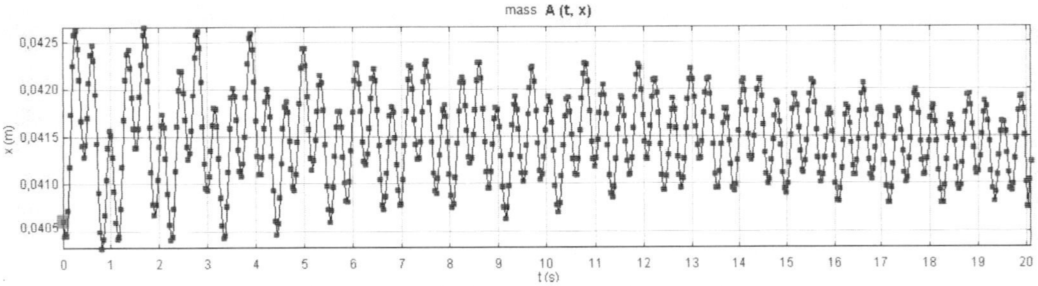

Figure 2. Displacement graphs mass A.

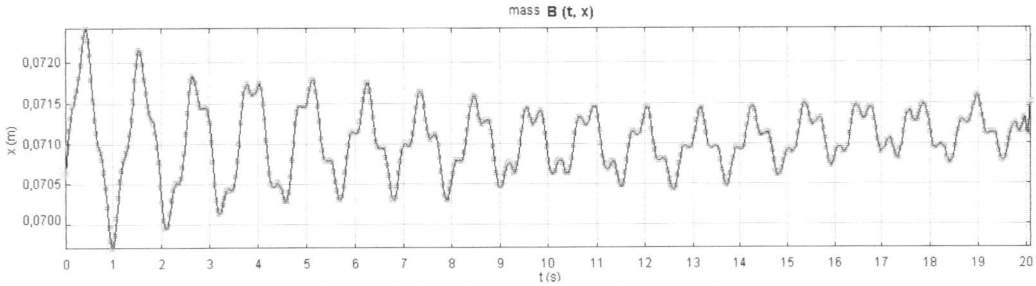

Figure 3. Displacement graphs mass B.

For the mathematical analysis, a diagram of the real situation is presented, see Figure 4.

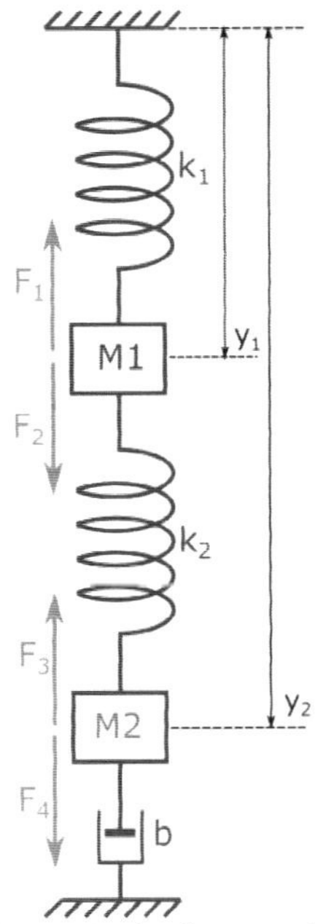

Figure 4. Analysis diagram free body.

Taking into account the second law of Newton, applied to the dynamic system of coupled masses. The mathematical model of the system is given in the following way:

$$m_1 * a_1 = F_1 - F_2 \tag{1}$$
$$m2 * a2 = F3 - F4 \tag{2}$$

Replacing by its derivatives in equations 1 and 2 it is observed that:

$$m_1 \frac{d^2 y_1}{dt^2} = -k_1 y_1 + k_2 \left(y_2 - y_1 \right) \tag{3}$$

$$m_2 \frac{d^2 y_2}{dt^2} = -k_2 \left(y_2 - y_1 \right) - b \frac{dy_2}{dt} \tag{4}$$

Reorganizing equations 3 and 4 we have:

$$m_1 \frac{d^2 y_1}{dt^2} + \left(k_1 + k_2 \right) y_1 - k_2 y_2 = 0 \tag{5}$$

$$m_2 \frac{d^2 y_2}{dt^2} + b \frac{dy_2}{dt} + k_2 y_2 - k_2 y_1 = 0 \tag{6}$$

If we make the following note in which:

$$\omega_{11}^2 = \frac{k_1}{m_1}, \omega_{22}^2 = \frac{k_2}{m_2}, \omega_{21}^2 = \frac{k_2}{m_1}, \lambda_b = \frac{b}{m_2} \tag{7}$$

The system becomes:

$$\frac{d^2 y_1}{dt^2} + \left(\omega_{11}^2 + \omega_{21}^2 \right) y_1 - \omega_{21}^2 y_2 = 0 \tag{8}$$

$$\frac{d^2 y_2}{dt^2} + \lambda_b \frac{dy_2}{dt} + \omega_{22}^2 y_2 - \omega_{22}^2 y_1 = 0 \tag{9}$$

Making a change of variables $x_1 = y_1$ $x_2 = y_2$ $\dot{x}_1 = v_1 = \dot{y}_1$ $\dot{x}_2 = v_2 = \dot{y}_2$ and replacing in equation 5 and 6 we have that:

$$\dot{v}_1 = -\left(\omega_{11}^2 + \omega_{21}^2 \right) x_1 - \omega_{21}^2 x_2 \tag{10}$$

$$\dot{v}_2 = -\lambda_b v_2 - \omega_{22}^2 x_2 - \omega_{22}^2 x_1 \tag{11}$$

Taking into account equations 7 and 8, the system can be written matrix as follows:

$$\begin{bmatrix} \dot{x}_1 \\ \dot{v}_1 \\ \dot{x}_2 \\ \dot{v}_2 \end{bmatrix} = \begin{bmatrix} 0 & 1 & 0 & 0 \\ -\left(\omega_{11}^2 + \omega_{21}^2\right) & 0 & \omega_{21}^2 & 0 \\ 0 & 0 & 0 & 1 \\ \omega_{22}^2 & 0 & -\omega_{22}^2 & -\lambda_b \end{bmatrix} \begin{bmatrix} x_1 \\ v_1 \\ x_2 \\ v_2 \end{bmatrix} \tag{12}$$

The system can be described as follows in equation 13:

$$\dot{x} = A_x \tag{13}$$

Where matrix A is the deterministic matrix of the system and x is the solution vector. The general solution can be written as follows:

$$x(t) = \sum_{i=1}^{n} C_i e^{\lambda i t} V_{0i} \tag{14}$$

Where λi are the proper values of A and $0i$ the eigenvectors of A .

To determine λi we must determine the roots of the n degree equation resulting from performing the following operation:

$$det\left(A - l\right) = 0 \tag{15}$$

Where l it is an identity matrix and to determine the eigenvalues it must be solved:

$$AV_0 = \lambda V_0 \tag{16}$$

Now taking into account our dynamic coupled mass system:

$$det\left(A - l\right) = \begin{bmatrix} -\lambda & 1 & 0 & 0 \\ -\left(\omega_{11}^2 + \omega_{21}^2\right) & -\lambda & \omega_{21}^2 & 0 \\ 0 & 0 & -\lambda & 1 \\ \omega_{22}^2 & 0 & -\omega_{22}^2 & -\lambda_b - \lambda \end{bmatrix} = 0 \tag{17}$$

By developing the following fourth degree polynomial is obtained:

$$\lambda^4 + \lambda_b \lambda^3 + \left(\omega_{11}^2 + \omega_{21}^2 + \omega_{22}^2\right)\lambda^2 + \lambda_b\left(\omega_{11}^2 + \omega_{21}^2\right)\lambda + \omega_{11}^2\omega_{22}^2 = 0 \tag{18}$$

IOP Conf. Series: Materials Science and Engineering **519** (2019) 012009 doi:10.1088/1757-899X/519/1/012009

Our System is described as: however, in order to determine the solution of our system we have to write it in canonical form. For this we make the following change of variable as it is presented:

$$X = PY \tag{19}$$

Where Y will be a system equivalent to X and P is a transformation matrix as follows:

$$X' = PY'$$
$$AX = PY'$$
$$APY = PY'$$
$$Y' = p^{-1}APY = JY$$

Where J is the canonical matrix of Jordan [29 - 30]. For our case with complex values of $\alpha_j \pm \beta_{ji}$

$$J = P^{-1}AP\begin{pmatrix} B_1 & \cdots & \\ \vdots & \ddots & \vdots \\ & \cdots & B_j \end{pmatrix} = o\text{, where } b_j = \begin{bmatrix} \alpha_j & \beta_j \\ -\beta_j & \alpha_j \end{bmatrix} \tag{20}$$

The eigenvectors V_j which corresponds to the j-esimo eigenvalue are also complex and the vector V_{2j} (pairs) is the complex conjugate V_{2j-1} (odd). Then P will be formed by the columns of eigenvectors such that:

$$P_{2j-1} = Re\left(V_{2j-1}\right) \tag{21}$$

And:

$$P_{2j} = lm\left(V_{2j-1}\right) \tag{22}$$

Then, the general solution of $X' = AX$ will be $X = PY(t)$ where:

$$Y(t) = \begin{pmatrix} a_1e^{\alpha_1 t}cos(\beta_1 t) + b_1e^{\alpha_1 t}sin(\beta_1 t) \\ -a_1e^{\alpha_1 t}sen(\beta_1 t) + b_1e^{\alpha_1 t}cos(\beta_1 t) \\ a_je^{\alpha_j t}cos(\beta_j t) + b_1e^{\alpha_j t}sin(\beta_j t) \\ -a_je^{\alpha_j t}sen(\beta_j t) + b_1e^{\alpha_j t}cos(\beta_j t) \end{pmatrix} \tag{23}$$

IOP Conf. Series: Materials Science and Engineering **519** (2019) 012009 doi:10.1088/1757-899X/519/1/012009

Taking into account the physical values of the system in Table 1. The initial values and values calculated in Table 2 can be observed.

Table 2. Real data of the system.

	Initials Value		Calculated Value
$x_1(0)$	-0.0099 (m)	λ_b	0.2462
$v_1(0)$	0.041 (m/s)	ω_{11}^2	36.05
$x_2(0)$	0.015 (m)	ω_{21}^2	64.45
$v_2(0)$	0.053 (m/s)	ω_{22}^2	39.66

Now our system is reflected in the following way:

$$\begin{bmatrix} \dot{x}_1 \\ \dot{v}_1 \\ \dot{x}_2 \\ \dot{v}_2 \end{bmatrix} = \begin{bmatrix} 0 & 1 & 0 & 0 \\ -(147.73) & 0 & 92.05 & 0 \\ 0 & 0 & 0 & 1 \\ 56.6462 & 0 & -56.6462 & -0.2462 \end{bmatrix} \begin{bmatrix} x_1 \\ v_1 \\ x_2 \\ v_2 \end{bmatrix} \tag{24}$$

Now our system with complex eigenvalues is shown as follows:

$$D = \begin{bmatrix} -0.0287 & + & 13.6948i \\ -0.0287 & + & -13.6948i \\ -0.0944 & + & 4.0998i \\ -0.0944 & + & -4.0998i \end{bmatrix} \begin{bmatrix} x_1 \\ v_1 \\ x_2 \\ v_2 \end{bmatrix} \tag{25}$$

And with complex eigenvectors is shown as follows:

$$V_0 = \begin{bmatrix} -0.0001-0.0668i & -0.0001+0.0668i & -0.0023-0.1362i & -0.0023+0.1362i \\ 0.9154+0.0000i & 0.9154+0.0000i & 0.5587+0.0033i & 0.5587+0.0033i \\ -0.0005+0.0289i & -0.0005-0.0289i & -0.0045-0.1938i & -0.0045+0.1938i \\ -0.3959-0.0078i & -0.3959+0.0078i & 0.7948+0.0000i & 0.7948+0.0000i \end{bmatrix} \tag{26}$$

The P matrix will be in this way:

$$P = \begin{bmatrix} -0.0001 & -0.0668 & -0.0023 & -0.1362 \\ 0.9154 & 0 & 0.5587 & 0.0033 \\ -0.0005 & 0.0289 & -0.0045 & -0.1938 \\ -0.3959 & -0.0078 & 0.7948 & 0 \end{bmatrix} \tag{27}$$

And the canonical Jordan form [28 - 30] will be in this way:

$$J = P^{-1}AP = \begin{bmatrix} -0.0287 & 13.6948 & 0 & 0 \\ -13.6948 & -0.0287 & 0 & 0 \\ 0 & 0 & -0.0944 & 4.0998 \\ 0 & 0 & -4.0998 & -0.0944 \end{bmatrix} \tag{28}$$

And the $Y(t)$ matrix will be:

$$Y(t) = \begin{bmatrix} e^{-0.0287t}\left(a_1 cos(13.6948t) + b_1 sin(13.6948t)\right) \\ e^{-0.0287t}\left(-a_1 sin(13.6948t) + b_1 cos(13.6948t)\right) \\ e^{-0.0944t}\left(a_2 cos(4.0998t) + b_2 sin(4.0998t)\right) \\ e^{-0.0944t}\left(-a_2 sin(4.0998t) + b_2 cos(4.0998t)\right) \end{bmatrix} \tag{29}$$

$$Y(0) = \begin{bmatrix} a_1 \\ b_1 \\ a_2 \\ b_2 \end{bmatrix} = P^{-1}X(0) = \begin{bmatrix} -0.01 \\ -0.008 \\ -0.0091 \\ 0.0048 \end{bmatrix} \tag{30}$$

The mathematical model of the mass, spring and damper system will be as follows:

$$x_1(t) = e^{-0.0287t}\left(0.0005cos(13.70t) - 0.0007sin(13.70t)\right) + e^{-0.09447t}\left(-0.0006cos(4.09980t) - 0.0012sin(4.0998t)\right) \tag{31}$$

$$v_1(t) = e^{-0.0287t}\left(-0.0091cos(13.70t) - 0.00074sin(13.70t)\right) + e^{-0.09447t}\left(-0.0051cos(4.09980t) - 0.0027sin(4.0998t)\right) \tag{32}$$

$$x_2(t) = e^{-0.0287t}\left(0.0040cos(13.70t) + 0.0003sin(13.70t)\right) + e^{-0.09447t}\left(-0.0009cos(4.09980t) - 0.0018sin(4.0998t)\right) \tag{33}$$

$$v_2(t) = e^{-0.0287t}\left(-0.0091cos(13.70t) - 0.00074sin(13.70t)\right) + e^{-0.09447t}\left(-0.0072cos(4.09980t) - 0.0038sin(4.0998t)\right) \tag{34}$$

IOP Conf. Series: Materials Science and Engineering **519** (2019) 012009 doi:10.1088/1757-899X/519/1/012009

2.1. Fitting and modelling in MATLAB

In this part of work, we presented the general model of fitting curve [31 - 34] using MATLAB software.

General mode movement of mass A:

$$x_1(x) = \exp(-a1.*x).*(a2*\cos(a3*x)+a4*\sin(a3*x)) + \exp(-a5.*x).* \\ (a6*\cos(a7*x)+a8*\sin(a7*x)) + a9 \tag{35}$$

Coefficients (with 95 % confidence bounds):

$$\begin{aligned} a1 &= 0.02774(0.02533, 0.03015) \\ a2 &= -0.003153(-0.003328, -0.002977) \\ a3 &= 17.29(17.28, 17.59) \\ a4 &= -0.006619(-0.006795, -0.006442) \\ a5 &= 0.09154(0.08601, 0.09707) \\ a6 &= -0.003745(-0.003971, -0.003519) \\ a7 &= 5.429(5.423, 5.434) \\ a8 &= 0.005036(0.004809, 0.005264) \\ a9 &= 0.4149(0.4148, 0.005264) \end{aligned} \tag{36}$$

Now we can see the adjustment of the movement of mass A. See Figure 5.

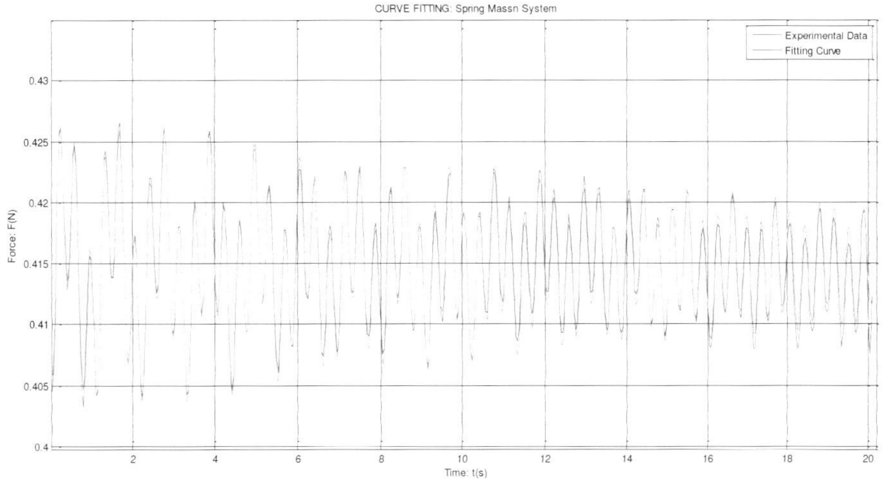

Figure 5. Adjustment of mass movement A.

General mode movement of mass B:

$$x_2(x) = a1*\exp(-a2*x).*\cos(a3*x+a4) + a5\exp(-a6*x).*\cos(a7*x+a8) + a9 \tag{37}$$

Coefficients (with 95 % confidence bounds):

$$\begin{aligned}
a1 &= -0.002426(-0.002682, -0.00217) \\
a2 &= -0.03434(0.02334, -0.04533) \\
a3 &= 17.28(17.27, 17.29) \\
a4 &= 2.084(1.979, 2.188) \\
a5 &= 00.01071(0.01039, 0.01103) \\
a6 &= -0.08843(0.08403, 0.09283) \\
a7 &= 5.426(5.421, 5.43) \\
a8 &= -2.168(-2.197, -2.139) \\
a9 &= 0.7104(0.7103, 0.7105)
\end{aligned} \tag{38}$$

Now we can see the adjustment of the movement of mass B. See Figure 6.

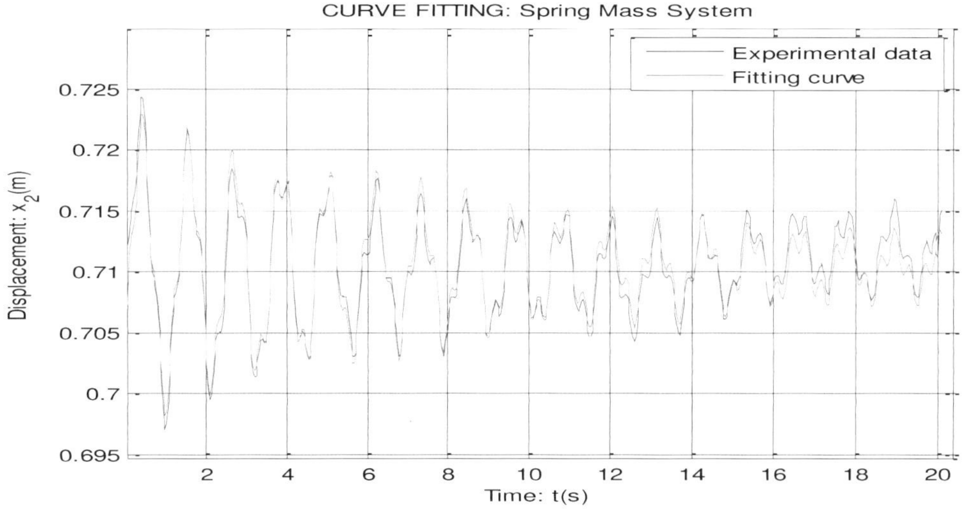

Figure 6. Adjustment of mass movement B.

Taking into account the calculation made for the general mode of movement of the mass B. It should be taken into account that in order to calculate the $\lambda_b = a2$, the general mode of the equation must be as shown in the equation:

$$FIT(x) = a1.*\exp(-a2.*x).*\cos(a3*x + a4) + a5 \tag{39}$$

Coefficients (with 95 % confidence bounds):

$$\begin{aligned} a1 &= 0.0106(0.008578, 0.01262) \\ a2 &= 0.2436(0.1817, 0.311) \\ a3 &= 17.08(16.94, 17.23) \\ a4 &= -0.3941(-0.9898, 0.2016) \\ a5 &= 0.7149(0.7144, 0.7154) \end{aligned} \tag{40}$$

Knowing that, $\lambda_b = \dfrac{b}{m2}$, we have to:

$$v_1(t) = e^{-0.0287t}\left(-0.0091\cos(13.70t) - 0.0074sen(13.70t)\right) + e^{-0.00944t}\left(-0.0006\cos(4.098t) - 0.0012sen(4.0998t)\right) \tag{41}$$

$$b = \lambda_b m_2 = 0.2463 * 0.325 = 80 + 10^{-3} \tag{42}$$

Now we can see the damped motion adjustment of mass B. See Figure 7.

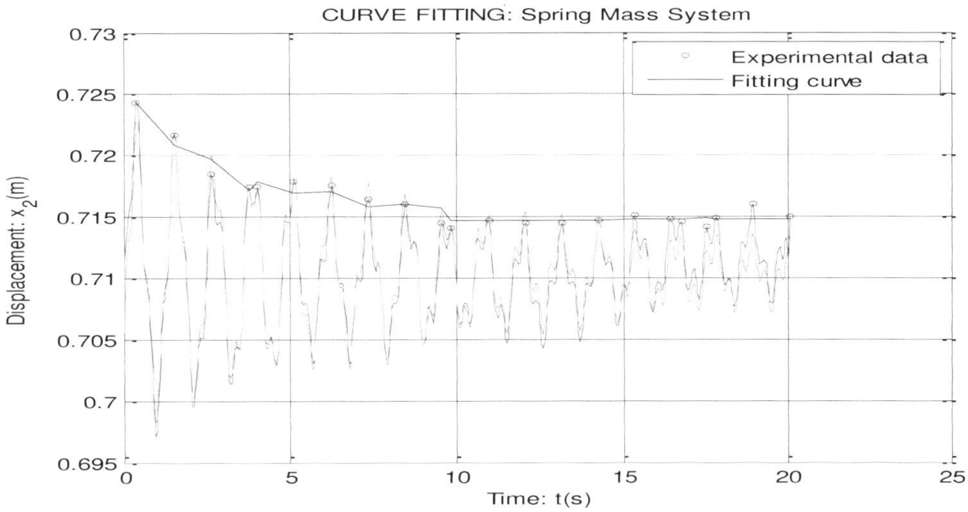

Figure 7. Damped motion adjustment of mass B.

2.2. Numerical Solution with, Runge Kutta 4 order

It is clarified that the 4-order Runge-Kutta Method [35 - 37] achieves the precision of the Taylor series, without having to determine the higher order derivatives. This is done to determine the solution of the initial value problem in the following way:

$$\frac{dy}{dt} = f(t, y), y(t_0) = y_0 \tag{43}$$

Through the general form:

$$w_{i+1} = w_i + \phi h \tag{44}$$

Where $\phi = a_1 k_1 + a_2 k_2 + ... + a_n k_n$

a_i Are constants.

$k_i = f\left(t_i + p_{i-1}h, y_i + q_{i-1,1}k_1 h + q_{i-1,2}k_2 h + ... + q_{i-1,i-1}k_{i-1}h\right)$

p_i Y q_i are constants.

Now we are looking to solve the initial value problem in this way:

$$\frac{dy}{dx} = f\left(t, y\right), y\left(t_0\right) = y_0 \tag{45}$$

The Runge-Kutta method of order 4 [35 - 37] is used based on iterations to approach the solution of $y(t)$ through w , using a step size h in this form:

$$w\left(i+1\right) = w\left(i\right) + \frac{1}{6}\left(k_1 + 2k_2 + 2k_3 + k_4\right)h \tag{46}$$

Where:

$k_1 = f\left(t_i, y_i\right)$

$k_2 = f\left(t_i + \frac{1}{2}h, y_i + \frac{1}{2}k_1 h\right)$

$k_3 = f\left(t_i + \frac{1}{2}h, y_i + \frac{1}{2}k_2 h\right)$

Taking into account the aforementioned equations, the systems of differential equations are represented as follows:

$$\left\{\begin{array}{l} \frac{dy}{dt} = f_1\left(t, y_1, y_2, ..., y_n\right) \\ \frac{dy}{dt} = f_2\left(t, y_1, y_2, ..., y_n\right) \\ . \\ . \\ . \\ \frac{dy}{dt} = f_n\left(t, y_1, y_2, ..., y_n\right) \end{array}\right\} \tag{47}$$

Now to find the solution of n differential equations, is required n initial values $y_1(t_0) = y_{0,1}, y_2(t_0) = y_{0,2}, \dots, y_n(t_0) = y_{0,n}$

It is clarified that the solution of the $j-esima$ equation, is found iteratively in the following way:

$$w(i+1,j) = w(i,j) + \frac{1}{6}\left(k_{ij} + 2k_{2j} + 2k_{3j} + k_{4j}\right)h \tag{48}$$

Where:

$$k_{ij} = f_j\left(t_i, y_{i,1}, \dots, y_{i,n}\right)$$
$$k_{2j} = f_j\left(t_i + \frac{1}{2}h, y_{i,1} + \frac{1}{2}k_{1,1}h, \dots, y_{i,n} + \frac{1}{2}k_{1,n}h\right)$$
$$k_{3j} = f_j\left(t_i + \frac{1}{2}h, y_{i,1} + \frac{1}{2}k_{2,1}h, \dots, y_{i,n} + \frac{1}{2}k_{2,n}h\right)$$
$$k_{4j} = f_j\left(t_i + h, y_{i,1} + k_{3,1}h, \dots, y_{i,n} + k_{3,n}h\right)$$

And $j = 1, 2, 3, \dots, n$

2.3. Numerical Solution with, Runge Kutta 4 order

In this section we present the mathematical model development of the mass coupled with spring and shock absorber with the Runge-Kutta method of 4 order [38 - 39]. It should be clarified that the method works with first-order differential equations and therefore the system of equations must be brought to such a level.

Taking into account the equation (1), (2), (3) and (4), the system can be written as follows:

$$\begin{Bmatrix} \dot{x}_1 = v_1 \\ \dot{v}_1 = -\left(w_{11}^2 + w_{21}^2\right)x_1 + w_{21}^2 x_2 \\ \dot{x}_2 = v_1 \\ \dot{v}_2 = w_{22}^2 x_1 - w_{22}^2 x_2 - \lambda_b v_2 \end{Bmatrix} \tag{49}$$

The system with values replaced will be:

$$\begin{Bmatrix} \dot{x}_1 = v_1 \\ \dot{v}_1 = -147.73x_1 + 92.05x_2 \\ \dot{x}_2 = v_1 \\ \dot{v}_2 = 56.6462x_1 - 56.6462x_2 - 0.2462v_2 \end{Bmatrix} \tag{50}$$

$$\begin{bmatrix} x_1(0) \\ v_1(0) \\ x_2(0) \\ v_2(0) \end{bmatrix} = \begin{bmatrix} 0.0099 \\ 0.041 \\ 0.0015 \\ 0.053 \end{bmatrix} \quad (51)$$

The analytical solution will look like this:

$$x_1(t) = e^{-0.0287t}\left(0.0005\cos(13.70t) - 0.0007sen(13.70t)\right) + e^{-0.00944t}\left(-0.0006\cos(4.0998t) - 0.0012sen(4.0998t)\right) \quad (52)$$

$$v_1(t) = e^{-0.0287t}\left(-0.0091\cos(13.70t) - 0.0074sen(13.70t)\right) + e^{-0.00944t}\left(-0.0051\cos(4.0998t) + 0.0027sen(4.0998t)\right) \quad (53)$$

$$x_2(t) = e^{-0.0287t}\left(0.0002\cos(13.70t) + 0.0003sen(13.70t)\right) + e^{-0.00944t}\left(-0.0009\cos(4.0998t) - 0.0018sen(4.0998t)\right) \quad (54)$$

$$v_2(t) = e^{-0.0287t}\left(0.040\cos(13.70t) + 0.0003sen(13.70t)\right) + e^{-0.00944t}\left(-0.0072\cos(4.0998t) + 0.0038sen(4.0998t)\right) \quad (55)$$

Finally, after executing the coding in MATLAB software with RK4 application we have the mass A position and velocity of mass A. See Figure 8 and Figure 9. And the mass B position and velocity of mass B. See Figure 10 and Figure 11.

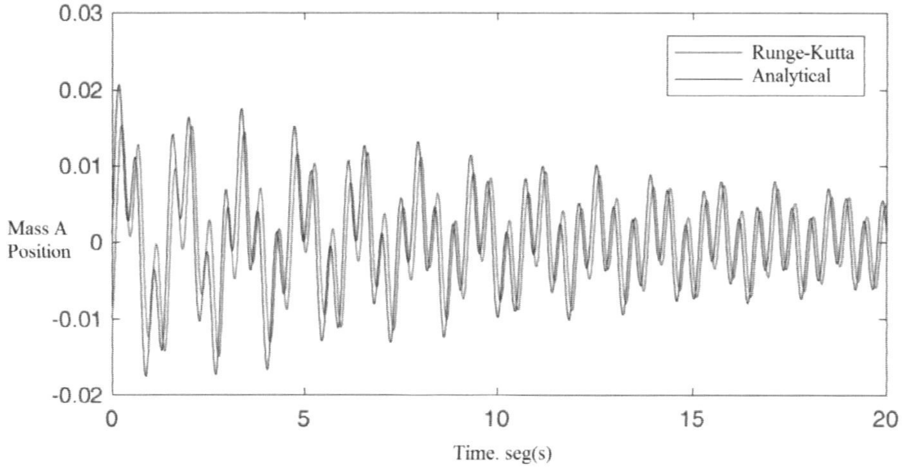

Figure 8. Mass A position with RK4.

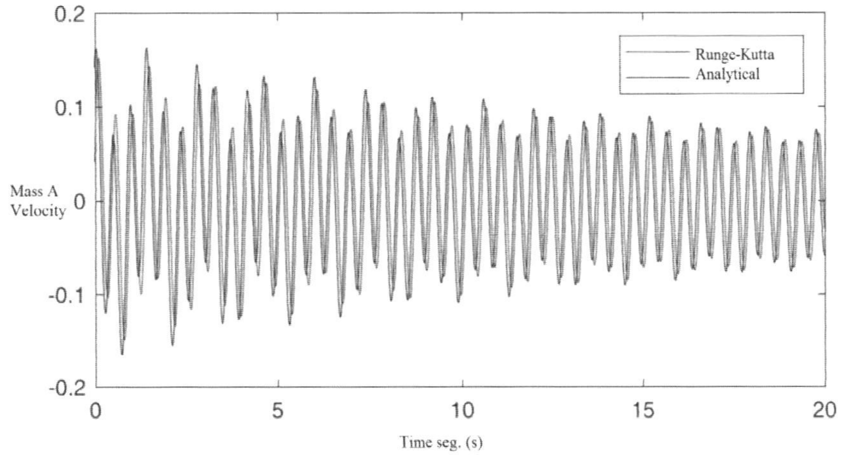

Figure 9. Mass A velocity with RK4

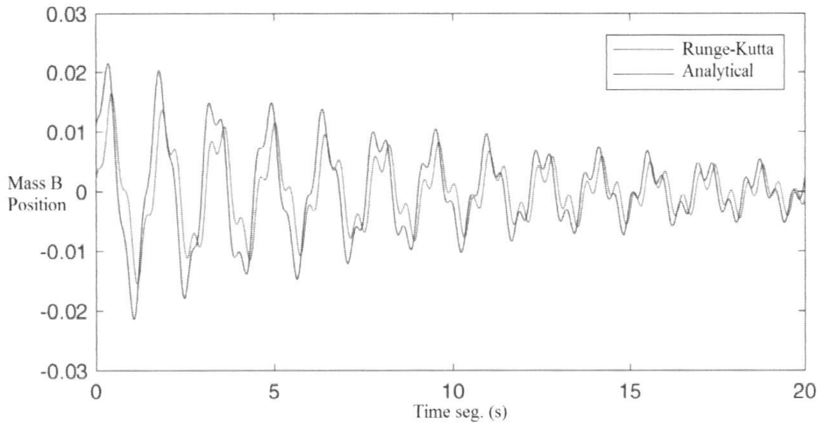

Figure 10. Mass B position with RK4.

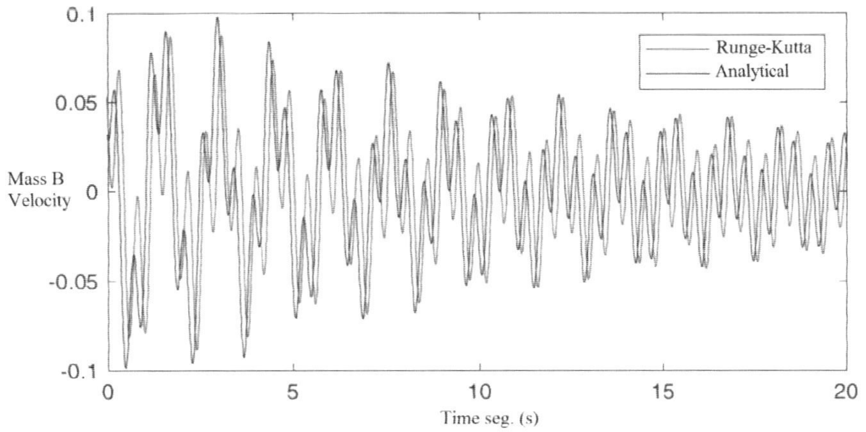

Figure 11. Mass B velocity with RK4

3. Conclusions

Making a comparison of the two analysis methods to adjust curves in modeling of second order dynamical systems through differential equations. It can be concluded that the RK4 method performs a better approximation to the initial state values. See Figures 5 and 6, in comparison with Figures 8 and 10.

The MSE between the data taken from Tracker and the data calculated using the Runge-Kutta fourth order method obtained for mass A, is 5.4005e-005 and for mass B, is 2.7258e-005, which were too small compared with their initial amplitudes.

It is concluded that the methods of analysis for modeling of coupled mass dynamic systems were ideal and are consistent with the movement and displacement capture evidenced in the Tracker software.

References

[1] Qi C, Zhao X, Gao F and Li H 2014 Modeling and identification of nonlinear distributed parameter dynamics of the micro-cantilever *Proceedings of the 33rd Chinese Control Conference* 5924-5929

[2] Chen Z , Qi Q and Zhou Y 2013 Mathematical Model and Simulation for Motion of Underwater Vehicle 2013 *Sixth International Symposium on Computational Intelligence and Design* 293-296

[3] Wang D, Turitsyn K and Chertkov M 2012 DistFlow ODE: Modeling, analyzing and controlling long distribution feeder *IEEE 51st IEEE Conference on Decision and Control (CDC)* 5613-5618

[4] Rath G, Harker M and Zaev E 2017 Direct numerical solution of stiff ODE systems in optimal control *6th Mediterranean Conference on Embedded Computing (MECO)* 1-5

[5] Lobão W, Dias D and Pacheco M 2016 Genetic programming and automatic differentiation algorithms applied to the solution of ordinary and partial differential equations *IEEE Congress on Evolutionary Computation (CEC)* 5286-5292.

[6] Korolova O, Popp M, Mathis W and Ponick B 2017 Performance of Runge-Kutta family numerical solvers for calculation of transient processes in AC machines *20th International Conference on Electrical Machines and Systems (ICEMS)* 1-6

[7] Krysko V, Saltykova O, Krysko A and Papkova I 2017 Nonlinear dynamics of contact interaction of MEMS beam elements accounting the Euler-Bernoulli hypothesis in a temperature field *Dynamics of Systems, Mechanisms and Machines (Dynamics)* 1-6

[8] Kern R and Gehring N 2017 Tracking control for a long pneumatic transmission line *22nd International Conference on Methods and Models in Automation and Robotics (MMAR)* pp 180-185

[9] Guangyao L, Yugang D, Chengning Z and Liwen P 2014 Numerical simulation and analysis of contact thermal resistance on motor controller based on Matlab *IEEE Conference and Expo Transportation Electrification Asia-Pacific (ITEC Asia-Pacific)* 1-4

[10] Swetha G, Priya M, Gupta R and Pandey A 2017 Implementation of chaotic oscillator using Matlab *International Conference on Recent Innovations in Signal processing and Embedded Systems (RISE)* 97-101

[11] Nenov H, Dimitrov B and Marinov A. 2013 Algorithms for computational procedure acceleration for systems differential equations in matlab *36th International Convention on Information and Communication Technology, Electronics and Microelectronics (MIPRO)* 238-242.

[12] Molina A. and Lazcano G. 2018 Matlab programmed method for the optical flow estimation based on the integral image *4th International Conference on Control, Automation and Robotics (ICCAR)* 394-399

[13] Bai Y and Zhang M 2009 Visual Teaching of Numerical Analysis Based on MATLAB *First International Conference on Information Science and Engineering* 3552-3555

[14] Trimbitas R and Trimbitas M 2007 Runge-Kutta Methods and Inverse Hermite Interpolation

Ninth International Symposium on Symbolic and Numeric Algorithms for Scientific Computing 118-123

[15] Ghassemi A, Fazel S, Maghsoud I and Farshad S 2014 Comprehensive study on the power rating of a railway power conditioner using thyristor switched capacitor *IET Electrical Systems in Transportation* **4** 4 97-106

[16] Kavitha S, Varuna S and Ramya R 2016 A comparative analysis on linear regression and support vector regression *Online International Conference on Green Engineering and Technologies (IC-GET)* 1-5

[17] Hirose H, Soejima Y and Hirose 2012 K NNRMLR: A Combined Method of Nearest Neighbor Regression and Multiple Linear Regression *IIAI International Conference on Advanced Applied Informatics* 351-356

[18] Feng X , Zhou Y, Hua T, Zou Y and Xiao J 2017 Contact temperature prediction of high voltage switchgear based on multiple linear regression model *32nd Youth Academic Annual Conference of Chinese Association of Automation (YAC)* 277-280

[19] Kong W, Zhou H, Zheng K, Mu X and Hong W 2017 FFT-Based Method With Near-Matrix Compression *IEEE Transactions on Antennas and Propagation* **65** 11 5975-5983

[20] Su T, Yang M, Jin T and Flesch R 2018 Power harmonic and interharmonic detection method in renewable power based on Nuttall double-window all-phase FFT algorithm *IET Renewable Power Generation* **12** 8 953-961

[21] Laih C, Harn L and Lee J 1989 On the design of a single-key-lock mechanism based on Newton's interpolating polynomial *IEEE Transactions on Software Engineering* **15** 9 1135-1137

[22] Van Beeumen R, Michiels W and Meerbergen K 2015 Linearization of Lagrange and Hermite interpolating matrix polynomials *IMA Journal of Numerical* Analysis **35** 2 909-930

[23] Lian Y, Juncheng L and Guohua C 2010 A Class of Algebraic Trigonometric Interpolation Splines and Applications *International Conference on Computational and Information Sciences* 1174-1177

[24] Pattanadech N and Yutthagowith P 2015 Fast curve fitting algorithm for parameter evaluation in lightning impulse test technique *IEEE Transactions on Dielectrics and Electrical Insulation* **22** 5 2931-2936

[25] Wang X 2017 Strong convergence rates of the linear implicit Euler method for the finite element discretization of SPDEs with additive noise *IMA Journal of Numerical Analysis* **37** 2 965-984

[26] Simos T 2001 A fourth algebraic order exponentially-fitted Runge–Kutta method for the numerical solution of the Schrödinger equation *IMA Journal of Numerical Analysis* **21** 4 919-931

[27] Haut T, Babb T, Martinsson P and Wingate B 2016 A high-order time-parallel scheme for solving wave propagation problems via the direct construction of an approximate time-evolution operator *IMA Journal of Numerical Analysis* **36** 2 688-716

[28] Zhou W and Bovik A 2009 Mean squared error: Love it or leave it? A new look at signal fidelity measures *IEEE signal processing magazine* **26** 1 98-117

[29] Zhu L and Rivera L 2002 A note on the dynamic and static displacements from a point source in multilayered media *Geophysical Journal International* **148** 3 619-627

[30] Szyld D and Xue F 2014 Several properties of invariant pairs of nonlinear algebraic eigenvalue problems *IMA Journal of Numerical Analysis* **34** 3 921-954

[31] Yoshida H and Tanaka T 2007 Positive Controllability Test for Continuous-Time Linear Systems *IEEE Transactions on Automatic Control* **52** 9 1685-1689

[32] Pattanadech N and Yutthagowith P 2015 Fast curve fitting algorithm for parameter evaluation in lightning impulse test technique *IEEE Transactions on Dielectrics and Electrical Insulation* **22** 5 2931-2936

[33] Bakhshizadeh M 2017 The Application of Vector Fitting to Eigenvalue-Based Harmonic Stability Analysis *IEEE Journal of Emerging and Selected Topics in Power Electronics* **5** 41487-1498

[34] Ding T, Li C, Huang C, Yang Y, Li F and Blaabjerg F 2018 A Hierarchical Modeling for Reactive

Power Optimization With Joint Transmission and Distribution Networks by Curve Fitting *IEEE Systems Journal* **12** 3 2739-2748

[35] Batzelis E, Kampitsis G and Papathanassiou S 2017 Power Reserves Control for PV Systems With Real-Time MPP Estimation via Curve Fitting *IEEE Transactions on Sustainable Energy* **8** 3 1269-1280

[36] Haut T, Babb T, Martinsson P and Wingate B 2016 A high-order time-parallel scheme for solving wave propagation problems via the direct construction of an approximate time-evolution operator *IMA Journal of Numerical Analysis* **36** 2 688-716

[37] Nicolet A and Delince F 1996 Implicit Runge-Kutta methods for transient magnetic field computation *IEEE Transactions on Magnetics* **32** 3 1405-1408

[38] Darkner S, Pai A, Liptrot M and Sporring J 2018 Collocation for Diffeomorphic Deformations in Medical Image Registration *IEEE Transactions on Pattern Analysis and Machine Intelligence*, **40** 7 1570-1583

[39] Abdelgawad H and Sood V 2016 Performance analysis of boost converter by using different integration algorithms *IEEE Canadian Conference on Electrical and Computer Engineering (CCECE)* 1-5

[40] Talbi S, Kassmi K and Malek R 2016 Modeling and simulation of a solar oven box-type with thermal storage *International Renewable and Sustainable Energy Conference (IRSEC)* 348-352

IOP Conf. Series: Materials Science and Engineering **519** (2019) 012010 doi:10.1088/1757-899X/519/1/012010

Identification of EMG activity with machine learning in patients with amputation of upper limbs for the development of mechanical prostheses

J N Zuleta[1], M Ferro[1], C Murillo[1], R A Franco-Luna[1,2]
[1]Tecnoacademia Risaralda, Servicio Nacional de Aprendizaje SENA, Carrera 21 con 73 bis, Dosquebradas, Colombia
[2]Facultad de Ingeniería, Universidad Tecnológica de Pereira, Carrera 27 # 10-02, Pereira, Colombia

rfranco@sena.edu.co

Abstract. A study of electromyographic signals (EMG) in subjects with partial hand amputation using machine learning techniques (ML) is presented in this document. The EMG were analyzed for five hand poses. We used the Fast Fourier Transform (FFT), and Wavelet transform as descriptors for the feature extraction, the identification and classification system was implemented based on Vector Support Machines (VSM). Percentages of accuracy greater than 90% were obtained in the cases of close hand, left hand, right hand and relax hand, while open hand obtained an acceptable performance with accuracy percentages lower than 90%.

1. Introduction

The use of EMG applied to the automatic identification of muscle activity is increasing. In medicine are essential in clinical diagnosis of neuromuscular and neurological disorders patients [1], in the field of biomedical instrumentation are important for the development of robotic prostheses, wheelchairs and devices physical rehabilitation as exoskeletons [2] . EMGs show electrical potentials associated with the level of muscle activity generated during contraction and relaxation. The potentials can be detected by means of intramuscular devices (iEMG) and superficial devices (sEMG) located on the muscle, this work considers only sEMG.

Different EMG features extraction techniques have been used, in 1975, Graupe and Cline classified EMG with a success rate of more than 75% using Mean Absolute Value (MAV), this approach was taken up in several works, among which stands out Bingken and Shiyou in 2016 [3], Khushaba et. to the. in the same year [4] and Bhattacharya and Sarkar in 2017 [5]. In 1995, W. Kang used Mel-Frequency Cepstral Coeficients (MFCC), obtaining an accuracy of 85% [6], a method adopted by Lee and Kim in 1996 [7] and recently by several authors [8,9,10]. The Fourier transform and its variations - fast Fourier transform (FFT) and fast short pulse Fourier transform (STFFT), are widely used in the characterization of EMG [11,12]. Techniques based on time-frequency combinations have also been used, such as the Wavelets transform [13,14], obtaining accuracies greater than 90%.

In this work we analyzed the EMG for five hand poses: close hand, open hand, left hand, right hand and relax hand. FFT and wavelet transform were used as descriptors for the features extraction. Vector

Content from this work may be used under the terms of the Creative Commons Attribution 3.0 licence. Any further distribution of this work must maintain attribution to the author(s) and the title of the work, journal citation and DOI.
Published under licence by IOP Publishing Ltd

IOP Conf. Series: Materials Science and Engineering **519** (2019) 012010 doi:10.1088/1757-899X/519/1/012010

support machines (VSM) were implemented as a machine learning technique (ML) for the identification and classification system.

2. Materials and methods

2.1. Subjetcs
Ten subjects aged between 15 and 48 years participated in this investigation. All subjects completed a document on consent to participate. There was also psychological accompaniment considering details related to the emergence of his disabling condition, life story collected based on precise clinical interview guidelines and the help of projective measuring instruments, this to assess the feasibility of work with the subject and possible risks of generating considerable levels of anxiety due to the expectation of a possible corrective intervention in later phases of the investigation.

2.2. Acquisition of EMG
The myo armband from Thalmic was used, it has eight Medical Grade Stainless Steel EMG sensors, highly sensitive nine-axis IMU containing three-axis gyroscope, three-axis accelerometer, three-axis magnetometer, and sampling frequency of 200Hz. Software was designed in Matlab to capture, visualize and store the EMG.

2.3. Hand poses
The myo armband sensor was located on the forearm. Five hand poses were analyzed: close hand, open hand, left hand, right hand and relax hand. Subjects repeated ten times each pose. Figure 1 shows the five hand poses.

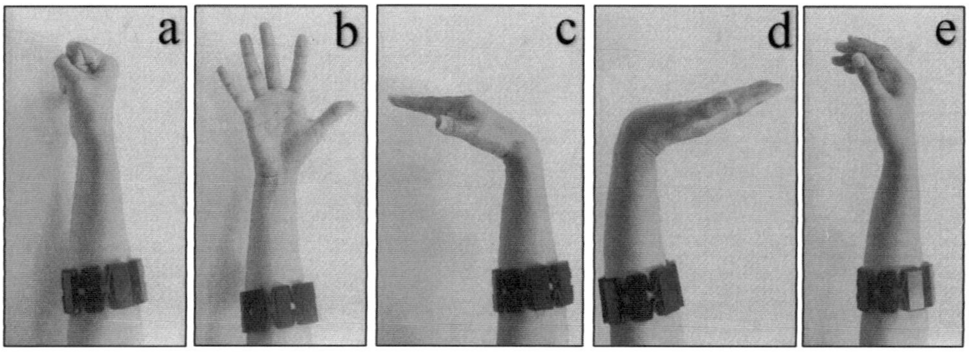

Figure 1. Hand poses: a) close hand, b) open hand, c) left hand, d) right hand, e) relax hand.

3. Data analysis
For the development of this document a pattern of three stages worked: the initial step is the detection signal, followed by the step of feature extraction and finally the classification stage. No dimensionality reduction is applied, since the aim is to identify the performance of the ML system based on raw data and compare two signal characterization methods: FFT and Wavelets.

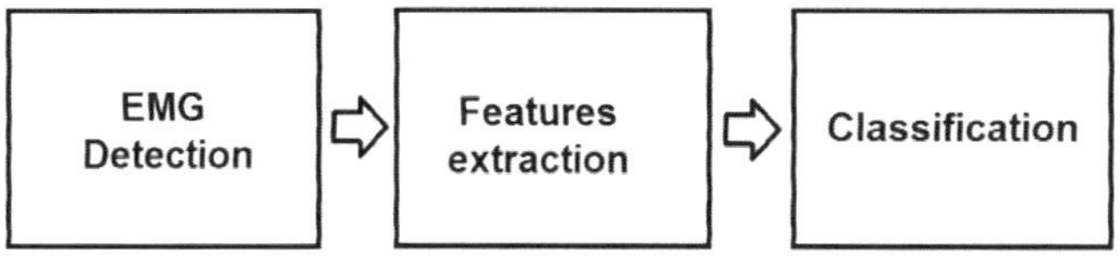

Figure 2. Scheme of 3 stages for ML implementation

3.1. Detection and adaptation of EMG.

After the configuration of the Myo Armband bracelet, the programming of a data acquisition system using the SDK of the development company and Matlab mathematical analysis software was carried out. A script was developed that allows collecting and organizing the information as shown in figure 3.

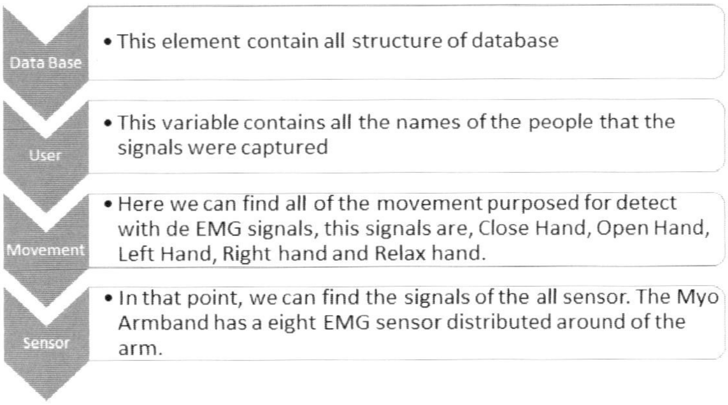

Figure 3. Sequential outline of the data structure.

3.2. Feature extraction

For the initial development and validation of the database, the mathematical methods to be compared are FFT and the transformed Wavelet haar. The equation (1) shows the discrete Fourier transform.

$$X_k = \sum_{n=0}^{N-1} x_n e^{-i2\pi k\frac{n}{N}} \quad k = 0,1,\dots,N-1 \tag{1}$$

From the FFT implementation, the frequency coefficients are obtained, which will be used as a descriptor of the signal. Figure 4 shows a set of 5 captures of the experiment for the pose close hand where the behavior of the signals in the time of the 8 sensors can be observed, and in Figure 5 the FFT corresponding to each sensor is observed.

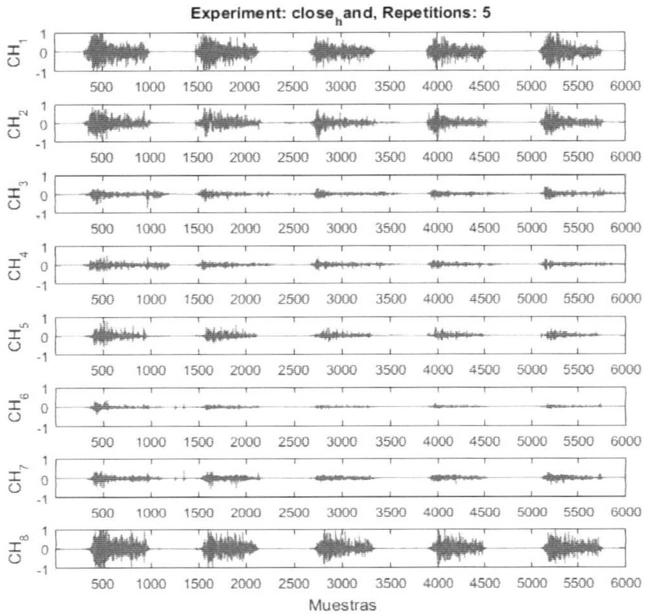

Figure 4. Pose close hand on the eight sensors

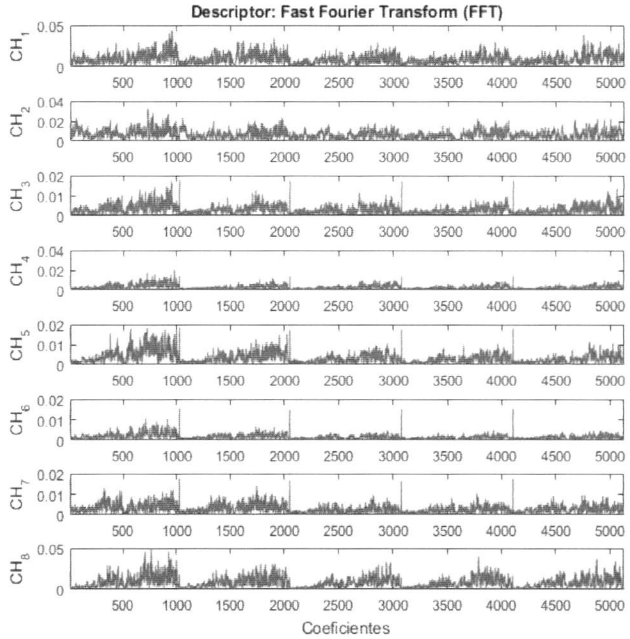

Figure 5. FFT descriptor

Another technique widely used in signal analysis is the Wavelet transform with a "haar" mother signal, for the implementation a decomposition of order 6 was used and a vector was constructed with the detail and approximation coefficients of the Wavelet haar transform, the descriptor obtained with the application of the wavelet transform can be seen in figure 6 (left).

In the experimental development a third descriptor was implemented from the Wavelet signal (WL-2), each of the groups of approximation and detail coefficients can be seen as a modal decomposition process. Each group was implemented a set of statistical moments composed of the average, standard deviation, kurtosis, median mean and autocorrelation, this descriptor is shown in Figure 6 (right).

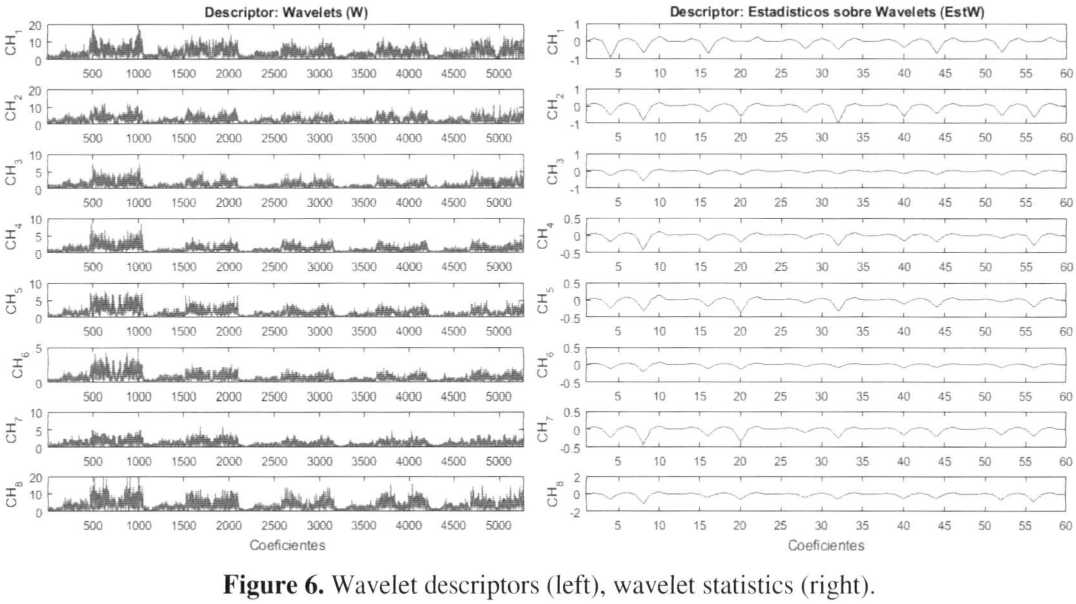

Figure 6. Wavelet descriptors (left), wavelet statistics (right).

3.3. Training and classification

After having built the database of the five poses to be identified, each pose composed of a set of signals from the 8 EMG sensors of the bracelet, the features extractors based on FFT, Wavelets haar and their statistics were applied. A one-to-one structure was implemented for each hand pose and the configuration of a binary SVM.

For the training the database was randomly divided 70% training and 30% validation in each iteration of Monte Carlo to guarantee statistical relevance. Figure 7 shows the structure of data processing for training using Monte Carlo iterations. Two experiments were performed with 1000 and 100 iterations of Monte Carlo with an approximate computation time of 95 minutes and 13 minutes respectively, the deviation between both experiments was less than 1%.

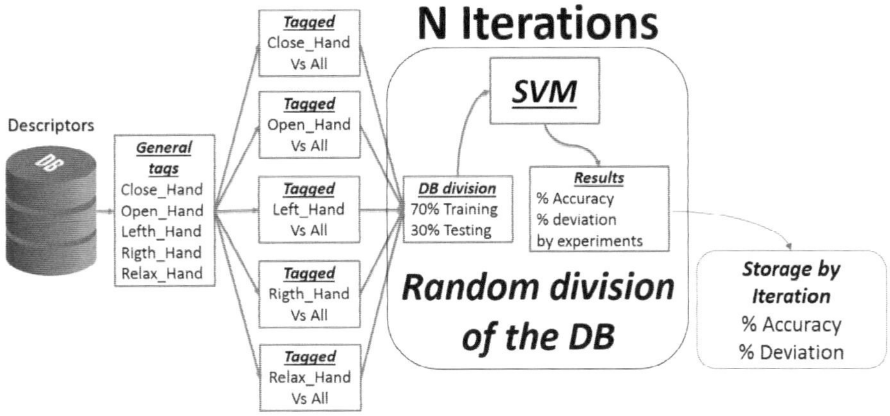

Figure 7. Structure of data processing for training using Monte Carlo iterations.

4. Results and discussion

After training, the performances of SVM for identifying poses with the use of the proposed descriptors were evaluated. Table 1 shows the percentage of accuracy and deviations obtained by the descriptors FFT, WLhaar and WL-2.

Table 1. Comparison of the FFT, WLhaar and WL-2 statistics descriptors of each hand pose. The first 3 columns correspond to the average of success in the classification process, and the next 3 columns correspond to the standard deviations of the average of success for each pose and each descriptor.

Pose	Accuracy (%)			Deviation (%)		
	FFT	WLhaar	WL-2	FFT	WLhaar	WL-2
close hand	97.33	94.38	95.04	0.85	1.39	1.16
open hand	87.50	79.56	84.30	2.19	2.88	2.40
left hand	94.64	91.97	94.44	1.22	1.54	1.37
right hand	92.75	88.55	93.68	1.40	1.89	1.44
relax hand	84.47	92.97	89.33	2.40	1.52	1.86

The FFT descriptor presented an outstanding performance in three of the poses (close hand, left hand and right hand), with accuracies over 92%, reaching 97.33% in close hand, this shows great separability in its frequency spectra. The lowest performances using FFT were obtained in open hand

and relax hand with accuracies of 87.50% and 84.47% respectively, which are considered acceptable for the identification process.

Moreover, the descriptor WLHarr presented an outstanding performance in the poses close hand, left hand and hand relax with accuracies above 90%. In the right hand and open hand poses the performance was lower with success percentages of 88.55% and 79.56% respectively. In turn, the descriptor WL-2 presented accuracies greater than 93% in the poses close hand, left hand and right hand, in addition, 89.33% and 84.30% in relax hand and open hand respectively.

Analyzing the performance by hand pose, the results of the FFT descriptor showed a greater accuracy in the poses close hand, open hand and left hand. The descriptor WLhaar classified with greater accuracy the pose relax hand while WL-2 did it in right hand. Additionally, open hand was the pose with the lowest percentage of acerts, being lower than 90% in the three descriptors, a percentage that is comparable with results obtained in other works [2,5,10]. In contrast, close hand and left hand were classified with accuracies greater than 90% by the three descriptors.

5. Conclusions

An identification and classification structure was obtained for the five proposed poses with accuracies higher to 90% in the cases close hand, left hand, right hand and relax hand, while open hand obtained an acceptable performance with accuracies less than 90 %.

The performance results of the descriptors suggest that some hand poses present combinations of muscle activities that could be confused by the identification system and be classified erroneously, especially in the open hand and relax hand cases. The implementation of techniques for eliminating redundant components such as PCA could improve the performance of the classifier.

The variations in the results of the three descriptors allowed to identify the importance of these in the performance of the classifier. It is concluded that the implementation of additional descriptors and the fusion of methodologies could allow a broader spectral separation and consequently better performances of the classifier.

6. References

[1] Gokgoz E and Subasi A 2015 Comparison of decision tree algorithms for EMG signal classification using DWT *Biomedical Signal Processing and Control* 138-144.

[2] Negi S, Kumar Y and Mishra V 2016 Feature extraction and classification for EMG signals using linear discriminant analysis Advances *in Computing, Communication, & Automation* 1-6

[3] Zhang B and Zhang S 2016 The estimation of grasping force based on the feature extracted from EMG signals. *IEEE Advanced Information Management, Communicates, Electronic and Automation Control Conference.*

[4] Khushaba R, Al-Timemy A, Al-Ani A and Al-Jumaily A 2016 Myoelectric feature extraction using temporal-spatial descriptors for multifunction prosthetic hand control *38th Annual International Conference of the IEEE Engineering in Medicine and Biology Society.*

[5] Bhattacharya A, Sarkar A and Basak P 2017 Time domain multi-feature extraction and classification of human hand movements using surface EMG *4th International Conference on Advanced Computing and Communication Systems* (ICACCS)

[6] Kang W, Shiu J and Cheng C 1995 The application of cepstral coefficients and maximum likelihood method in EMG pattern recognition [movements classification]. *Biomedical Engineering, IEEE Transactions* 777-785

[7] Lee S, Kim J and Park H 1994 An enhanced feature extraction algorithm for EMG pattern classification *IEEE Transactions on Rehabilitation Engineering.*

[8] Hazarika A, Dutta L and Barthakur M 2016 Fusion of projected feature for classification of EMG patterns *International Conference on Accessibility to Digital* World (ICADW).

[9] Mishra V, Bajaj V and Kumar A 2016 Classification of normal, ALS, and myopathy EMG signals using ELM classier *2nd International Conference on Advances in Electrical, Electronics, Information, Communication and Bio-Informatics* (AEEICB).

[10] Khushaba R, Al-Timemy A, Al-Ani A and Al-Jumaily 2016 Myoelectric feature extraction using temporal-spatial descriptors for multifunction prosthetic hand control *38th Annual International Conference of the IEEE Engineering in Medicine and Biology Society* (EMBC).

[11] Garg C, Narayan and and Mathew L 2015 Development of a software module for feature extraction and classification of EMG signals *Communication, Control and Intelligent Systems*

[12] AL-Quraishi M 2015 Impact of feature extraction techniques on classification accuracy for EMG based ankle joint movements *10th Asian Control Conference* (ASCC).

[13] Pomboza-Junez G and Terriza J 2016 Hand gesture recognition based on sEMG signals using Support Vector Machines *IEEE 6th International Conference on Consumer Electronics - Berlin* (ICCE-Berlin).

[14] Mannini A and Sabatini A 2011 Machine Learning Methods for Classifying Human Physical Activities from on-body sensors *Pattern Recognition and Image Analysis: 5th Iberian Conference.*

Development of a prototype of an electronic module that allows to calculate the frequency of spawning in laying hens

S A Sanchez[1], A A Aguilar[1], H J Romero[1] and A D Morales[1]

[1]Corporación Universitaria Antonio José de Sucre, Sincelejo, Colombia, Engineering science faculty, Grupo de investigación GINTEING

sergio_sanchez@corposucre.edu.co

Abstract. Poultry farming is an industry recognized worldwide, being one of the main activities in different countries. From the national level, poultry farming is a source of income dedicated to the production of eggs and poultry meat, which has had a constant growth since the mid-twentieth century. This market in Colombia is one of the few economic sectors with permanent growth, this is evidenced by the increase in the per capita figures for this type of product. Nowadays in the market there is no evidence of a low cost system that allows to measure the hens' position in real time, this necessity being the main objective to develop an electronic module that allows to calculate the frequency of spawning in laying hens, with the purpose that this information can be used by poultry farmers in order to determine what food treatments and environmental conditions improve egg productivity. The materials used for the prototype were; a module cell (independent cage for each bird), an Arduino UNO card, infrared proximity sensor, 12x2 LCD display, MicroSD card module and an SD memory. This investigation is framed in a technological development of experimental type, for the construction of the prototype 3 stages were taken into account; prototype design and its characteristics, final assembly of the prototype and tests and results. In the results found there is evidence of a versatile tool that allows to inform in real time in a graphic way the postures. The tests carried out demonstrate the potential of a low- cost, lightweight and state-of-the-art prototype.

1. Introduction

The development of technologies is advancing rapidly in the world, due to factors that are changing their leading role, this is how we can appreciate the change that has been generated in different sectors and the great progress that the sector has made in recent years. Poultry In the poultry farms technified egg producers, different techniques are implemented to improve the productivity of these. Therefore, the breeders carry out tests with manual and inaccurate data on the laying time of the birds; to determine which treatment works best.

Currently in some municipalities of the department of Sucre, poultry farmers carry out this activity from the empiricism, there are no certified processes that allow improving the quality of the products and achieve a positioning in the national and international market. For this reason, the national government from different programs has been encouraging and financing the public and private sector (companies, universities, industries, among others) to generate projects that help position and generate competitiveness in these markets [1]. It should be noted that poultry farming is one of the economic activities that take place in this area.

Content from this work may be used under the terms of the Creative Commons Attribution 3.0 licence. Any further distribution of this work must maintain attribution to the author(s) and the title of the work, journal citation and DOI.
Published under licence by IOP Publishing Ltd

According to the National Service of Learning, SENA, "poultry farming worldwide and nationally, thanks to advances in genetics, nutrition and animal management, currently presents a rapid growth, improving the supply and facilitating the access to egg consumption as one of the most complete foods in human nutrition" [2].

According to the National Federation of Poultry Farmers of Colombia, Fenavi; Nowadays, most laying hens are raised in intensive systems with a controlled environment, where parameters such as light intensity and duration are monitored by man, in order to improve the egg position, modify the behavior or simply save on energy costs. Birds are characterized by having a greater number of cones than rods in the retina, so they have better daytime vision than nighttime. As for laying hens, they are bred with light intensities ranging between 5-10 lux, enough to maintain the physiological behavior of the laying, through the stimulation of the optic nerve. In this context, electric light plays an important role, since it is used to keep the laying birds active and influence the production of eggs. Initially, the work of regulating the lighting period was done manually by a manager, who turned on and / or turned off the light according to the convention held in the house [3].

With the advancement of technology, electronic devices were designed, which must be programmed to perform this function. However, to prevent the light from being turned on and off abruptly, the agricultural research company, Embrapa Suínos e Aves, linked to the Ministry of Agriculture, Livestock, and Supply of Brazil, developed a new adjustable device. An automatic luminosity controller, with an approximate dimension of 25x25 cm, which promises to simulate gradually the beginning of the day or the arrival of the night. In addition, the regulator also stores in memory everything that happens during the batch process. Another interesting aspect is the light compensation work that this system implements in relation to external light.

According to Saatkamp, M. analyst in the area of technology transfer of Embrapa. "This automatic controller optimizes the light inside the house. The light is turned on at the programmed time gradually, simulating the start of the day" [4].

By the time the end of the light period is programmed, it decreases until it is turned off, simulating the end of the day. "He had already announced that the management of light / dark / light illumination should be a gradual change, to avoid a greater stress in the birds. These researchers analyzed the behavior of the hens in both cage and ground, subjected to two different lighting management practices: in the first, turning on and off the lights was instantaneous, while in the second, the passage from light to dark or vice versa was gradual, simulating a sunrise or natural sunset. In both situations, an increase in food intake was observed before the darkness, to later go to the resting areas, when the light was suddenly extinguished, some s hens, especially in the system of accommodation in soil, did not have enough time to look for the slats, having it done in the dark, experienced stress. This did not happen when simulating a sunset with a gradual reduction of light. Regarding the ignition, when it was sudden, it caused in the hens a situation of alarm and disorientation, reaching, even, to cause some stereotyping (they revolved around themselves). These authors concluded that a gradual ignition and / or shutdown would increase the comfort of the hens. However, they pointed out that chickens must be accustomed to sudden changes of light / darkness or vice versa, as this way, in case of any failure in the electrical system, a situation of generalized panic in the house would not be triggered. The first precaution that should be considered with this brightness regulating device is the installation, which must be carried out by technicians. Another important aspect is the regulation, where it must be assigned to a professional for the operation of the device, since the lighting program introduced should not be altered. In addition, you should have the normal basic care of the equipment in bird plants such as protect the electric discharges and prevent water from falling into the equipment.

It highlights the research conducted by San Lucas et al. where they implement an automatic system with sensors to measure some variables within the house, the data taken are sent to a website to be observed via the internet [5]. Focusing on a chicken farm for meat production, where variables such as humidity and temperature are controlled by means of combustion heaters and curtains that rise or descend automatically allowing the air to flow inside the house depending on the alteration in the variables before mentioned, in addition, depending on the age of the chickens their food and water consumption varies, for this they carry out an automatic system that supplies a precise quantity.

Coturniculture is a branch of the poultry sector, whose objective is the production of quails for obtaining eggs, meat and other derivatives. For Vásquez et al. in the exploitation of the quail, several processes are implicit, such as egg harvesting, food irrigation and cleaning the house [6]. The main cause of dirt is the codornaza. (manure produced by the quail), this produces gases such as ammonia and increases humidity, raising the temperature inside the house and affecting production.

The quail is a bird very sensitive to the environment, this can get sick because of the poor conditions of this. To carry out the cleaning in the house, the codornaza must be extracted from the deposit sheets, this process requires between two and three hours per man for each thousand quails, as it requires physical contact and exposure to toxic gases. Which the coturnicultor is susceptible to diseases and infections. They address the problem of codornaza extraction by developing a prototype on a fully automated scale, which will comply with the process at a frequency determined by the coturniculturist through an HMI (Human Machine Interface) interface or when the ammonia gas level exceeds the maximum allowed value. With the objective of guaranteeing which environment is the most suitable for the exploitation of the quail, measuring and controlling the temperature variable [7].

These new technological tools that are offered in the poultry industry, specifically for the posture sector, provide favorable measures to reduce bird stress and therefore improve the welfare of the birds. This type of products help to increase egg productivity as well as allow energy savings, consequently improving the final profitability. The electronic devices used for the development of the remote monitoring prototype are specified below:

1.1 Arduino programmable card

According to the information consulted, the Arduino development board is a free hardware and software microcontroller, designed to favor the indispensable operation of the electronics in different projections. One of the peculiarities of this, is that the hardware resides on a board with a small controller. Which allows the interconnection with other shields or modules. Nowadays it is very used in the digital world, since the electronic circuit assemblies are made easier. Its applications can be found in biomedical equipment, unmanned aircraft, 3D printers, among others [8].

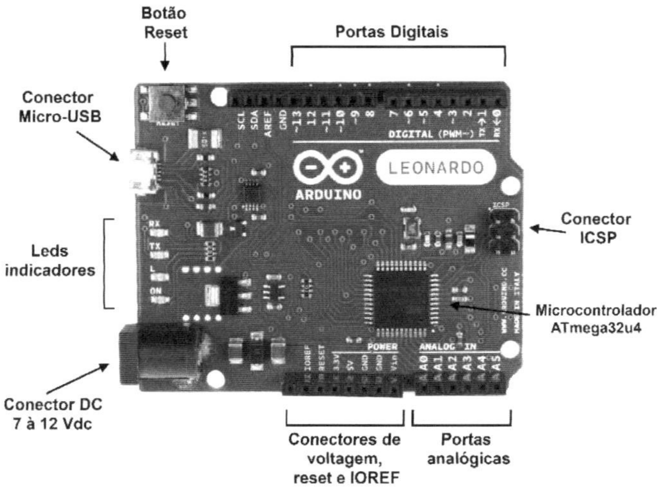

Figure 1. Programable Arduino Card Uno

1.2 Lithium-ion battery

It is a device designed for storage of electrical energy that uses a lithium salt as electrolyte that obtains the necessary ions for the reversible electrochemical reaction that takes place between the cathode and the anode.

Figure 2. Lithium-ion battery

1.3 Display LCD (Liquid Crystal Display)

It is a passive display of low consumption, based on a substance trapped between two plates of glass. This electronic component serves to display messages by means of characters such as letters, numbers or symbols.

Figure 3. Display LCD

1.4 Micro SD card interface module

This device allows an SD memory card to be connected to a microcontroller, so that information can be stored. This is ideal for carrying out different types of projects.

Figure 4. Module interface Micro SD Card

1.5 Infrared proximity sensor

It is a distance measurement sensor, which is based on a system of emission / reception of light radiation in the infrared spectrum [9].

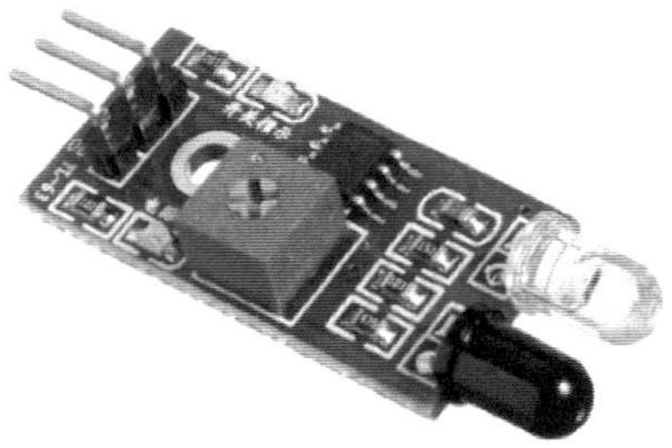

Figure 5. Infrared proximity sensor

The terms presented below are concepts collected from the project, these describe important aspects within the research process:

- *Poultry Farms:* A poultry farm is an agricultural establishment for the breeding of poultry such as chickens, turkeys, ducks, and geese, with the purpose of using them as a food base either by killing them for meat or by collecting their eggs.
- *Egg Posture:* Action of the birds, to lay eggs in the nest.
- *Module Cells:* Individual plastic or metal cages for laying hens.
- *Oviduct:* Conduit of the female reproductive system of the mammals that communicates each ovary with the uterus. In the reproduction of the birds, the ovule, when mature, penetrates inside the oviduct and after being fertilized by a spermatozoon it is surrounded by the clear and the shell, to finally form the egg.
- *Lighting pattern:* are the different characteristics of the lighting (type of light, intensity, exposure time, etc.), to adapt to specific needs [10].

1.6 Reasons why chickens stop laying eggs

a) *Shorter days.* Chickens are sensitive to the length of daylight, and particularly to the direction in which the length of the day is changing, when it comes to laying eggs.
b) *Inadequate nutrition.* Chickens need a balanced and adequate diet to maintain egg production. Each egg contains large amounts of protein and energy, which the hen must first consume as part of its daily food intake.
c) *Age.* A hen can live for many years. It is not unusual for a backyard flock owner to have several generations of birds and loses the notion of the age of some chickens. As in other species, an old chicken will eventually lose its ability to be active for reproduction and stop producing eggs.
d) *Illness.* Many bird diseases will affect the production of eggs. Often birds will show symptoms of disease, but sometimes they will not.

2. Methodological framework of the investigation.

2.1 Kind of investigation.

This investigation is framed in a technological development, of EXPERIMENTAL type, where it is tried to develop a practical prototype and of low cost that will be put under different types of tests to find the most optimal operation.

2.2 Research method.

The focus of this research is quantitative, according to Sampieri this type of approach is based on analyzing an objective reality from numerical measurements and statistical analysis to determine predictions or patterns of behavior of the phenomenon or problem posed, also employing experimentations and cause-effect analysis.

2.3 Technological resources.

The technological means with which the objectives of the project will be achieved are:

- Computer.
- Internet connection.
- Printer.
- Digital camera.
- Academic database.
- Office automation applications.
- CAD software for electronic design.

- Programming software.
- Procedures of the investigation.

2.4 Procedures of the investigation.
In order to successfully carry out the present project, it will be carried out in chronological fulfillment of the specific objectives and three stages were defined:

2.4.1 Design of the prototype and its characteristics.
In this stage the functional and non-functional requirements of the prototype will be defined, be they technical, mechanical, electronic, among others.

2.4.2 Final assembly of the prototype.
For the development of this stage it is necessary to validate all the requirements to proceed to the construction of the prototype, using the elements and suitable devices to achieve efficient and effective operation of the module.

2.4.3 Tests and results.
For this stage, the module undergoes different types of tests in order to meet the basic characteristics of a prototype, whatever it may be; complete, reachable and verifiable.

3. Results

For the development of the prototype, the productive chain of commercial poultry farming was validated, which has six production processes; interdependent, technified and demanding in aspects of genetics, nutrition, health, biosecurity and environment.

Specifically, the process taken into account for the development of the electronic module is carried out in the production farms of laying hens, where the birds arrive a day after birth and are vaccinated according to the risk factors of the area. The birds are generally housed in modules, floor, rice husk bed, chip, among others. The complete cycle in the laying hens lasts approximately 72 weeks, divided in turn in the breeding, raising and laying phases.

For the development of the prototype, a scaled module cells were designed, to which a proximity sensor module (consisting of an infrared LED and a phototransistor) was adapted. This module fulfills the function of detecting the passage of the eggs deposited by the hen laying The module cells are inclined at an angle of 30 degrees, making the eggs go down a ramp made of wood, passing through an area where the sensor is located, in which a beam of light is emitted and, if interrupted, a signal is sent. electrical signal to a microcontroller (Arduino Card), this system processes the information, then goes through a micro SD card module, which stores the data in a flat file of .txt format, later this information displayed on an LCD display, so that the Farmer poultryman can observe in real time the frequency of posture of the hen at any instant of time. In the figures 9 the final prototype is observed.

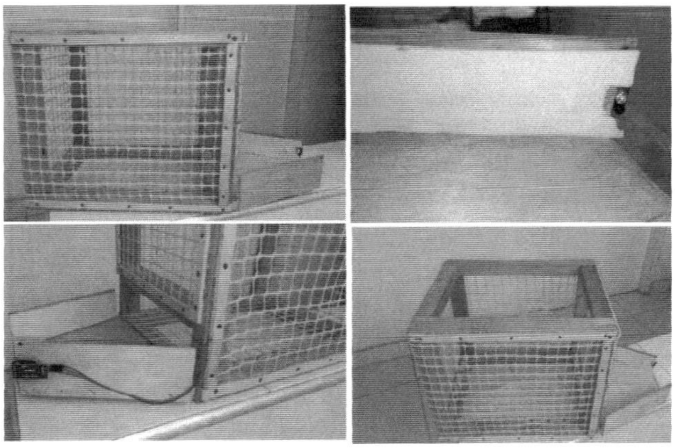

Figure 6. Electronic Module allows to measure the pawning frequency of laying hens.

In the figure 10 and 11, the tests and validations that were carried out with several angles of inclination of the module to achieve the suitable location for the postures of the hens are evidenced, it was validated that the system made count of postures and showed them in a LCD screen. It was found that it is necessary that poultry farms are divided into lots, to determine what factors can increase or decrease production. All the information collected and stored, the poultry farmer can take advantage of it to perform a thorough analysis in order to optimize the processes. The size and comfort of the module were also evaluated, important factors to avoid chicken stress and to reduce egg production.

For the feeding stage of the prototype it was determined that the consumption per hour is 1W, for this case 5 rechargeable batteries of 3.7 volts were placed at 1 ampere, in order to achieve a time of autonomy of the module of approximately 18 hours in case of failure of the electrical supply.

This tool can be of great importance for poultry farms, since all these processes are carried out by many poultry farmers through the observation method, which can generate a lot of uncertainty. With this device you can make projections and draw conclusions about the behavior of the hens.

Figure 7. Final Prototype to measure the pawning frequency of laying hens.

Figure 8. Technology used for final prototype

In the pilot tests of the prototype, the following technical characteristics were found, which are evidenced in Tables 1, 2 and 3. As future work, the investigation should be conducted in tests on poultry farms of the Department of Sucre.

Table 1. Technical tests of the Prototype.

Test	Average
Security	100%
Functionality of the Prototype	100%
Interaction with the user	95 %
Easy to use	95 %
Response time	99 %

Table 2. Characteristics of the final prototype vs. market prototypes

Device	Sensors	Size	Weight	Warning mode	Market price (Approx)
Module to measure egg frequency (created)	Infrared	Adjustable	[200-550 gr]	Tone and warning by light	20,2 €
Module to measure egg frequency (In the market)	Without sensors	Adjustable	[300-1000] gr	By observation	[25,00 - 100] €

Table 3. Characteristics of the final prototype

Elements	Infrared proximity sensor	LCD display	Battery	Arduino card
Working range	0.1 – 3 V	0.5 –5 V	0.5 –12 V	0-5V
Immunity external factors	No	Si	Si	Si
Energy consumption	0.03 W	0,15 W	0.21 W	0.25 W

For this first phase of the investigation, it only focuses on the design of the module. For phase 2, it is intended to test poultry farms in the department of Sucre to perform these processes manually and by the technique of observation to ensure that these are optimized.

The important aspects that must be had in the poultry industry are the following:

- The birds must be in well-designed module, clean, disinfected and with high levels of biosecurity.
- Birds must have good ventilation, air quality, temperature and space.
- The spread of diseases should be controlled by housing birds of one age and one origin.
- The homogeneity of the birds should be controlled.
- Prevent, detect and treat diseases.
- Cover nutritional needs, make good management of food and water supply.
- Waste collection system (excreta, corpses, etc.).

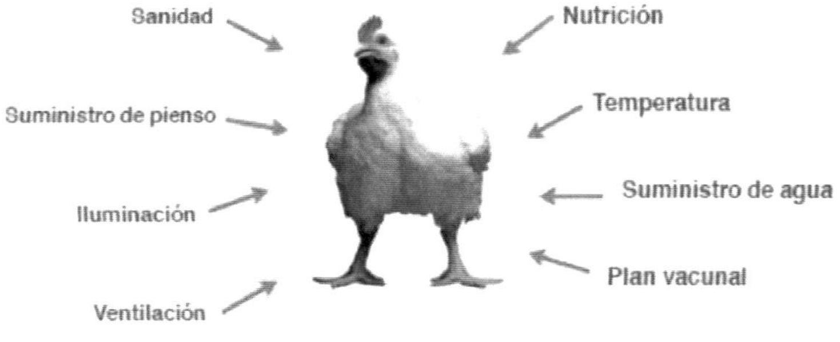

Figure 9. Important factors in poultry farming.
Source: Ross, 2010 Handbook.

4. Conclusion.

After carrying out this research work, it was concluded that it is possible to develop an electronic device that is economical, functional and versatile. A module was generated that allows to collect egg position data in real time to achieve the optimization of the productive processes of the poultry farmers; and with good results. This project aims to generate greater profitability and encourage quality processes that improve the competitiveness of the poultry sector in the department of Sucre, also providing the opportunity to use technology and human talent. The prototype was subjected to technical tests of operation, response time, use and safety where successful results were found and demonstrate the versatility of the tool. Finally, the aim is to encourage poultry farmers to invest in more competitive technologies in national and international markets. As future work, it is intended to generate a mobile application that allows to visualize in a graphic way the pawning of the chickens and generate historical records.

5. Recommendations

- The module cells must be properly inclined so that the egg has a free movement towards sensor detection.
- It is very important to check that the position detector sensors are not obstructed by foreign bodies, in order to avoid erroneous captures of the data.
- In the event that the electrical supply to the house fails, it is essential to restore it within 18 hours, in order to prevent the device from shutting down and the continuity of the process to be lost.

6. References

[1] Plan Nacional de Desarrollo *"Todos por un Nuevo País"* de Colombia (2014-2018)
[2] Servicio Nacional de Aprendizaje, SENA 2013. Manual de gallinas ponedoras. Recuperado en septiembre de 2017.
[3] Federación Nacional de Avicultores de Colombia (FENAVI) 2017 Cifras estadísticas. Recuperado en septiembre: http://www.fenavi.org/index.php?option=com_content&view=article&id=2472&It emid=133 0
[4] Marcio Saatkamp 2014 Automatización de iluminación en galpones de ponedoras Recuperado en noviembre de: https://www.wattagnet.com/articles/21351-automatizacion-de-iluminacion-en-galpones-de-ponedoras
[5] San Lucas I, y Garzón, D 2009 Control y Monitoria de un criadero Avícola Controlado por microcontrolador desde un sitio Web dinámico.
[6] Vásquez Rodrigo y Ballesteros 2007 Ambiente, Construcciones, Instalaciones y Equipos. En: La Cría de Codornices. Bogotá, DC.: Produmedios.
[7] Andrez Jhair Segovia Cruz 2014 Diseño e implementación de prototipo para sistema automatizado de recolección y extracción de codornaza en módulos de explotación de codorniz. Recuperado en abril de 2018.
[8] UCA 2014 Comenzando con Arduino Recuperado en febrero de 2018 de http://www.uca.es/recursos/doc/Unidades/Unidad_Innovacion/Innovacion_Docente /ANEXO S_2011_2012/22232441_310201212102.pdf
[9] Maloney Timothy J 2006 Sensor de proximidad. Recuperado en Marzo de 2018.
[10] Articulo Universidad de Georgia 2005 Razones por las cuales las gallinas dejan de poner recuperado en septiembre de 2017de http://www.elsitioavicola.com/articles/2484/razones-por-las-cuales-las-gallinas-dejan-de- poner-huevos/.
[11] FAO Alojamiento y manejo de las aves de corral. Recuperado en septiembre de 2017de www.fao.org/docrep/106/a1740s/a1740s00.pdf.
[12] GMBH Ponedoras. Guía de manejo, Lohmann Brown-Classic. Recuperado en septiembre de 2017 de http://www.morrishatchery.com/docs/Brow_spanisch_A4,_Endv.pdf
[13] Control y Monitoria de un criadero Avícola Controlado por microcontrolador desde un sitio Web dinámico.
[14] DNP 2013 *Plan estratégico departamental de ciencia, tecnología e innovación de Sucre.* Recuperado en abril de 2018 de: http://www.colciencias.gov.co/sites/default/files/upload/paginas/pedcti-sucre.pdf.

IOP Conf. Series: Materials Science and Engineering **519** (2019) 012012 doi:10.1088/1757-899X/519/1/012012

Analysis of adaptability estimation algorithms in dynamic systems

J N Zuleta[1] and R A Franco L.[1,2]
[1]Tecnoacademia Risaralda, Servicio Nacional de Aprendizaje SENA, Carrera 21 con 73 bis, Dosquebradas, Colombia
[2]Facultad de Ingeniería, Universidad Tecnológica de Pereira, La julita, Pereira, Colombia

jdnanez@sena.edu.co

Abstract. In this paper we show the adaptability analysis of estimation algorithms based on mobile average autoregressive models (ARMA) in an easy implementation system composed of a network of operational amplifiers (OpAmp) known as the Sallen- Key. Parameters are adjusted by means of projection algorithms, projection algorithm with covariance matrix, Recursive Least Squares (RLS) and Normalized RLS. The comparison of the performance of the estimation algorithms shows that those based on RLS and Normalized RLS quickly stabilize their values before disturbances in the input signals of the dynamic system.

1. Introduction

The implementation of control algorithms requires the construction of *a priori* mathematical models that describe the evolution of a system according to initial conditions and parameters. However, in the case of dynamic systems, developing a mathematical model that determines their behaviour, outside of ideal conditions, requires considering a large number of parameters whose behaviour against perturbations can't always be approximated to a linear and homogeneous relationship [1, 2], increasing the complexity of the model and the cost in computation time with implications in the control of processes in real time [3,4].

The algorithms of adaptive estimation in dynamic systems allow to establish conditions on the model that facilitate, before disturbances, adjust the parameters relatively quickly and stabilize the response of the system [5], thus decreasing the response time of the control algorithms [6, 7].

In this paper we show the adaptability analysis of estimation algorithms based on mobile average autoregressive models (ARMA) [8,9] in an easy implementation system composed of a network of operational amplifiers (OpAmp) known as the Sallen- Key.

2. Materials and methods

2.1. *Dynamic system*

The simulation of a second order system based on a network of operational amplifiers (OpAmp) known as the Sallen-Key network was performed, the topology is shown in figure 1. An easy-to-implement system composed of an acquisition card was designed ARDUINO ONE, and an interface in MATLAB (Figure 2).

Content from this work may be used under the terms of the Creative Commons Attribution 3.0 licence. Any further distribution of this work must maintain attribution to the author(s) and the title of the work, journal citation and DOI.
Published under licence by IOP Publishing Ltd

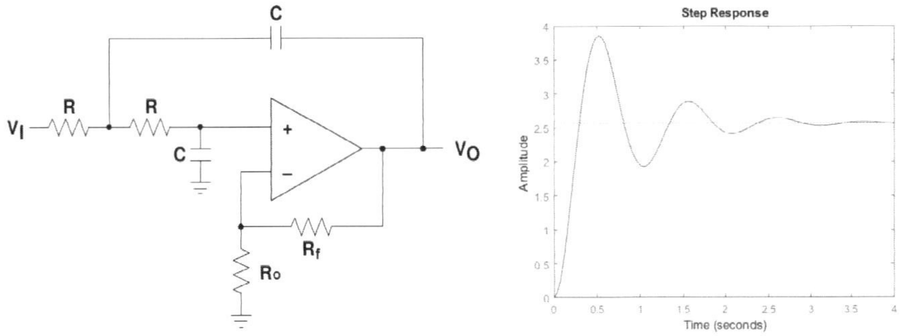

Figure 1. Topology of the Sallen-Key network and its step response design.

The design of the system is shown in Figure 1, which can be obtained the response of the dynamic system, was designed with the criterion of a maximum impulse (mp) of 50%, an establishment time (ts) of 3 seconds and a stable state gain μ of the Sallen-Key network (see equation 2) shown in Figure 1, its transfer function as a function of the circuit parameters (R and C) is shown in equation 1.

$$H(s) = \frac{\mu\left(\frac{1}{RC}\right)^2}{s^2 + \frac{(3-\mu)}{RC}s + \left(\frac{1}{RC}\right)^2} \qquad (1)$$

$$\mu = 1 + \frac{R_f}{R_o} \qquad (2)$$

Where, $R = 16.16k\Omega$, $C = 10\mu F$, $R_f = 15.69k\Omega$ and $R_o = 10k\Omega$

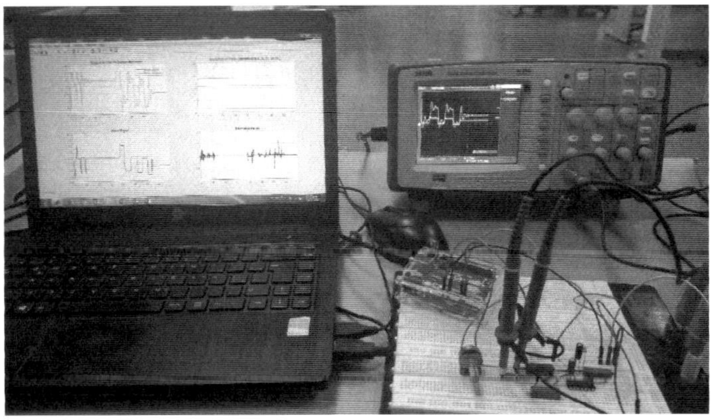

Figure 2. System assembly

2.2. *Estimation algorithms*

In this work an online estimation process is carried out. The estimation of parameters is based on autoregressive models of moving average (ARMA) [8]. This process can be done online if you have only one data at a time or offline if you have a data set. For the application of estimation algorithms in the identification of dynamic systems, it is necessary to capture the input signal $u\ (k)$ and output of the system $y(k)$ and construct a discrete linear model in the time domain (Eq.3) and in differences (Eq. 4). The system can be carried to a reduced model for regression (Eqs. 5 and 6).

$$H(z) = \frac{y(z)}{u(z)} = \frac{b_1 z + b_2}{z^2 + a_1 z + a_2} \qquad (3)$$

$$y(k) = -a_1 y(k-1) - a_2 y(k-2) + b_1 u(k-1) + b_2 u(k-2) \quad (4)$$

$$y(k) = [-y(k-1) \quad -y(k-2) \quad u(k-1) \quad u(k-2)] \begin{bmatrix} a_1 \\ a_2 \\ b_1 \\ b2 \end{bmatrix} \quad (5)$$

$$y(k) = \varphi^T \theta_o \quad (6)$$

The parameters to be estimated are a1, a2, b1 and b2, which belong to the dynamic model of the system as shown in equation 1. These parameters are adjusted by means of projection algorithms, projection algorithm with covariance matrix, Recursive Least Squares (RLS) and Normalized RLS.

The estimation algorithms that are compared in the development of this document are the following:

a. Projection Algorithm (PA)

$$\theta_k = \theta_{k-1} + \frac{\varphi}{\varphi^T \varphi}(y - \varphi^T \theta_{k-1}) \quad (7)$$

b. Projection Algorithm with Covariance Matrix (PA CM)

$$P_k = P_{k-1} - \frac{P_{k-1} \varphi \varphi^T P_{k-1}}{1 + \varphi^T P_{k-1} \varphi} \quad (8)$$

$$\theta_k = \theta_{k-1} + \frac{P_k \varphi}{\varphi^T P_k \varphi}(y - \varphi^T \theta_{k-1}) \quad (9)$$

c. Recursive Least Squares (RLS)

$$\theta_k = \theta_{k-1} + P_k \varphi (y - \varphi^T \theta_{k-1}) \quad (10)$$

d. Recursive Least Squares Normalized (RLS N)

$$\theta_k = \theta_{k-1} + \frac{P_k \varphi}{1 + \varphi^T P_k \varphi}(y - \varphi^T \theta_{k-1}) \quad (11)$$

With the equation (8) the calculation of the covariance matrix is performed, and in the same way for the algorithms shown in equations 9, 10 and 11.

3. Methodology

The methodology proposed for the development of the experimentation was carried out in the following way:

a. To the capture of the set of input and output signals of the dynamic system was performed, although an off-line estimate could be made because the whole set of samples is available, the process will be performed simulating the implementation recursively.
b. During the calculation of the estimation recursively by testing the 4 proposed algorithms, the calculation of the RMS error will be made for each algorithm, thus identifying those with the least error, this procedure will be carried out by means of a Monte Carlo experiment with a stop criterion based on the calculation of the minimum variance.

$$E_{rms} = \sqrt{\frac{1}{N}\sum_{i=1}^{N}\left(y_i - \phi_i^T\theta_{i-1}\right)^2} \qquad (12)$$

c. From the calculation of the estimate, we count on the signals generated from the parametric evolution during the estimation, to these signals obtained from the calculation of the average value and the minimum variance adjusting the RANSAC method to guarantee a constant as a result and the minimum variance this procedure was performed on the same Montecarlo experiment with the stopping criterion based on the minimum variance of the RMS error.
d. The criterion for the identification of the adaptability and the estimation capacity is given by the algorithms that present the minimum mean squared error (RMS), and identifying those that present lower variance and therefore a static behavior in the parametric evolution during the process of estimation.

4. Results and discussions

For the estimation of the stable parameters of the dynamic system, the evolution of the parameters a2, a1, b2 and b1 was identified, and the parametric evolution of 4 recursive identification algorithms was compared. The comparison of the algorithms will be made using the same set of input and output signals (Figure 3).

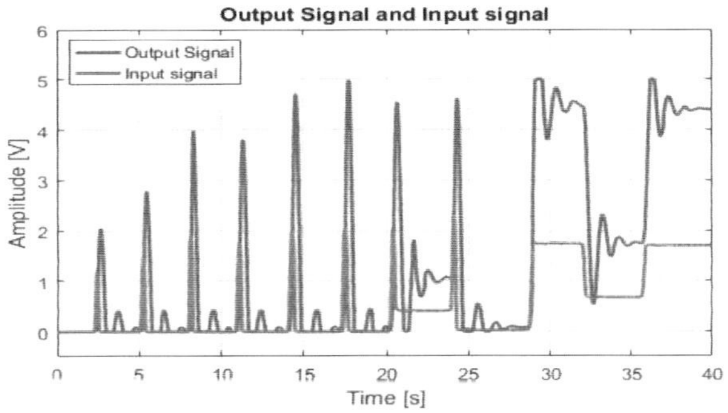

Figure 3. Set of input and output signals.

Figure 4 (Left) and (right) show the parametric evolution and the error, respectively, of the implementation of projection algorithms on the set of input and output signals shown in Figure 3. This implementation readjusts the parameters before changes of the input signal, however, does not allow to obtain a stable behavior.

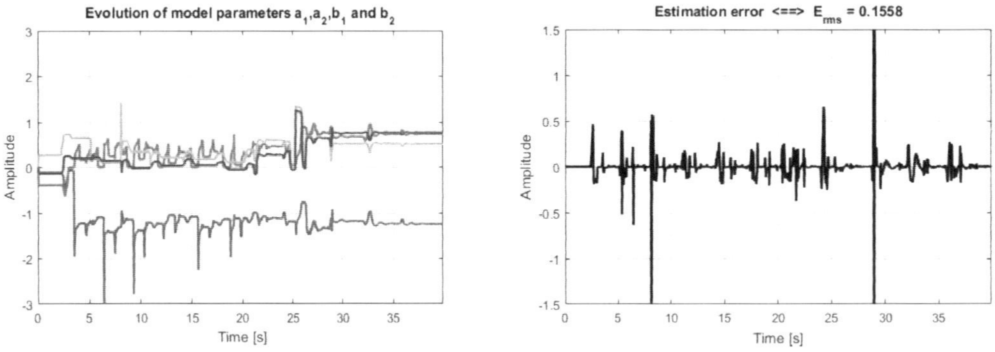

Figure 4. Parametric evolution projection algorithm

Figure 5 shows the parametric evolution obtained by the projection algorithm with covariance matrix (left) and the associated error (right). As in the case of projection algorithms, the parameters are sensitive to disturbances but do not exhibit stability.

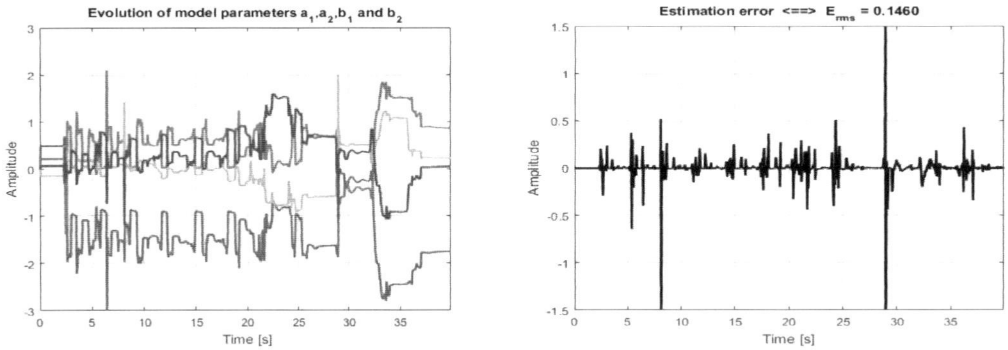

Figure 5. Projection algorithm with covariance matrix.

Figures 6 and 7 show the evolution of the parameters before disturbances of the input signal (left) and the estimated error (right). These algorithms present the most optimal performances, minimizing the error and exhibiting a stable readjustment of the parameters against variations of the input signal.

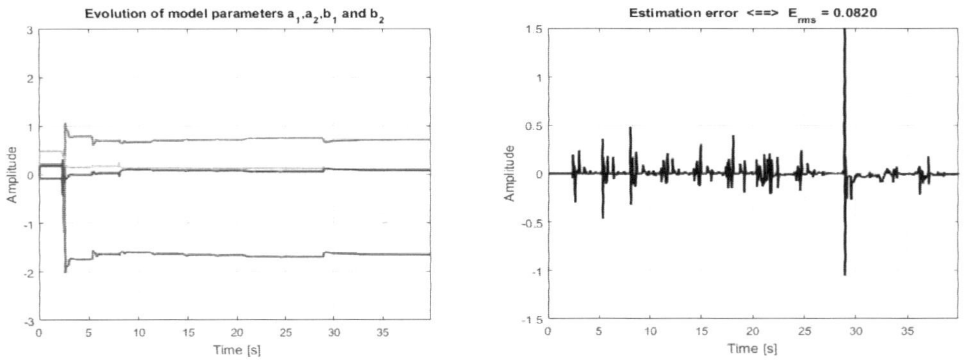

Figure 6. Evolution of RLS algorithms.

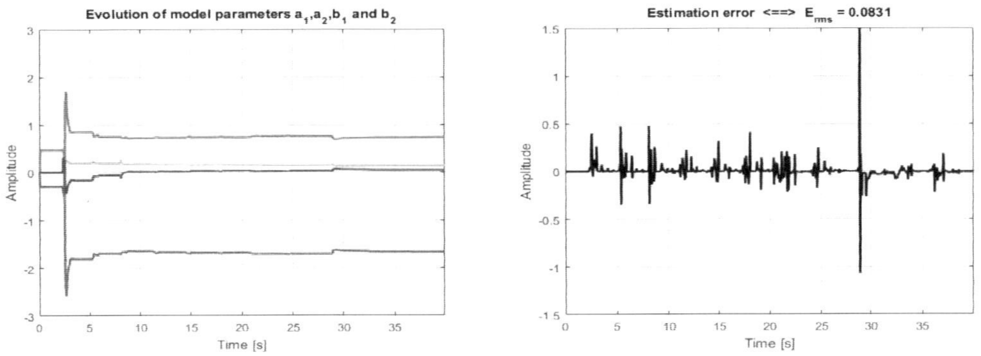

Figure 7. Evolution of Normalized RLS algorithms

The table 1 shows the relationship of the RMS errors obtained through the Montecarlo iterations, and in Figure 8 the behavior of the RMS error during the implementation of the Montecarlo experiment can be observed.

Table 1. Comparison of the RMS error of the estimation algorithms using statistical moments, Montecarlo experiments approx. 1000 iterations.

Estimation Algorithm	Error RMS mean	Error RMS deviation
Projection Algorithm (PA)	0.1591	$2973.50x10^{-6}$
PA with Covariance Matrix (PA CM)	0.1459	$31.41x10^{-6}$
Recursive Least Squares (RLS)	0.0819	$56.09x10^{-6}$
RLS Normalized (RLS N)	0.0885	$5994.90x10^{-6}$

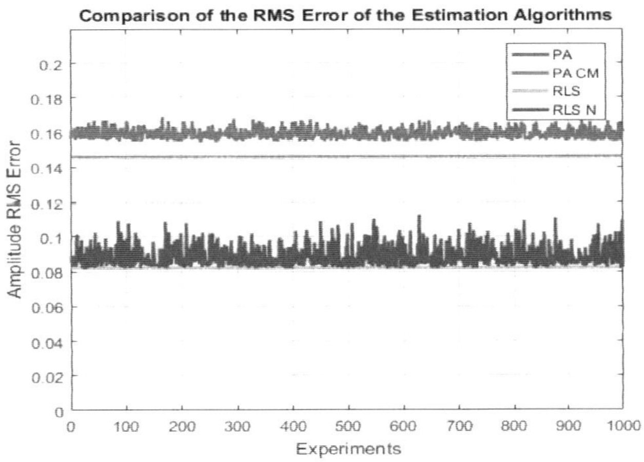

Figure 8. Comparison of the all algorithms using RMS error.

Table 2 shows the standard deviations obtained by the Montecarlo experiment, using in each iteration the RANSAC method to calculate the constant value and the minimum variance.

Table 2. Comparison of the standard deviations of the parameters Montecarlo experiments approx. 1000 iterations.

Estimation Algorithm	Parameter a_1 deviation	Parameter a_2 deviation	Parameter b_1 deviation	Parameter b_2 deviation
Projection Algorithm (PA)	0.1202	0.1175	0.0927	0.1685
PA with Covariance Matrix (PA CM)	0.2202	0.1669	0.1465	0.1660
Recursive Least Squares (RLS)	0.1458	0.0713	0.0284	0.0316
RLS Normalized (RLS N)	0.1554	0.0955	0.0360	0.0376

5. Conclusions

An easy-to-implement system was built that allowed analysing the performance of the estimation algorithms defined in terms of their response to variations in signals and ability to estimate and adapt.

The comparison of the performance of the estimation algorithms shows that those based on RLS and Normalized RLS quickly stabilize their values before disturbances in the input signals of the dynamic system, making it possible to approximate the transfer function in discrete time.

From Table 1, it can be identified that the RLS and RLS N algorithms have the lowest mean squared error of estimation in comparison with the projection algorithms. However, the RLS algorithm not only has the lowest RMS error, but also the minimum standard deviation. In Figure 8 you can observe the behavior of the errors during the evolution of the Monte Carlo experiment.

Finally, using table 2, we could identify that for each parameter of the estimation model the minimum possible variance was found, where for parameter a_1, the PA algorithm showed the least deviation, however, the RLS algorithm is in second place, and for the rest of the parameters (a_2, b_1 and b_2) the algorithm has the least deviation.

From the results obtained from the experimentation it can be concluded that the best estimation algorithm with parametric stability is the RLS algorithm, since it presents a low mean squared error of estimation, and the parametric evolution presents the minimum variance and stability in most of the estimated parameters.

6. References

[1] Safaei A and Mahyuddin M 2018 Adaptive Model-Free Control Based on an Ultra-Local Model With Model-Free Parameter Estimations for a Generic SISO System *IEEE Access* **6** 4266-4275

[2] Wang X, Zhao J, Wang P and Tang Y 2015 Finite-time parameter estimation of switched systems *The 27th Chinese Control and Decision Conference (2015 CCDC)* 392-396

[3] Majeed F 2018 Model based estimation approach to the missing data problem *Advances in Science and Engineering Technology International Conferences (ASET)* 1-6

[4] Yang Y 2017 A neural network model reference adaptive system for speed estimation of sensorless induction motor *29th Chinese Control And Decision Conference (CCDC)* 6864-6868

[5] Sihag S and Tajer A 2018 Power System State Estimation Under Model Uncertainty in *IEEE Journal of Selected Topics in Signal Processing* **12** 4 593-606

[6] Lee H and Kim H 2015 Cost Estimation of Hybrid System Models in Simulation Based Acquisition *4th International Conference on Advanced Information Technology and Sensor Application (AITS)* 55-58.

[7] Liang X, Li W, Lin J, Su L and Li H 2010 Model Identification for Autonomous Underwater Vehicles Based on Maximum Likelihood Relaxation Algorithm *2010 Second International Conference on Computer Modeling and Simulation* 128-132.

[8] El-Sinawi A, El-Baz H and Amer N 2013 Multi-input single-output (MISO) random system modeling using methods of system identification *5th International Conference on Modeling, Simulation and Applied Optimization (ICMSAO)* 1-6.
[9] Lin J and Li S 2012 Parameter estimation of continuous model based on Marquardt algorithm *24th Chinese Control and Decision Conference (CCDC)* 903-906.

IOP Conf. Series: Materials Science and Engineering **519** (2019) 012013 doi:10.1088/1757-899X/519/1/012013

Software-Defined Optical Network (SDON), new generation networks

G V Leytón[1]
[1] Engineering Faculty, University Institution CESMAG, Pasto COLOMBIA.

ginnaleytony@gmail.com

Abstract. At the time of establishing communications, the configuration of network resources, storage and computer systems worked independently. With the innovation and the new communication alternatives, the management was initiated so that all the parts mentioned above work together. This is how the idea of virtualization emerged, a process by which virtual versions of computers, operating systems and applications are created. For all network devices and servers to communicate with each other and function correctly there must be a centrally functioning controller, this is when a promising SDN paradigm emerges; networks defined by software considered as a promising method to rebuild the Internet architecture and improve the implementation and distribution of applications in communication networks. In addition, we will discuss how SDN networks perform their inclusion in optical networks, study these advanced architectures, in order to facilitate flexibility and resource optimization in optical transport.

1. Introduction

The increase in communication networks, the thousands of data transmitted and the constant use of the network, require improvements in service and resources that are part of it, in addition to avoiding the various causes of performance impact such as packet loss, latency, bit rate and network congestion [1].

The main characteristic of software-defined networks is the separation of the control plane from the data plane, the first is responsible for managing the necessary resources of the network with their respective traffic and the second is composed of different network devices, transmission media and different services.

On the other hand, SDN creates networks that work with the centralized control of a software application, which eliminates the need to use specifically designed network devices; in this way a network administrator can shape traffic from an SDN controller without having to touch individual network devices.

The decisions on the transmission path of the network flow, in addition to the configuration of the network infrastructure from a centralized application, are made by the SDN, which is responsible for responding efficiently to the data plane, taking into account the status and requests of the application plane. These decisions taken from the control plane are communicated to the data plane with the necessary actions to be carried out on the network infrastructure. Each plane of the SDN has its specific functions and communication protocols [2].

The Open Networking Foundation (ONF) is in charge of developing protocols or open standards that allow the implementation of SDN. Within the control protocols implemented in software defined networks, the Borden Gateway Protocol (BGP) used to exchange routing information between gateway terminals in a network of autonomous systems is identified. Each network has a router and its respective network information such as routing tables that store the routes of the different connected routers, the ip addresses that can be reached, in addition to a cost metric associated with the route to choose the best

Content from this work may be used under the terms of the Creative Commons Attribution 3.0 licence. Any further distribution of this work must maintain attribution to the author(s) and the title of the work, journal citation and DOI.
Published under licence by IOP Publishing Ltd

IOP Conf. Series: Materials Science and Engineering **519** (2019) 012013 doi:10.1088/1757-899X/519/1/012013

and available [3]. Also, there is the Netconf protocol used for the secure administration of the network, opting for the remote procedure call that is used to solve the different inconveniences present in the configuration of the network devices. Additionally, there is the MPLS Transport profile protocol (MPLS-TP) which was designed to be used as a network layer technology, as a connection-oriented packet switching application (CO-PS,) dedicated to eliminating features that are not relevant in CO-PS applications and add devices for critical transport. The purpose of the ONF is to develop an architecture that allows full control from a software controller and manages the operation of the network [4].

The OpenFlow protocol makes it possible to coordinate the routes for sending packets from one terminal to another, in addition to administering and programming the network as a whole, since in the control plane the decisions for migrating packets in the data plane are centralized. We chose this protocol because its use extends to optical networks, has been developed to allow the virtual allocation of network resources from various levels of granularity. In the application plane can require or dynamically control network resources wavelength, circuit, flow in granularity to seamlessly reconfigure resources based on the requirements of the application plane [5].

2. Software Defined Networking

Before entering SDN, let's remember the traditional networks. A data network is the set of network devices organized in a hierarchical way and interconnected with each other through a means of transmission that allows the exchange of information and resources. Each network and terminal device that is connected to the network, manages its own configuration and preset policies that define the tasks of each device, as well as what it must do with the data or information that passes through it, this configuration is done before the network operates.

SDN, is a paradigm that allows the control of the network infrastructure located in the data plane intelligently, implements the use of centralized software with which manages the network functions, carries out a centralized and distributed control of the network, for this reason the control plane and the data plane must operate independently [6].

In the past, the network administrator had to configure each independent device belonging to the network, a rather complicated and inflexible task for companies with large networks. It is now in charge of traffic management by means of controllers with which an efficient flow of the network can be obtained, which will be able to respond to all the needs that arise during its operation [7].

In SDN the packet forwarding is controlled by an interface, the way of processing the packets is no longer fixed, each packet can be handled separately and managed by the control plane, by a controller that is responsible for meeting the requirements of performance, scalability, flexibility and agility necessary for Internet communications [8].

To summarize the elements that characterize SDN are:

- The control and packet switching functions are separated.
- The functions of the control plane are centralized in a single node, in which the data traffic routes are configured.
- Network traffic control, SDN can share network traffic between several servers without having to create a system that is capable of supporting this traffic.
- The functions of the control plane are implemented in a controller or software, which manages the separation and coordination of network traffic.
- Any type of software can be developed for the network.
- The network administrator can make modifications or verify novelties in an efficient way, since it performs permanent monitoring of the network.
- It enables and promotes the development and use of routing algorithms to exploit network capacity.

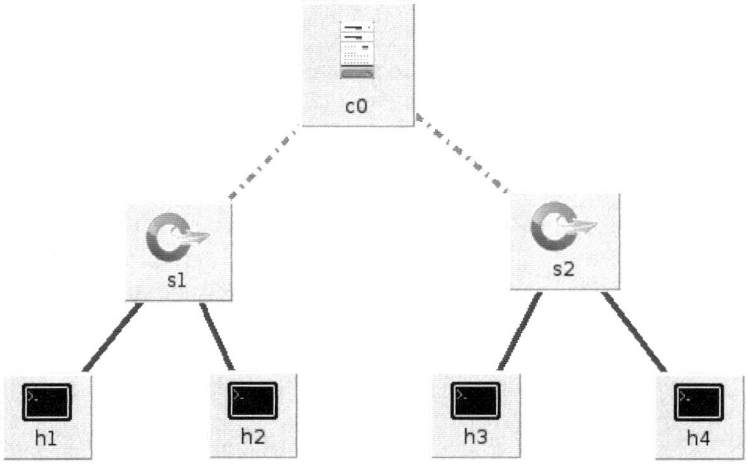

Figure 1. Software defined network topology.

2.1 Data Plan

This plane consists of network devices or terminals that form a data network and perform certain tasks depending on the functions defined by the logical components of control and data.

In the control is the state of the network that is communicated to the controller, which simultaneously indicates the routes for packet forwarding, these routes can be optimized by implementing routing algorithms. In the data, the network devices forward the packets according to the decisions made by the control plane. The first packet of a data stream is sent to the controller for processing of the packet, in order to update the routing tables and send the packets according to it [9].

2.2 Control Plan

Within this plane the most important thing is the controller, since it is in charge of managing the application and infrastructure plane, centralizing the monitoring of all the devices of the network. On the other hand, this plane is in charge of the communication with the SDN applications and the data or infrastructure plane.

The control plane communicates with the application plane and translates the requested requirements, as well as collecting all the information concerning the data plane and displaying it in the application plane.

At the same time, as previously mentioned, the control plane through communication with the infrastructure plane, is the one that collects the state of the network and according to the requirements of the applications, the control plane updates the route of sending packets in the devices [9].

2.3 Application Plan

As mentioned above, one of the main features of SDN applications is to communicate requirements to the network through the application programming interface (API), which is located between the application plane and the control plane, and is designed to facilitate user interaction and meet the needs of users [10].

Separation of two planes: data plane and control plane.

The data plane performs a sequence of tests and verifications on the datagrams that travel at the link level and are responsible for verifying that the data is adequate. If the datagram is incorrect, the package is delivered to the controller to decide what to do with it. If it is correct, the FIB (Forwading information base) operates to see what the destination of this data is, since the FIB is used in the forwarding of packets.

The control plane consists of several local data that allow you to create the packet forwarding table called RIB (Routing information base), which are used by the data plane to manage network devices.

The RIB exchanges information all the time to keep up to date. After the RIB we have the FIB, but this comes into operation when the RIB is stable. The RIB is used in the control plane and the FIB in the data plane [11].

This type of searches improve the performance in the forwarding of packets, searches that are very desired for services with a large bandwidth. To facilitate this sending of packets, the datagram has a header, thanks to it can also facilitate the search time, identifying much faster the destination of the packet.

2.4 OpenFlow Protocol

This protocol emerged at Stanford University and seeks the separation of the control plane and the data plane, the standardization of protocols between network devices and controllers, from a centralized API we can promote the programmability of the network.

For this protocol to work, this architecture proposes the existence of OpenFlow switches. This is formed by flow tables that are managed from the controller and an external channel that connects to the controller. The controller manages the behavior of the switch through the secure channel using the OpenFlow protocol [11].

The controller can add, update, and delete information from the flow table, both reactively (in response to packets) and proactively (generating actions).

Each flow entry works in the following ways:

- Send the packet flow to a respective port.
- Encapsulate and send the packet stream to the controller for him to decide what to do with it.
- Override the packet flow for security.

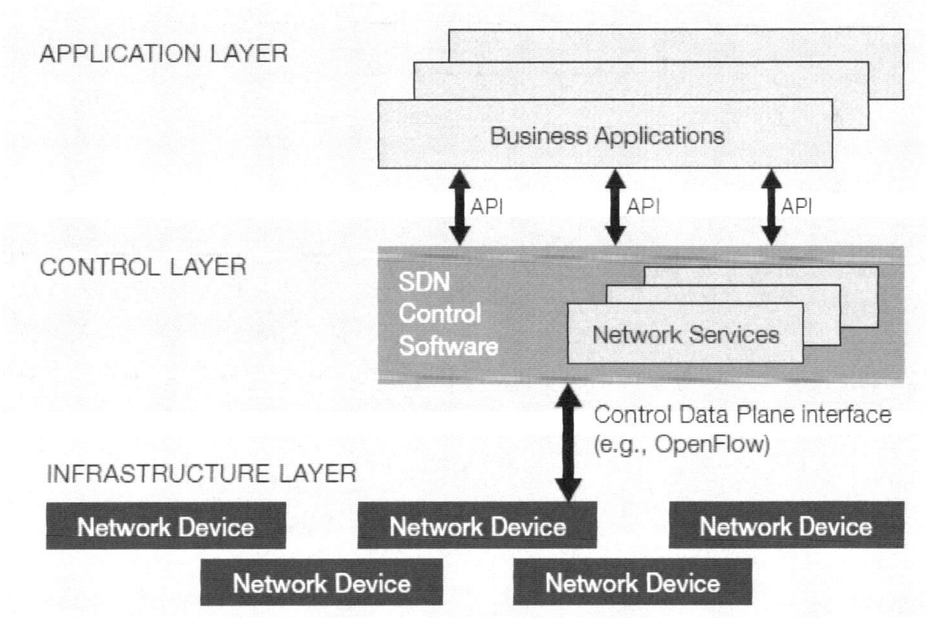

Figure 2. Software Defined Network Architecture [9]

3. Software Defined optical network

We have already carried out a brief review of the networks defined by software. Now they are, because they are included in optical networks. Optical networks offer too many advantages over conventional networks, since they offer greater capacity, high scalability and low power consumption. The single fibre transmission capacity has exceeded 100 Tb/s, the 100 Gb/s per channel wavelength multiplexing (WDM) system is being implemented and the superchannels handle a data rate of 400 Gb/s or 1Tb/s [12].

Optical networks handle data traffic, which allows SDNs and particularly the OpenFlow protocol to support and integrate multiple network technologies. In 2013 the Open Network Foundation (ONF) created the Optical Transport Working Group (OTGW) teaches the benefits of extending software-defined networks and the Open Flow protocol into optical transport, including flexibility in network management or administration, improved control of optical transport network, implementing new virtualization services and systems control [13].

Packet switching in conventional networks is less scalable than switching optical circuits that can change to higher granularity, the optical circuit consumes much less energy than the packet switch that is growing by the volume of traffic, in addition the combination and filtering of basic operations of optical signals are completely passive. Software defined optical networks have similar characteristics to SDNs, such as: network management and administration from centralized control, and network devices in the infrastructure plane offer high flexibility to support different network conditions.

The structure of the optical networks defined by software, exposes 3 main elements. The application plane, the highest plane of the structure, allows the end user from a control application to apply changes to the network without any inconvenience. This plane communicates with the next plane, through an API that facilitates interaction with the control plane, which is characterized by having a centralized controller that manages the network, improvement and custom through software that allows optimizing energy, resource allocation and protection.

And finally we have the data plane, where there is the programmable optical hardware, including transmitters, receivers, modulators, couplers, filters, amplifiers, switches, among others. These devices perform operations such as transmission, modulation, elastic bandwidth commutation, allow to establish the optimization and management of SDON. As in optical transport the main functions are transmission and switching, it is important to have two important optical network devices in SDON [14].

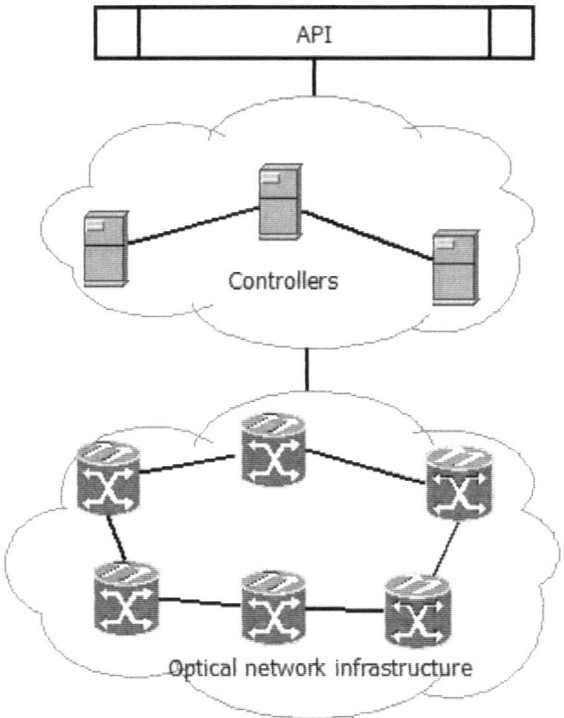

Figure 3. SDON Architecture.

The Flexible Switch is considered a reconfigurable optical insertion and extraction multiplexer (ROADM). It is responsible for switching any output port, has multiple ports, in addition, have the wavelength conversion function and the ability to fully switch data streams in the optical domain. ROADM is applied to SDON providing flexible switching, which allows you to adapt to different types of signals without additional problems in costs or signal quality. As described The main features are [15]:

No filter: The transponder does not need an optical demultiplexer or filter to dynamically select the WDM channel from the flex-grid.

No Grid: The channel bandwidth can be dynamically varied to accommodate variable transponder signals.

No channelization: Adjacent WDM channels can be grouped into wavebands that can be switched.

No connection: Signals with the same wavelength can use different simultaneous inputs and outputs.

Colorless: ROADM can extract or add any wavelength from the transponder.

No direction: The transponder receives an extracted input signal or adds an output signal.

Another important device is the variable transponder, a device used in telecommunications as a receiver and transmitter of signals. Signal characteristics such as data rate, modulation format and correction coding can be dynamically changed for WDM channels from the transponder. Various schemes are found to achieve the above-mentioned features [15]:

Digital transmitter: high-speed digital signal processing is used in real time for data modulation format, then the optical signals are modulated in phase and quadrature through electro-optical modulators. Additionally, it has a coding of errors that allows you to correct them are in need of more resources.

Variable format transmitter: An O/E/O signal generator and cascade optical modulators are used to select between QPSK, QAM, taking into account spectral efficiency.

Variable Transmitter: Optical orthogonal frequency division multiplexing is used to create high capacity superchannels.

We use flexgrid and elastic rates to supplement the above-mentioned features. Flexgrid allows adapting the bandwidth in a heterogeneous way according to the demand; in addition, the ITU must establish a new recommendation for the thinnest frequency slots, where the spectrum of an optical channel includes several frequency slots [16].

The Elastic Rates dynamically adjust data rates by adjusting modulation formats and the number of subcarriers. The elastic transponders used are defined by software, allowing the flexible allocation of sufficient optical bandwidth according to the current traffic demand, which is an advantage, since it speeds up the transport in optical networks [17].

It is as we see here, the joint work between optical networks and SDN, where centralized control based on SDN, is necessary for adaptation to the characteristics of the flexible optical layer, where some specific functions are included:

Flexible allocation of network resources: from the SDON controller the allocation of network resources is programmed in a flexible way to achieve performance optimization [18].

Spectrum optimization: SDON uses routing algorithms for wavelength assignment and spectrum assignment to optimize spectrum utilization taking into account demand, network status, required transmission speed, and the appropriate light path for transmission [19].

Network defragmentation: Several researches are known to develop network defragmentation algorithms, which have the capacity to consolidate the spectrum and reoptimise the use of network resources [20].

4. Viability and relevance

At present, several routing and wavelength assignment (RWA) algorithms have been proposed in optical network studies, but they have the disadvantages of assigning a specific wavelength and a fixed bandwidth for each connection. Meanwhile, in the routing and assignment of spectrum, a central frequency and a variable bandwidth are assigned depending on the requirements of the service.

Due to the above, is motivating the study of routing and spectrum allocation (RSA), which has emerged as a way of offering an efficient use of available optical resources, hand in hand with networks defined by software. Which can solve the problem of routing; the calculation of routes between the source origin and destination through a centralized controller, and additionally the allocation of spectrum could be worked through artificial intelligence techniques.

Bearing in mind that the implementation of optical networks defined by software is a clearly new subject, a feasibility study was carried out identifying the limitations, opportunities and assumptions of the same. With regard to the limitations in the development of the solution presented, the limitation was found in the testing and implementation of an optical network defined by software in real environments, since there is no technological network infrastructure to perform such tests, therefore, has opted for the use of simulators and network emulators such as Omnet + + and Mininet. The first, allows the modular simulation of discrete events in networks, programming is object-oriented, the second allows creating a realistic virtual network on a single machine, running real Kernel, since it is done through CLI. The opportunities of the study, are the development of algorithms of assignment of wavelength, spectrum or routing that allows to speed up the configuration of the optical networks and likewise to solve the main disadvantages as the speed of bit, loss of packets and latency.

5. Future jobs

This is the beginning of the development of the end-of-master project "To develop a dynamic spectrum assignment method for flexible optical networks defined by software, using artificial intelligence and performance quantification techniques, in order to optimize the use of spectral characteristics". Bearing in mind the increase in network traffic and the demand for new telecommunications services that require high bandwidth, the aim is to avoid wasting bandwidth by increasing network capacity, reduce the complexity of network elements and improve network reconfigurability with software-defined optical networks (SDON), since this way large volumes of information can be transported quickly and securely but with relatively small computational efforts. The algorithms will be tested in the Omnet++ network simulator.

6. Conclusion

In this work a partial description was made of the networks defined by software and their development process and incorporation into optical networks. Since they are considered new generation networks, since they have design features with the separation of control functions, transport and services, they are considered open interfaces, they have planning and integration of services and maintenance and integration with current networks. The most relevant characteristic in optical networks are the centralized controllers based on SDN and the development of routing algorithms for the assignment of spectrum and wavelength, since from them an intelligent control can be carried out that allows network administrators to improve the management and distribution processes of the network. Additionally, from the control plane a protection scheme can be centralized and thus guarantee the stability of the network. From another angle, we have optimization in optical transport, given that there is flexible allocation of bandwidth depending on traffic demand. SDON therefore has great potential and a promising future in the development of telecommunication networks.

References

[1] Gringeri S, Bitar N and Xia T 2013 Extending software defined network principles to include optical transport *IEEE Xplore* ***51*** *3* 32-40

[2] Channegowda M, Nejabati R and Simeonidou D 2013 Software-defined optical networks technology and infrastructure: Enabling software-defined optical network operations *IEEE Xplore* ***5*** *10* A274 - A282

[3] Wessing K 2015 Future network architectures *Deliverable d12.3*

[4] McNickle M 2014 Cinco protocolos SDN que no son OpenFlow *Techtarget*

[5] McKeown N, Anderson T, Balakrishnan H, Parulkar G, Peterson L, Rexford J, Shenker S and Turner J 2018 OpenFlow: enabling innovation in campus networks *ACM SIGCOMM Computer Communication Revie 38 2* 69-74

[6] Khondoker K 2014 Feature-based comparison and selection of Software Defined Networking (SDN) controllers *Computer Applications and Information Systems (WCCAIS)* 1-7

[7] Vargas W 2014 Emulación de una red definida por software utilizando Mininet *Escuela Técnica Superior de Ingenieros en Telecomunicaciones (ETSIT - UPM)*

[8] Nunes B, Mendonca M, Nguyen X, Obraczka K and Turletti T 2014 A Survey of Software-Defined Networking: Past, Present, and Future of Programmable Networks *IEEE Xplore* **16** 3 1617-1634

[9] Thyagaturu A, Mercian A, McGarry M, Reisslein M and Kellerer W 2016 Software Defined Optical Networks (SDONs): A Comprehensive Survey *IEEE Xplore* 2738-2786

[10] Foundation O 2012 Software-Defined Networking: The New Norm for Networks *ONF White Paper*

[11] Nadeau T and Gray K 2013 SDN: Software Defined Networks. An Authoritative Review of Network Programmability Technologies *O'REILLY Media*

[12] Xia G 2011 Field Experiment with Mixed Line-Rate Transmission (112-Gb/s, 450-Gb/s, and 1.15-Tb/s) over 3,560 km of Installed Fiber Using Filterless Coherent Receiver and EDFAs Only *IEEE Xplore*

[13] O. N. F. (ONF) 2015 Optical Transport Protocol Extensions *ONF TS-022* **1** 01-33

[14] Rouskas G, Dutta I and Baldine I 2008 A new internet architecture to enable software defined optics and evolving optical switching models *Broadnets* 71-76

[15] Philip N 2013 Software Defined Optical Network *IEEE Xplore*

[16] Zhang S, Martel C and Mukherjee B 2012 Dynamic traffic grooming in elastic optical networks *IEEE Xplore* 4-12

[17] Bhaumik P, Zhang S, Chowdhury P, Lee S, Lee J and Mukherjee B 2014 Software-defined optical networks (SDONs): a survey *Springer Science*

[18] Patel A, Ji P, Jue J and Wang T 2012 Routing, Wavelength Assignment, and Spectrum Allocation in Wavelength-Convertible Flexible Optical WDM (WC-FWDM) Networks *OSA Technical Digest*

[19] Patel A, Ji P, Jue J and Wang T 2012 Routing, wavelength assignment, and spectrum allocation algorithms in transparent flexible optical WDM networks *ONDM* **9** 191-204

[20] Ankikumar P, Ji P, Jue J and Wang T 2011 Defragmentation of transparent Flexible optical WDM (FWDM) networks *IEEE Xplore*

IOP Conf. Series: Materials Science and Engineering **519** (2019) 012014 doi:10.1088/1757-899X/519/1/012014

Model for the interoperability between health service providers (IPS), based on the electronic clinical information standard (HL7) and architecture to design and develop distributed systems (SOA)

J G Arévalo[1, 2], A Gonzales[1] and L Arévalo[1]
[1]Universidad de La Guajira, Riohacha, Colombia.
[2]ENAPROT, grupo de invetigscion Energía alternativa, productividad y la tecnología, Universidad de La Guajira, Riohacha, Colombia.

jarevalo@uniguajira.edu.co

Abstract. This study addresses the need to exchange clinical records, efficiently and uniform among health institutions (IPS Spanish acronym for Instituciones Prestadoras de Salud), which is generated in clinical care users from using the Hospital Information Systems (HIS). From defining a model for interoperability, with reference to the regulations governing the General System of Social Security in Health in Colombia (SGSSS Spanish acronym for Sistema General de Seguridad Social en Salud), the standards to facilitate electronic exchange of clinical information Health Level Seven (HL7) and service oriented architecture (SOA). The model it wasdesigned in two dimensions, on the one hand horizontally would facilitate electronic exchange of clinical information between IPS and on the other side vertically was projected was obtained as they settled components for the exchange of healthcare and administrative records the SGSSS different actors, in this case with the health Promotion Companies (EPS Spanish acronym for Empresas Promotoras de Salud) and the Territorial Entities through health secretaries.

1. Introduction

During the provision of health services to users, a large amount of data is generated in a fragmented manner, due to the fact that the SGSSS actors in Colombia (hospitals, clinics, clinics, pharmacies, laboratories, IPS, EPS, among others) operate usually in an isolated manner using HIS, which partially facilitate the generation of documents necessary for the exchange of information between institutions, in order to guarantee accessibility, opportunity, continuity and quality in the care of services, such as treatments and medical authorizations to other complexity levels [1] and other regulatory standards.

In various investigations and regulations of the national government [2][2][3][4] and international level [5][6][7], Alternatives have been postulated for the exchange of clinical information and architectures that promise to integrate different information systems, in order to provide a solution to the need to interoperate care and administrative records between the institutions providing the health services. Investigative projects of reference, guide this study, to reach a level of interoperability among the HIS of the different IPS in Colombia, which allows the use of information between them, through the design of an architectural model in which involve, the regulations the SGSSS, the HL7 [8][9] development framework and the SOA architecture.

Content from this work may be used under the terms of the Creative Commons Attribution 3.0 licence. Any further distribution of this work must maintain attribution to the author(s) and the title of the work, journal citation and DOI.
Published under licence by IOP Publishing Ltd

IOP Conf. Series: Materials Science and Engineering **519** (2019) 012014 doi:10.1088/1757-899X/519/1/012014

2. Interoperability and health services

The term interoperability can be defined as the ability of systems to exchange and share information between two or more heterogeneous systems and then be processed independently. The interoperability requirements are defined taking into account the scope and subjectively consider the skills, knowledge, socio-political and cultural restrictions present. All this, with the iterations necessary to achieve business objectives [10]. The HIS, health information systems, or hospital information systemns, contain the demographic and general information of the patient, the medical agenda and the clinical history, it also stores and organizes all the specific information of the diagnoses and treatments carried out in a health institution and allows access to the treatment information to the staff doctor to obtain a broad knowledge of the patient's condition.

SOA is an architecture that establishes a frame of reference for the integration of systems, which manage to communicate and share functionalities, generally through the network with the implementation of web services, managing to establish functional services in the form of modules for the integration of monolithic systems. In order for the HIS interoperability to be implemented, the main HL7 standards are Messaging HL7, Clinical Document Architecture (CDA), Structured Product Labeling (SPL), HL7 Medical Records, GELLO, Arden Sintax and Clinical Context Object Workgroup (CCOW).). Prepared by the HL7 organization, created in 1987 and accredited by the American National Standards Institute – ANSI.

3. Methodology HIS-DF

HIS-DF, integrates the Unified development processes of Rational (RUP), architectural styles such as the Reference Model for Open and Distributed Processing (RM-ODP), Architecture Directed by Models (MDA), Service Oriented Architecture (SOA), and HL7 frameworks. Dividing in 5 phases the process for the development of an architectonic model of interoperability from the implementation of RUP, the HL7 standards and the SOA architecture, described below: I), identification of the system; II), Identification of domains; III), complexity management; IV), Establishment of the reference architecture, V) Establishment of the development process.

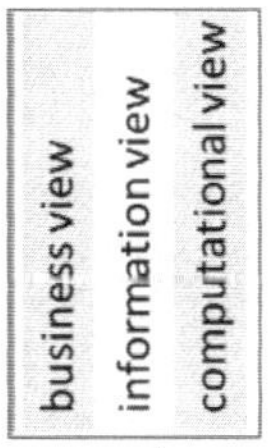

Figure 1. Reference views of RM-ODP

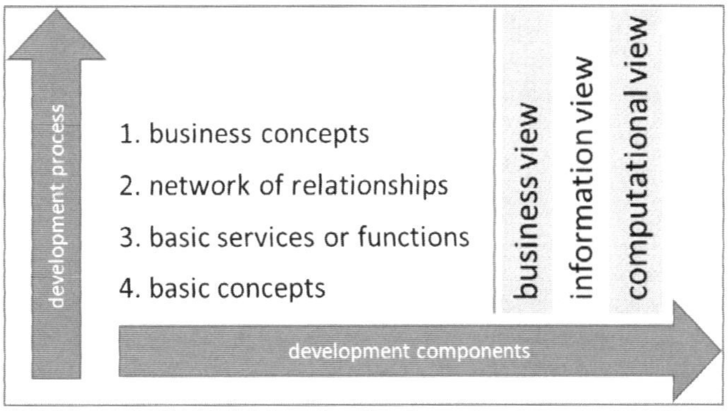

Figure 2. Establishing the development process (RUP).

IOP Conf. Series: Materials Science and Engineering **519** (2019) 012014 doi:10.1088/1757-899X/519/1/012014

From the use of the methodology, the components of the RM-ODP that refer to business, information and computational views. Subsequently, the development processes that integrate RUP are established, to establish the stages of the design of the interoperability model. In this context, based on the establishment of the development process as mentioned in phase 5 of the RM-ODP model, the architecture of the Hospital Information Interoperability Model is designed. Defining a correlation between them, for the reference of packages, the evolution between them and the integration with UML to facilitate the development and specifications of the system. Is presented below High-level diagram of the business view, the main objective of the business view is to identify the requirements established for the design of the interoperability model, supported by the specification of Requirements according to the IEEE 830 standard and the actions that must be implemented in it, which allow achieving the scope of the functionalities indicated in the general description and based on the identified needs in the characterization stage. The main purpose of the information view, is to understand and describe the type, structure and content of data to be handled, in relation to the types of information used in the business view, taking into account the semantics and use of information for each request.

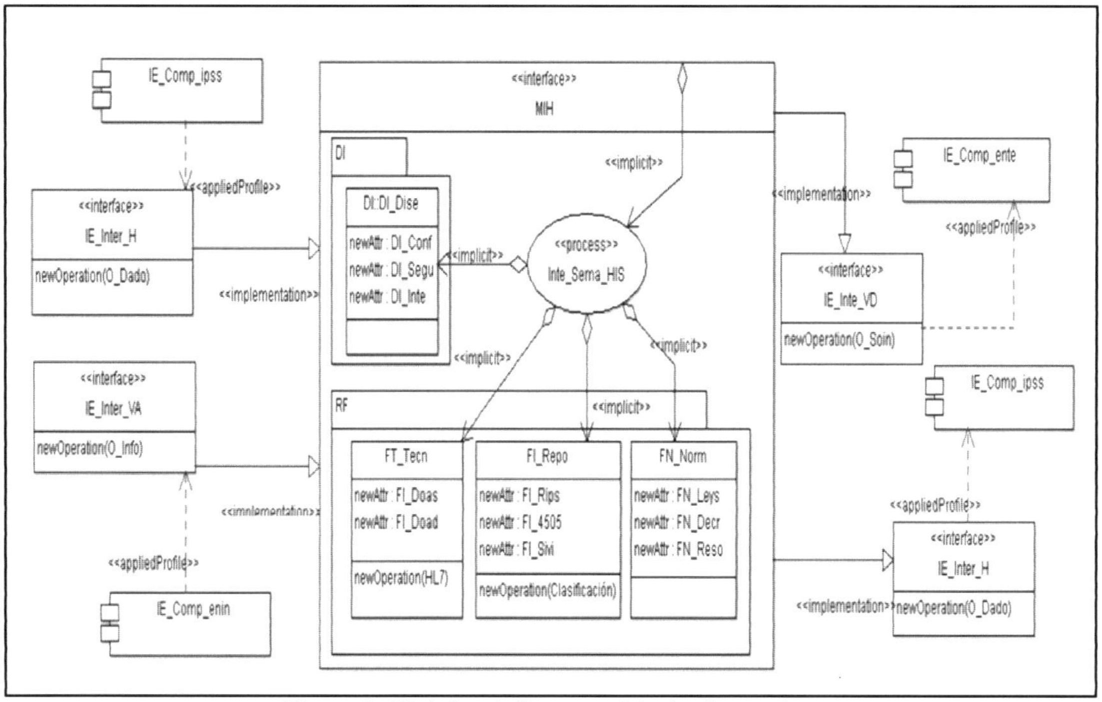

Figure 3. High-level diagram of the business view

Then, in the computational view and in its structures of logical components, we seek to define the logical components, through the representation of a use case diagram, where the design elements and the relationship between them are identified, to generate the architecture of the interoperability model from of the use of SOA. The services are interconnected through a BUS solution (ESB), through which access is allowed from the HIS for interoperability between the IPS, grouped according to the message formats defined in HL7.

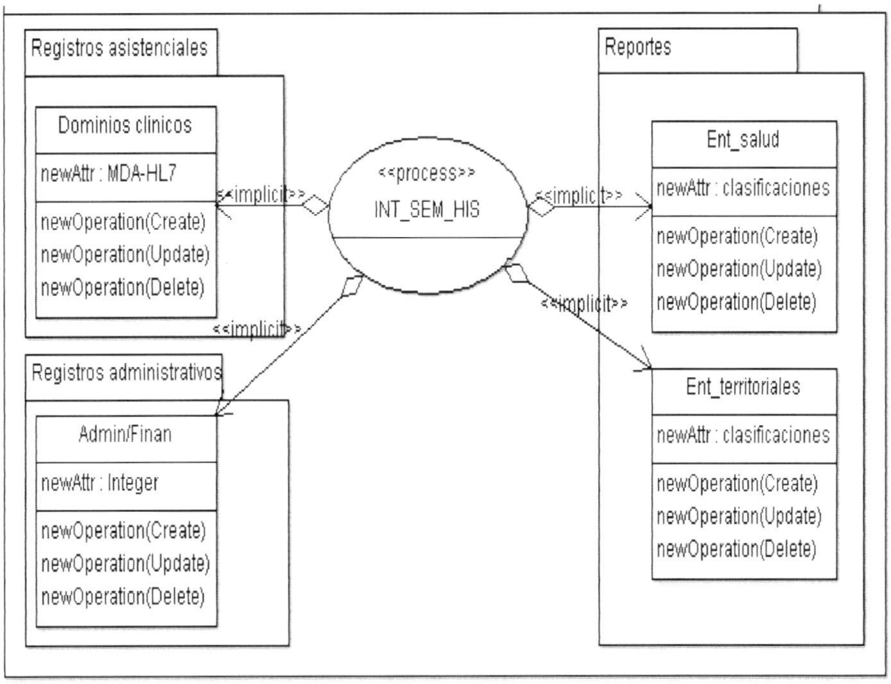

Figure 4. High-level information view diagram

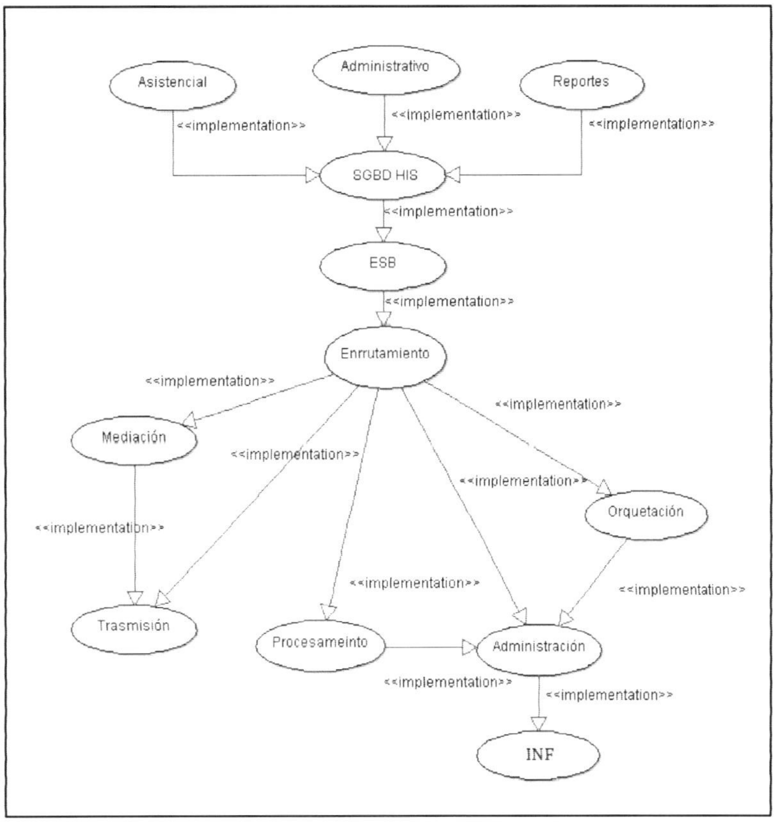

Figure 5. Structures of logical components of the computational view.

4. Results and discussion

From the development of the components, for the architectural representation of the interoperability model, business, information and computational views were obtained, by means of which the requirements of the model, the data structure, and the definition of the components are related. Logical respectively. Based on the above, through the implementation of the proposed model, a series of steps are carried out in various interoperability scenarios, to validate compliance with the requirements, which allows to determine the suitability of the design in the horizontal and vertical dimensions. In addition, the steps for the validation procedure of the interoperability model among the IPSs, corresponding to the implementation of the proposed architectural design in an HIS, are presented, describing the requirements, structure of data and implicit logical components, and the analysis of prototype tests implemented in one of the various tested interoperability scenarios.

For the implementation of the horizontal interface implementation, it was developed under the design for the scope of information exchange of the HIS modules among the IPS. Taking as reference the functional requirements, for the records of service requests, which would allow to verify the sender and the requested information, access control, for security through an encrypted code for each IPS. Where they were involved, the actions for requests for exams and diagnostic images, transfer of electronic medical records, user records and administrative records. In the figure 6, the horizontal interface for the exchange of information between the IPS is presented, the interoperability through this interface implies the classifications single service delivery code (CUPS), International Classification of Diseases version 10 (ICD-10), unique code of medicines (CUM), and the message structure CDA of HL7 that can be used by the IPS. From the use of the horizontal interface, from the IPS, it is possible to make requests for the processes and activities that guarantee the quality, accessibility, opportunity, continuity and integrality of the health services to the users. For this process, it is necessary to implement connections between the HIS, through Internet services that allow the transition of data.

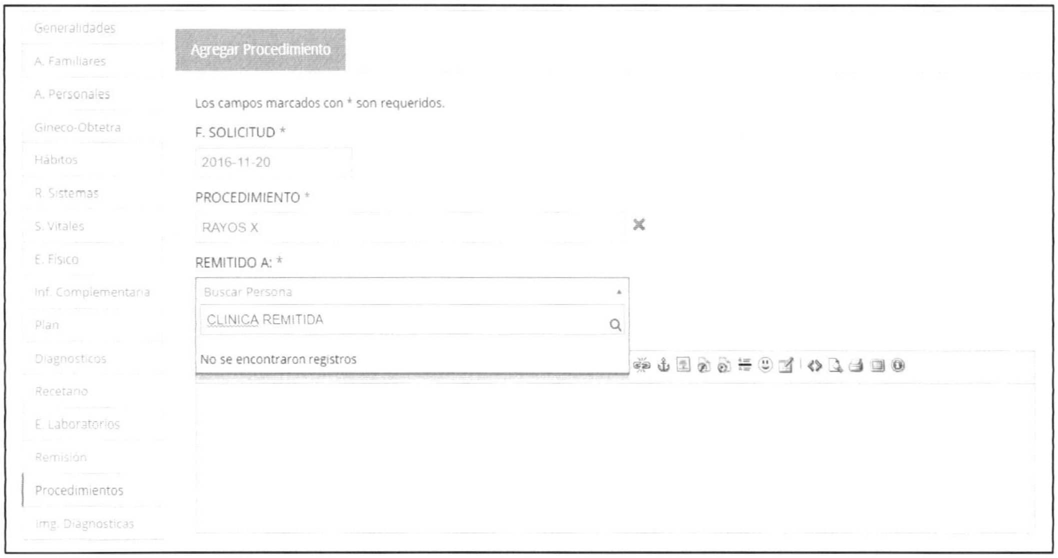

Figure 6. Implementation of the horizontal interface.

The execution of stored procedures corresponds to the set of instructions developed to consult the database of the HIS, in order to have the data structures defined in the information view of the interoperability model, corresponding to the exchange of information with other IPS, transfers of administrative level records, and issuance of reports to the EPS. Figure 7 shows a stored procedure of the database management system for the structure of the mandatory compliance report based on the provisions of resolution 4505 of the Ministry of Health and Social Protection since 2012, and in figure 8, the development of the basic structure of a CDA clinical document of HL7 is presented, where methods for the construction of the CDA message were implemented from the Information Reference Model (RIM) of HL7, with the attributes of stored procedures, obtaining as a result the main characteristics of electronic documents structured in HL7, such as the use of a header, attributes of the header and the body of the document.

Figure 7. Implementation stored Procedures.

```
<ClinicalDocument xmlns="urn:hl7-org:v3" xmlns:mif=
"urn:hl7-org:v3/mif"
xmlns:xsi="http://www.w3.org/2001/XMLSchema-instance"
xsi:schemaLocation="urn:hl7org:v3
CDA.xsd">
<!-- ENCABEZADO DEL DOCUMENTO ELECTRONIO (HL7-CDA)-->
<!-- Atributos del encabezado -->
<typeId root="2.16.840.1.113883.1.3" extension="POCD_HD000040"/>
<id root="..." extension="..."/>
<code code="..." codeSystem="..." codeSystemName="..." displayName=
"..."/>
<effectiveTime value="..."/>
<confidentialityCode code="..." codeSystem="..."/>
<!-- Fin de los atributos del encabezado -->
<component>
<!-- CUERPO DEL DOCUMENTO ELECTRONIO (HL7-CDA)-->
</component>
</ClinicalDocument>
```

Figure 8. Basic structure of the HL7 CDA.

After developing the Basic structure of the HL7 CDA, the Participations in the structure of the header are developed, related acts in the structure of the header, structure of the body of the CDA document of HL7, the sections and the data entries and the structure of the component, sections, data entry and clinical acts. Finally, it is required to establish a set of connections to the HIS database, through a server applications, to enable the exchange of information through the Internet. The configuration of the connections must be parameterized according to the services contracted between the IPS. In this context, a "pool of connections" was established through the glassfish server open source edition. GlassFish is a free software application server, compatible with Java EE, developed by the GlassFish and Oracle community. GlassFish integrates an Application Programming Interface (API, for its acronym in English of Application Programming Interface). JDBC, is the API for the execution of SQL statements with Java programming language. The set of functions and procedures can be configured from the glassfish web interface or for development purposes through an Integrated Development Environment (IDE), such as NetBeans, which is a free development environment with no restrictions on use, mainly for the Java programming language, tools of the Oracle Corporation. The GlassFish wizard allows you to create a new connection pool (New JDBC Connection Pool), through which the *java.sql.DataSource* resource is established, which corresponds to the extension package that provides data source connection and grouping functionality, and also, indicate the database management system used by the HIS. Finally the properties of the available database connections will be displayed, allowing to have a logical architecture, through which, through validations, messaging, transformation and routing are integrated, establishing a logical layer that allows virtual exchange of the information among the IPS, from the use of the HIS.

The scenarios developed in this context, allow the use of an HIS, and validate the compliance of the requirements for interoperability in the horizontal and vertical dimensions, as a basis to determine the suitability of the model for interoperability between the IPS. With reference to the above, the development of three (3) scenarios corresponding to: i) electronic transfer of the clinical records of a user of an IPS, for the request of diagnostic images in another IPS, from the use of the horizontal interface, ii) import of the records with the information of the users, of the affiliation database that is reported from the EPS, and iii) validation of the report in the ERP of an EPS, with the record of the activities of Specific Protection, Early Detection of mandatory compliance (resolution 4505 of 2012). Finally, a summary of the compliance with the architectural interoperability model requirements is presented and one of the validated scenarios is presented in this study.

5. Conclusions

In this research, the design of a model for interoperability between IPS, based on the standards for the electronic exchange of clinical information HL7, and in the architecture to design and develop distributed SOA systems was developed in an effective manner. The interoperability model focused on two dimensions, on the one hand the horizontal dimension was projected, through which the architecture was designed, based on the required components, for the exchange of clinical information of the users among the HIS modules of the IPS, and on the other hand, the vertical dimension was projected, through the architectural design, to exchange information vertically backwards, with actors of the health system that provide information to the HIS, and vertically forward, with entities that receive information from the IPS, processed from the use of the HIS.

References

[1] Ministerio de Salud y Protección Social 2007 *Decreto 4747* (Bogota D.C.).
[2] Castrillón H 2013 *Modelo arquitectónico de interoperabilidad entre instituciones prestadoras de salud en Colombia* (Popayán).
[3] Presidencia de la república de Colombia 2015 *Plan vive digital* (Bogota D.C.).
[4] Ministerio de Tecnologías de la Información y Comunicaciones 2008 *GEL-XML, Lenguaje de intercambio de información* (Bogota D.C.)
[5] Canada Government 2014 *Canada Health infoway, International Standard from Infoway* (Toronto).
[6] Abar F 2006 *Project National Health Information Infrastructure – NHII* (Taiwan).
[7] Jorcin L and Marichal E 2011 *Historia Clínica Electrónica HCE, NHS National Programme for IT – NpfIT* 10-14
[8] Bolaños R 2009 *HL7 Implementation Guide for Public Health Reporting Systems in Colombia* (Popayán) **7** 14 13-32
[9] Organización HL7 2000 *Estandares HL7.*

[10] González C, Blobel B and López D 2015 *Ontology-based interoperability service for HL7 interfaces implementation* (Popayán).

Acknowledgments
The authors thank the help of colleagues from the research group ENAPROT of the University of La Guajira and the organizing members of EXPOTECNOLOGÍA 2018 of the Antonio de Arévalo Technological Foundation.

IOP Conf. Series: Materials Science and Engineering **519** (2019) 012015 doi:10.1088/1757-899X/519/1/012015

Design of a WSN to determine the levels of CO and noise emitted in the parking of the CORPORACIÓN UNIVERSITARIA ANTONIO JOSÉ DE SUCRE of the city of Sincelejo-Sucre

A Barrios-Ulloa[1], E Martinez-Diaz[1], A Medina Medina[1], D Sánchez Hernández[1], C Mercado Montes[1], A Arrieta Hernadez[1], J Julio Jiménez[1] and J Serpa Serpa[1]

[1] Faculty of Engineering, Corporación Universitaria Antonio José de Sucre.

alexis_barrios@corposucre.edu.co

Abstract. Wireless sensor networks (WSN) have gained great importance in recent years in different investigations and different fields, due to its wide deployment, cost and utility advantages. The constant exposure to different gases produced by combustion engines and excessive levels of noise can cause alterations in human health, especially in medium and large cities in which the vehicle fleet increases year by year. In this project a real-time contamination monitoring system is developed using wireless sensor networks for the monitoring of carbon monoxide (CO) and noise pollution in the parking of the headquarters of the Corporación Universitaria Antonio José de Sucre (CORPOSUCRE) in Sincelejo, Sucre. The data collected by the sensor nodes are sent to a computer where they can be viewed on the Serial Plotter platform. The design of the system was carried out through a site study in order to obtain the data of the variables used to get the results of the present study. The monitoring system is based on the Arduinos and XBee modules, which results in a simple design and low cost due to the characteristics of the elements used. The data obtained will allow to determine if the CO emissions and the noise levels in the parking of CORPOSUCRE exceed or not the maximum limits allowed.

1. Introduction.

In Colombia, air pollution has become one of the main environmental problems, and the degradation of air quality has caused an increase in negative effects on human health and in the same environment [1]. According to the analysis carried out in 2005 and presented in the CONPES document 3344, fossil fuels mainly used by mobile sources (cars, airplanes, trains and others) and fixed sources (mainly industrial and thermoelectric) are the largest contributors to air pollution [2]. With respect to auditory noise, the Ministry of Health of Colombia has also launched alerts with the purpose of reducing the number of cases of hearing loss that occur by the involuntary exposure to high levels of noise generated in the environment. By March of the year 2015 it is estimated that about 11% of the total of Colombian population suffered from hearing problems and among the active working population from 25 to 50 years, the prevalence of hearing loss due to noise exposure was 14% [3].

In recent years wireless sensor networks (WSN) and data acquisition systems have evolved rapidly, allowing in the new circuits greater computational power, longer transmission distances, and more accurate measurements with lower power consumption [4]. In general, the constant advances of the

Content from this work may be used under the terms of the Creative Commons Attribution 3.0 licence. Any further distribution of this work must maintain attribution to the author(s) and the title of the work, journal citation and DOI.
Published under licence by IOP Publishing Ltd

IOP Conf. Series: Materials Science and Engineering **519** (2019) 012015 doi:10.1088/1757-899X/519/1/012015

WSN allowed to solve problems in various fields and today is applied to a large extent in everything that refers to intelligent environments such as smart cities, smart homes and others.

Wireless sensor networks can also be used for the monitoring of environmental variables that in values beyond permits, can cause harm to the health of human beings. In addition, these systems allow the compilation of data that later can be used to establish policies or actions that help to preserve human health. This paper presents the development and implementation of a prototype of a wireless sensor network for the monitoring of carbon monoxide (CO) and noise in the parking of the headquarters of the Corporación Universitaria Antonio José de Sucre (CORPOSUCRE) located in the La Maria neighborhood in Sincelejo, Sucre, Colombia; in which there is a high vehicular occupation especially in the night shift, due to the large number of students attending in motorcycles. Figure 1 shows the high congestion in the parking of headquarters A of CORPOSUCRE.

Figure 1. Top view of the parking of the main headquarters of CORPOSUCRE. Source: Authors

This work proposed the design and implementation of a prototype of measuring system that allows to determine the levels of CO and auditory noise that occur in the parking of the main headquarters of CORPOSUCRE, because these exceed the maximum levels recommended by the World Health Organization (WHO) and adopted by Colombian laws. For the development it has been decided to use the Arduino Uno modules that allow an implementation in a low cost, and the XBee S2B modules that use the IEEE 802.15.4 protocol and provide good coverage, energy efficiency and the possibility of deploying different topologies. The data will be sent to a computer to be stored and later analyzed.

The rest of the article is structured as follows: Unit 2 presents a theoretical framework with the definitions of the concepts used in the research and a description of the elements used in the implementation of the prototype. Unit 3 shows the research background and related works that served as a guide for structuring the proposal. Unit 4 presents the technological resources and the methodology used for the development of the measurement system and the communication of the data. In units 5 and 6 the results and conclusions are presented respectively; and finally the conclusions and recommendations are presented.

2. Theoretical framework.

By resolution 2254 of the year 2017, the Ministry of the Environment of Colombia established six criteria of pollution: particulate material less than 10 micrometers (PM10) and less than 2.5 micrometers

(PM2.5), sulfur dioxide (SO2), dioxide nitrogen (NO2), ozone (O3) and carbon monoxide (CO) [5]; these are the ones that cause greater risks in the population, and therefore, must have control of maximum permissible levels. For this project measurements of the concentration of CO will be made, this can be produced by the combustion of natural gas, gasoline, kerosene and other substances; and for which the Ministry of Environment establishes maximum values of 5000 µg / m3 for an exposure time of eight hours and 35000 µg / m3 for a one-hour exposure. In the same resolution, the Air Quality Index (AQI) was adopted for the implementation of a system to report the state of air quality based on a colour code, which must be taken into account to reduce the exposure to high concentrations of air polluted into the population. Table 1 shows the general description of the AQI index.

Table 1. General description of the Air Quality index. Source: [5].

Range	Colour	State of the air quality	Effects
0 - 50	Green	Good	Air pollution supposes a low risk to health.
51 - 100	Yellow	Acceptable	Possible respiratory symptoms in sensitive population groups.
101 - 150	Orange	Harm to the health of sensitive groups	Sensitive groups may have effects on their health.
151 - 200	Red	Harmful to health	All individuals can experience effects on their health. Sensitive groups may experience more serious health effects.
201 - 300	Purple	Very harmful to health	Alertness means that everyone can experience more serious effects on their health.
301 - 500	Brown	Dangerous	Health warning. The entire population can have serious adverse effects on their health and they are prone to be affected by serious health effects.

The same resolution also presents the cuts of the air quality index with respect to the pollutants criteria, which can be seen in table 2.

Table 2. AQI cut-off points for carbon monoxide. Source: [5].

Air quality index			AQI cut-off points for CO
AQI	**Colour**	**Category**	**CO µg/m3 8 hours**
0 - 50	Green	Good	0 -5094
51 - 100	Yellow	Acceptable	5095 - 10819
101 - 150	Orange	Harm to the health of sensitive groups	10820 - 14254
151 - 200	Red	Harmful to health	14255- 17688
201 - 300	Purple	Very harmful to health	17689 - 34862
301 - 500	Brown	Dangerous	34863 - 57703

For the calculation of the AQI, equation 1 is used:

$$AQUIP = \frac{I_{High} - I_{Low}}{PC_{High} - PC_{Low}} * (C_p - PC_{Low}) + I_{Low} \qquad (1)$$

Where:

AQIP = Air quality index for the pollutant P.

CP = 'Measured concentration for the pollutant P.

PCHigh = cut point greater than or equal to CP.

PCLow = Cut point less than or equal to CP.

IHigh = Value of the AQI corresponding to the PC High.

ILow = Value of the AQI corresponding to the PC Low.

With regard to noise levels, the exposure of humans to very high levels of sound pressure can lead to loss of health and other problems such as sleep disorders, anxiety, depression and changes in behavior [3]. Resolution 0627 of April 2006 of the Ministry of Environment defined the maximum allowable levels of noise emission expressed in dBA, and it categorizes the universities as sector B for tranquillity and moderate noise, for which the maximum permissible emission standards of noise are 65 dBA in the daytime (7:01 to 21:00 hours) and 50 dBA in the night time (21:01 to 7:00 hours) [6].

3. Related works.

Air pollution and excess of noise levels are problematic that occur in different parts of the world and for this several measurement and monitoring systems have been implemented. To continuation are some publications and reports that are related to the development of the measurement system proposed to CORPOSUCRE's parking and where are shown the different tools used - in Colombia and abroad - to address the problem of measuring air pollution and of noise levels.

3.1. Prototype system for the wireless monitoring of air pollutant gases developed under free hardware and software platforms.

Students of the Electronic Engineering and Telecommunications program of the Escuela Politécnica Nacional de Ecuador designed and implemented a prototype system for the wireless monitoring of carbon monoxide (CO), Ozone (O3) and nitrogen dioxide gases in order to monitor the quality of the air in some sectors of the city of Quito. The prototype has a global positioning system (GPS) that allows georeferencing with latitude, longitude, time and date every measurement made. The information of the variables measured at each monitoring point are received through an Arduino UNO module, which performs functions of acquisition and processing of the data; which are subsequently consolidated into a set of data, for which a Single Board Microcomputer Raspberry Pi is used. Finally, the information is sent through a cellular network to a reception system for processing and storage. Among the main conclusions, the authors indicate that the system allows the wireless monitoring of gases following the pre-established requirements in the project [7].

3.2. Design and implementation of a basic gas analyzer for vehicle emissions (HC, CO and CO2), developed under the Android platform

Researchers from the Universidad Surcolombiana de Neiva, Colombia; developed and implemented a vehicle gas analyzer (CO, CO2 and HC) with wireless communication through a Wifi Wasp Mote Pro 1.2 module that operates under the Arduino platform. For real-time monitoring via a smartphone, the Android operating system was adapted to the embedded Libelium system board. Among the positives results of this project, the main was that the values obtained by the basic gas analyzer are stable, reliable and reproducible. However, the authors indicate that it was not possible to make comparisons between the data obtained and those thrown by the measurement system such as those used in revisions on automotive diagnostic centers. [8].

3.3. Design of a wireless sensor network for the monitoring of the vehicular traffic and CO2 pollution in an urban sector.

Students of the electronic engineering program of the Universidad Politécnica Salesiana de Cuenca of Ecuador developed a vehicular behavior monitor system to measure concentrations of carbon dioxide (CO_2). The developed prototype consists of a wireless network based on the Zigbee standard with XBee modules with four sensor nodes and a coordinator. The sensors are responsible for sending the data via wireless to the coordinator, who through a graphical interface developed in JAVA, presents the behavior described by the measurement curves. The information is stored in a web server linked to MySQL. In conclusion, the authors affirm that the implemented wireless sensor network is reliable, since there was no interference or loss of information taking into account the maximum link distance and line of sight conditions; this made the network register efficiently the concentrations of CO_2 [9].

3.4. Implementation of the API mode in a WSN sensor network for the measurement of auditory contamination

The student Milton Vinicio Cacuango of the Universidad de las Américas, designed and implemented a system for measuring the levels of auditory contamination in some of the most affected areas of the city of Quito, Ecuador, for which he used the API mode in a WSN network. The system is composed of a sonometer that is responsible for measuring the noise level at a specific point, a GPS to determine the position of the node and an XBee wireless device for transmit the data at an approximate distance of 1.2 km from the Gateway to the node and the API frame, for the transmission of data from the node to the coordinator and vice versa. The results presented show that the developed system was able to measure noise levels in intervals of 48 dB to 80 dB [10].

3.5. Design of a noise measurement tool based on Arduino-Rasperry PI technologies.

This system is one of the phases of a macro project, of the Universidad de la Corporación Universitaria Americana, called "Study on the Use of Free Software in the MSMEs of the Colombian Caribbean Region" in which a technological tool was designed to measure the levels of auditory contamination in closed spaces. For the development of this system was worked with free hardware technologies such as Arduino and Raspberry, and for the sound measurement we used a DfRobot sensor 0034. Finally, the administration of the information was done with the MySQL database. Although the article doesn't show definitive results, the authors indicate that the tool will allow the community to control, monitor and analyze auditory contamination in enclosed spaces. So in a future will allow to design strategies to minimize the impact on human health caused by exceed the maximum values marked by the law [11].

3.6. Evaluación de la calidad del aire en la ciudad de Sincelejo, departamento de Sucre, Colombia.

In the case of the city of Sincelejo, Sucre; only the results presented by the Corporación Autonoma Regional de Sucre (CARSUCRE) about measurements made in August 2014 from four stations of measure of air quality, specifically particulate material, located in four points on the capital of the department. The measurement of auditory noise is not included. The results indicate that PM measurements weren't obtained above the maximum allowable. For the study PM10 equipment was used, located in two branches of the University of Sucre, another in the CARSUCRE headquarters and another one in a high vehicular traffic lane. [12].

IOP Conf. Series: Materials Science and Engineering **519** (2019) 012015 doi:10.1088/1757-899X/519/1/012015

4. Devices and tools used in the measurement system.

This project is framed in Colombian legislation. For the measurement of CO the resolution 2254 of the year 2017 of the Ministry of Environment and Sustainable Development was taken into account, and for the measurement of noise, Resolution 0627 of 2006 of the Ministry of Environment [5]; Housing and Territorial Development was considered. In the design of this system free hardware technologies were used due to the facilities offered by these tools in terms of costs and use. Below the different devices used in the carbon monoxide and noise level measurement system installed and tested in the parking of the main headquarter CORPOSUCRE are presented.

4.1. Arduino Card

Arduino is an open source platform that was designed to facilitate electronic projects. It can be used to develop interactive objects that work without being connected to a computer, in addition to taking information from the environment through sensors connected to its analog and digital inputs, controlling lights, motors and others solenoids in a direct way or from controlling signals generated by their outputs. It has a graphical development environment that uses a processing/wiring programming language and a boot manager. Another features of Arduino are: low cost, multi-platform, open source and extensible hardware [13]. In terms of hardware, it consists of a microcontroller and input and output ports. In the design of the system for this project, an Arduino One board was used as shown in figure 2, which has the following characteristics: ATmega328 microcontroller, input voltage 7-12V, 14 digital I/O pins (6 outputs) PWM), 6 analog inputs, 32k of Flash memory and 16MHz speed clock.

Figure 2. Arduino One programmable card. Source: [13]

4.2. MODULE MQ-9

It is a carbon monoxide (CO) and fuel gas detector sensor. Among its main features are its adjustable sensitivity, digital and analog output and voltage of 5V. The conductivity of the sensor is higher along with the increasing gas concentration. It has a high sensitivity to carbon monoxide, methane and liquefied petroleum gas (LPG). The sensor could be used to detect different gases that contain CO and combustible gases, and its sensitivity can be adjusted through a potentiometer [14]. In figure 3 can be seen the MQ-9 sensor and its sensitivity curve.

IOP Conf. Series: Materials Science and Engineering **519** (2019) 012015 doi:10.1088/1757-899X/519/1/012015

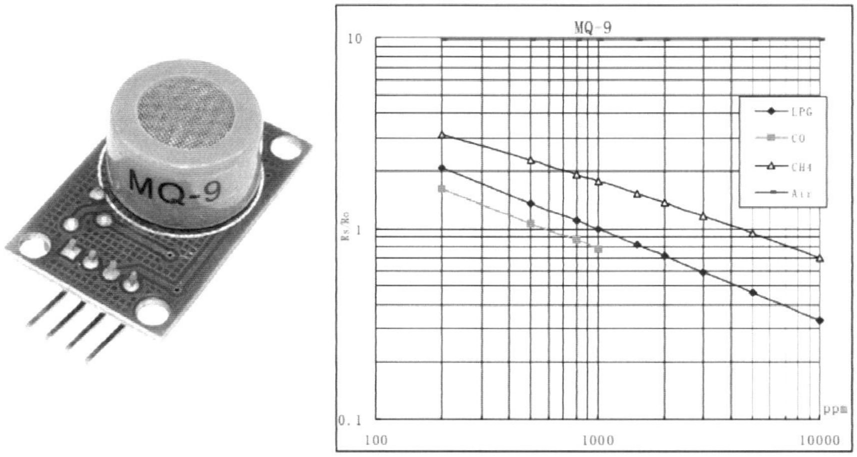

Figure 3. MQ9 module for gas measurement. Source: [14]

4.3. MAX9814 MODULE.

The MAX9814 microphone, like the one presented in figure 4, is a microphone amplifier with automatic gain control. It also has a low noise preamplifier, variable gain amplifier, output amplifier and a bias voltage generator. Automatic gain control reduces "loud" sounds that are close to the microphone in order to avoid saturation in the audio, and amplifies sounds that are generated at a greater distance [15].

Figure 4. Microphone MAX9814 for measuring audio levels. Source: [15].

4.4. XBEE PRO S2B

Xbee chips are wireless communication modules that work under the IEEE 802.15.4 standard and are manufactured by Digi International. This type of devices allows the creation of point-to-point and point-to-multipoint networks for applications where high data traffic and low latency are required. In addition, XBee modules offer the possibility to configure saving modes to reduce energy consumption [16]. For this investigation, modules of the series 2 were used, such as those shown in figure 5, and among its main characteristics are the maximum transmission rate of 250 Kbps, a coverage range of 1.6 km, output of 63mW, 8 digital I/O pins and local or wireless configuration.

IOP Conf. Series: Materials Science and Engineering **519** (2019) 012015 doi:10.1088/1757-899X/519/1/012015

Figure 5. XBee Pro S2B module. Source: [16].

4.5. SHIELD XBEE - ARDUINO
To use an XBee module with Arduino is needed a Shield or an adapter to connect the serial port of the XBee chip with the one of the Arduino module. To use or configure an XBee module with a computer you need an adapter that usually has an FTDI chip that acts as a gateway between the serial port and the USB [17]. Figure 6 shows the image of an adapter similar to the one used in the project.

Figure 6. Adapter from XBee to Arduino. Source: [17].

5. Methodological framework
The development of this project falls into the category of "technological development project" according to the typology established by Colciencias, because the result is a prototype for his first validation of a process in which is looking to reduce the initial uncertainty raised with respect to carbon monoxide and auditory noise concentration levels in the parking of the headquarters A of the Corporación Universitaria Antonio José de Sucre of Sincelejo, Sucre. The research is an experimental type, since different tests and trials were carried out in order to optimize the functioning of the measurer system.

5.1. Technological resources
The technological resources with which it is expected to fulfil the project's objectives are:

- Laptop.
- Smartphone.

- Broadband internet connection.
- Academic and/or scientific database.
- Search engines for scientific and/or academic information.
- Ofimatic applications.
- CAD Software for electronic desing.
- Programming software.

5.2. Research procedure.

Five stages were established in the methodology in order to achieve the objectives proposed in this project. Next, each of the phases is detailed:

5.2.1. Selection of design features of the tool. In the initial phase of the development of the carbon monoxide and noise monitoring system design, the following problems were considered: device selection, range and precision of the measured variables, processing and data transmission. The ease of use and acquisition of tools in the market was also considered.

5.2.2. Recognition of the performance perimeter of the wireless sensor network. The identification of the place where the measurements are carried out helps to identify the propagation characteristics of the environment and to clarify if the type of link is with line of sight (LOS) or without line of sight (NLOS). This step will support the selection of the communications standard and the components for communication. The area where the system operates is 669 square meters (m^2) and it turns out to be a scenario with little or almost no vegetation. The distance between the data measuring point and the receiver's location is 48.09 meters with no line of sight.

5.2.3. Identification of the electronic elements and materials for the construction of the tool. Among the general characteristics for the identification and selection of devices, low cost, low energy consumption, easy handling, easy assembly and that they were easily configurable were considered. In the case of the sensors, the accuracy of the measurements and the compatibility with the other elements were taken into account, in addition to the range. Regarding of the communication stage, the distance between transmitter and receiver and the data rate was mainly considered, but also the scalability so that in the future the number of sensor nodes could be increased. Finally, in the case of data processing, the selection features include the number of I/O ports, the ease of handling and robustness.

5.2.4. Selection of the topology. Taking into account the physical characteristics and disposition (pedestrian passages and parking spaces), of the parking of the A headquarters of Corposucre, the installation of a single measurement point located at the entrance of the institution was determined, which also allows obtaining data of the concentration of CO and noise level of the motor vehicles that circulate in the street that gives access to the University. Although a point-to-point topology is sufficient for the proposed system, a star configuration was decided in order to allow more measurement points to be added in the future and thereby ensure the scalability of the network.

5.2.5. Assembly and testing of the system. The elements mentioned in the third stage are connected with the arrangement shown in figure 7. The operation of the prototype is as follows: the MQ-9 sensor performs CO detection and measurement, and the Max-9814 sensor is used to detect sound changes. The first one has the adjustable sensitivity to have an adequate reading to the system. The MQ-9 has a digital output through a comparator with adjustable range and an analog output that can be measured with any microcontroller or development card with ADC. The second one includes an electret microphone from 20 to 20Khz, the center of this sensor is the MAX9814 amplifier which has a series of configurable options. By default, the maximum gain level is 60dB but it can be varied to 40 and 50 dB bypassing the Gain, VCC and GND pins. The output from the amplifier is around 2Vpp max on a base of 1.25V DC.

The signals obtained by the sensors are transmitted wirelessly to a PC through the XBee modules. The acquired signal enters into the Arduino card and is amplified so that it reaches the final user's visualization for its respective interpretation of the data and decision making.

To perform tests, the carbon monoxide measurement system was installed 48 hours in advance because the MQ-9 sensor must be subjected to a preheating process for its calibration. The tests were always carried out from 6:00 p.m. to 9:15 p.m. because it is the time zone with the highest occupation on the parking, more vehicular traffic and also increases the student population and the teaching staff. In this stage, the results obtained allow to determine in an initial form the utility of the system.

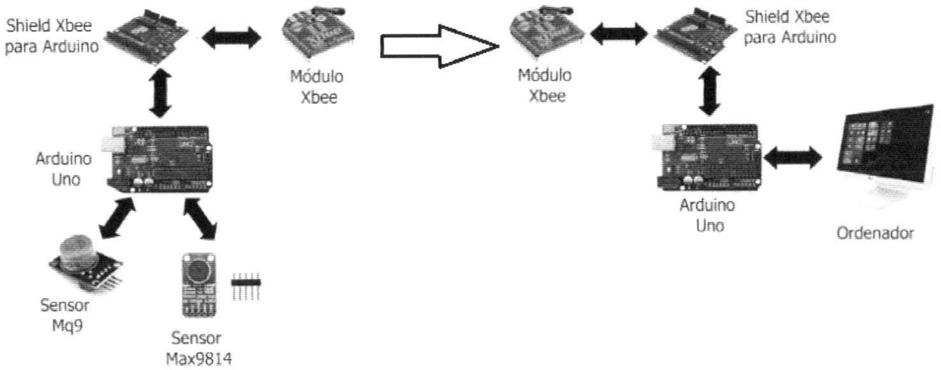

Figure 7. Diagram of the final prototype.

6. Results.

The calibration method for each of the sensors was by trial-error. For the case of the gas sensor, the characteristic curve delivered by the manufacturer was used as reference, that curve can be seen in figure #3. For the measurement of dB, a recommended ratio was used for this device. This relationship is determined by the manufacturer.

The prototype was configured with a star topology with simplex communication in coordinator-slave mode and is capable of recording information of carbon monoxide concentration (in ppm) and sound power (in dBA). In addition, each measure made by the node is recorded with the time it was made, in order to identify the times in which there is a considerable increase or when the maximum permissible values are exceeded.

Tests were performed with the X-CTU software to measure the signal strength indicator received by RSSI (for its acronym in English of Received Signal Strength Indicator) in the scenario considered. The results indicated that for a line of sight link, the RSSI value didn't decrease below -66 dBm. However, when placing the receiver in its final location, the link turns out to be NLOS and the obstacles (walls, furniture and others) cause a reduction of the RSSI to -79 dBm. In the different tests, no packet losses were registered.

6.1. Discussion.

In this section highlights the main limitations of the prototype that affect the accuracy of the results:

- The calibration of the MQ-9 sensor requires a 48-hour preheat by technical specifications of the manufacturer for optimal operation of the device. In the event of a disconnection in the power system, the preheating must be re-started.

- The system at been a prototype requires more evidence to help to determine the validity of the results. No comparisons have been made with certified equipment or certified reference points.

- For measuring ambient noise, the system must be adapted to the standardized measurement conditions in the resolution 0627 of 2006 in order to validate the data thrown by the measurement.

- The levels of RSSI obtained, and the number of packages sent and received indicate that the coverage level of XBee equipment is optimal for the characteristics of the scenario and propagation object of study. However, it is necessary to study another location for the receiving system to determine if the received signal strength level can be improved.

Conclusions and recommendations

At the end of this prototype, a low cost noise and CO monitoring system was achieved, with wireless communication to obtain contamination levels of the variables mentioned above through a network of wireless sensors working with the ZigBee standard.

Considering the data delivered by the manufacturers of the sensors in the data sheets. The prototype provides favorable results, so that the measurement and monitoring of CO and noise levels have been successful. In order to improve the accuracy levels of the measurement system, in the future comparison tests will be carried out with certified polluting gas measurement equipment such as those found in automotive diagnostic centers. During the measurement intervals no CO and noise levels above the maximum values allowed by Colombian regulations were detected. However, it is recommended to conduct a more extensive measurement campaign, including all the hours of the academic day. So that more concrete results can be obtained.

It can be evidenced that the communication by Xbee is stable and reliable according to the distance to which the coordinator and the slave are connected. It can also be noted that it is an accessible device compared to others devices on the market and also allows working in real time with very low delays. Although RSSI levels are satisfactory and no packet loss was registered, it is recommended to consider other location points for receivers as well as a range test for the purpose of increasing system performance and signal strength in order to provide greater stability to data transmission.

As future work, it is recommended to increase the number of measurement points that include not only the CORPOSUCRE facilities, but also the access roads to the institution which present a vehicular flow at different times of the day. It is also recommended to measure other pollutants including in the law and that may affect the health of people and the environment.

References

[1] Ministerio de Ambiente, 2010, *Vivienda y Desarrollo Territorial, Politica de Prevención y Control de la Contaminación del Aire, Ministerio de Ambiente, Vivienda y Desarrollo Territorial.*

[2] CONPES, 2005, *Lineamientos para la formulación de la política de prevención y control de la contaminación del aire*., Bogotá D.C.: CONPES.

[3] Ministerior de Salud, 2 Marzo 2015, «https://www.minsalud.gov.co». [Online]. Available: https://www.minsalud.gov.co/Paginas/Ruido-excesivo-en-entornos-una-de-las-principales-causas-para-perdida-auditiva.aspx. [Último acceso: 15 Agosto 2018].

[4] D. Trancă, D. Rosner, R. Curatu, A. Surpăteanu, M. Mocanu, S. Pardău y P. A., 2017, *«Industrial WSN node extension and measurement systems for air, water and*

environmental monitoring: IoT enabled environment monitoring using NI WSN nodes.,» IEEE, pp. 1-6.

[5] Minambiente 2017 *Ministerio de Ambiente y Desarrollo Sostenible 2017 Resolución 2254 de 01 de noviembre.*

[6] Minambiente 2017 *Ministerio de Ambiente, Vivienda y Desarrollo Territorial 2006 Resolución 0627 de 2006.*

[7] J. Acero, R. Larco y C. David, 2016, *«Sistema prototipo para el monitoreo inalámbrico de gases contaminantes del aire desarrollado bajo plataformas de hardware y software libre,»* Escuela Politécnica Nacional, Quito.

[8] J. Molina, A. Gutiérrez y P. Muñoz, 2016, *«Diseño e implementación de un analizador básico de gases para emisiones vehiculares (HC, CO y CO2), desarrollado bajo la plataforma android,» Ingeniería y Región (15),* pp. 57-64.

[9] B. Bravo, B. Belduma y A. Alberto, 2017, *«Diseño de una red de sensores inalámbricos para el monitoreo del tránsito vehicular y la contaminación CO2 dentro de un sector urbano,»* Universidad Politécnica Saleciana, Cuenca.

[10] C. Cacuango y M. Vinicio, 2017, *«Implementación del modo API en una red de sensores WSN para la medición de contaminación auditiva,»* Universidad de las Américas, Quito.

[11] D. López, R. Espinosa, I. Gutierrez y L. Oliveira, 2017, *«Diseño de una herramienta de medición de ruidos basados en tecnologías Arduino-Rasperry PI,» Producción+ Limpia,* vol. 12, nº 1, pp. 81-87.

[12] T. Ruiz, 2015, *«Evaluacion de la calidad del aire,»* Corporación Autónoma Regional de Sucre, Sincelejo.

[13] ARDUINO, «https://www.arduino.cc,» [Online]. Available: https://www.arduino.cc/en/Guide/Introduction. [Último acceso: 15 Agosto 2018].

[14] seeed, «http://wiki.seeedstudio.com,» [Online]. Available: http://wiki.seeedstudio.com/Grove-Gas_Sensor-MQ9/. [Último acceso: 15 Agosto 2018].

[15] Maxim Integrated, «https://datasheets.maximintegrated.com,» [Online]. Available: https://datasheets.maximintegrated.com/en/ds/MAX9814. [Último acceso: 15 Agosto 2018].

[16] XBEE, «http://xbee.cl,» [Online]. Available: http://xbee.cl/que-es-xbee/. [Último acceso: 15 Agosto 2018].

[17] Aprendiendo Arduino, «https://aprendiendoarduino.wordpress.com,» [Online]. Available: https://aprendiendoarduino.wordpress.com/tag/xbee/. [Último acceso: 15 Agosto 2018].

IOP Conf. Series: Materials Science and Engineering **519** (2019) 012016 doi:10.1088/1757-899X/519/1/012016

Position control module of myoelectric prostheses for people with upper limb loss at forearm level

C A Arenas[1], R E Monterroza[2], A J Marrugo[2] and L Rojas[1]

[1] Facultad de Diseño e Ingeniería, Fundación Tecnológica Antonio de Arévalo, Cartagena, Colombia.
[2] Escuela de Ingenieria de Sistemas, Universidad del Sinu seccional Cartagena, Cartagena, Colombia

carlos.arenas@tecnar.edu.co

Abstract. This project emerges as a solution to the problem present in the Colombian population with upper limb movement disability, related to access to prostheses that provide them with a better performance in their daily and work activities, thus offering them a better quality of life. The objective of this project will be to build prostheses with state-of-the-art technology at relatively lower costs than similar versions in other countries, as well as strategies for people with limited resources to access them easily.

The scope of the project is limited to the construction of a control module, that will be divided into two phases, the first phase covers the design of the EMG signal acquisition system; this will focus on the establishment of the design criteria necessary for the acquisition of the myoelectric signal, model for obtaining the characteristics of the EMGs signal in the time domain and, parameters for the identification of the movements. And a second phase, where the acquisition module is developed, for performing functional tests with EMGs signals in real time, in order to be able to calculate the percentage of error between the expected value and the real value.

1. Introduction

In Colombia, there are currently 413,269 people with disabilities that have alterations in the function or body structure related to the movement of the body, hands, arms or legs [1], of which at least 29,727 require upper limb prostheses. Despite the aforementioned and the advantages shown by myoelectric prosthetics to restore functionality to the subjects, this type of prosthesis has not yet been developed commercially in Colombia, this situation extends to the rest of Latin America, with the exception of Argentina where there are advances in such matter. Now, the process of manufacturing low cost prosthetics involves the use of durable and low cost materials, as well as the development of control models that can be implemented in open hardware systems of sufficient performance to obtain good results, which implies a challenge of fairly large design, considering that the processing of the EMGs signal requires a large calculation capacity that directly affects the performance of the hardware, and therefore the manufacturing costs. In addition to this, patients, follows-up is often complicated, since when faults occur in the equipment, vital information cannot be accessed at that precise moment, where variables such as energy consumption, movement and time of use can be related. These variables are important during the monitoring process to detect early failures and improve the models used.

Content from this work may be used under the terms of the Creative Commons Attribution 3.0 licence. Any further distribution of this work must maintain attribution to the author(s) and the title of the work, journal citation and DOI.
Published under licence by IOP Publishing Ltd

2. Methodology

2.1. Muscles selection

For the signal EMGs acquisition, two antagonist muscles of the forearm were selected, considering that the level of amputation or loss of the forearm is in the distal third, these muscles were: the long supinator and pronator round. Figure 1 shows both muscles and how they intervene in the movements of supination and pronation of the forearm.

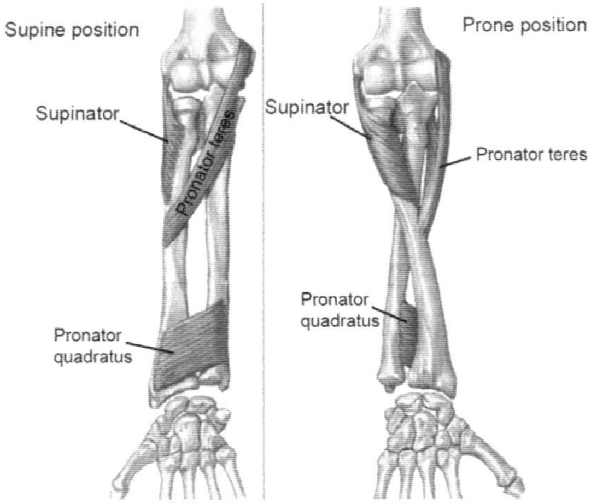

Figure 1. Supine position and Prone position

Taking into account that the best location to record EMG signal is in the central area of each muscle. It should be noted that we worked with healthy muscles in people who were not amputated [2].

Figure 2 shows the electrodes located in the muscles, where the blue electrode corresponds to the main entrance of the instrumentation amplifier, the red electrode to the secondary entrance and the black electrode to the reference.

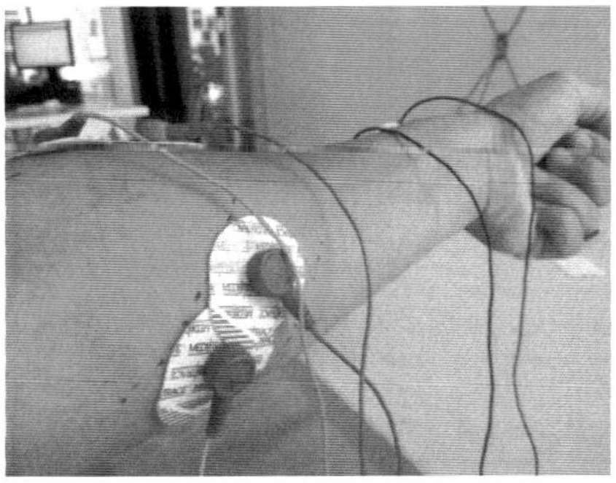

Figure 2. Location of the electrodes (two channels)

2.2. Signal Acquisition

For the acquisition of the signal, two Olimex Ekg-Emg modules were used, connected to an Arduino Uno module. Due to each module (shield) has a channel. See figure 3.

Figure 3. Shield EKG-EMG (Olimex)

This acquisition module is based on the INA321, which has a CMRR of 94 db, which is above the recommended value, which is 90 db minimum [3].

The recording of the electrical activity of the muscles was made by the analog-digital converter of an Arduino UNO. Figure 4 shows the signal EMGs corresponding to the long supinator muscle (channel 0) during an isometric contraction.

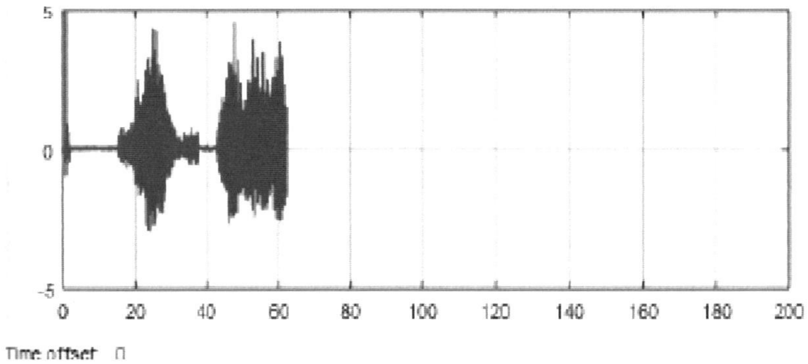

Figure 4. Isometric contraction of the long supinator muscle

In order to evaluate the signal in the time domain, during the acquisition, the implementation of a vector of 10 elements was performed, as shown below.

```
// data capture up to 10 samples
// channel A0 - muscle 1 (m1)

        valorSensor0 = analogRead(A0);
        valorEmg0 = map(valorSensor0, 0, 1023, 0, 255);

// Rectification of channel A0
        v_rec0[num] = abs(valorEmg0-127);

// channel A1 — muscle 2 (m2)
        valorSensor1 = analogRead(A1);
        valorEmgl = map(valorSensor1, 0, 1023, 255);
```

IOP Conf. Series: Materials Science and Engineering **519** (2019) 012016 doi:10.1088/1757-899X/519/1/012016

```
// Rectification of channel A1
        v_recl[num] = abs(valorEmg1-127);

        if(num = (muestras – 1)){
        //cálculo de RMS y VAR en cada muestra
                emg_feature ( ) ;
                num = O;
        }
num = num + 1;
```

As we can see in the previous code, the channels A0 and A1 of the converter are read, and then the rectification of the signal is carried out. Having the rectified signal, the function emg_feature () is called for the calculation of the characteristics.

2.3. *Characterization of the EMG signal*

Characterization on time domain is done because the control module must be accessible, which is achieved, in part, making its development using open source hardware platforms like Arduino, that don't have high data processing needs so it will not be expensive to obtain the final product.

The extracted features: Root Mean Square and the amplitude variance. These were calculated on each channel vectors. The formulas are the following [4]:

$$RMS = \sqrt{\frac{1}{N}\sum_{n=1}^{N} x_n^2} \qquad (1)$$

$$VAR = \frac{1}{N-1}\sum_{n=1}^{N} (x_n - \bar{x})^2 \qquad (2)$$

Then, it's shown a piece of the developed function code to calculate those values:

```
float emg_feature( ){
        for (int i=0; i<muestras; i++){
          suma0 = suma0 + v_rec0[i];
          suma1=suma1+v_rec1[i];
        }
        promedio = suma0/muestras;
        promedio=suma1/muestras;
      for (int i=0;i<muestras;i++){
      //canal 0
          volt_sqr0 = v_rec0[i]^*v_rec0[i];
        sumasqr0 = suma_sqr0 + volt_sqr0;
      //
        des0 = v_rec0[i]-promedio0;
        suma_des0 = suma_des0 + des0;
      //canal 1
        suma_sqr1 = suma_sqr1 + volt_sqr1;
```

IOP Conf. Series: Materials Science and Engineering **519** (2019) 012016 doi:10.1088/1757-899X/519/1/012016

```
        volt_sqr1 = v_rec1[i]* v_rec1[i];
    //
        des1 = v_rec1[i] – promedio1;
        suma_des1 = suma_des1 + des1;
    }

media_sqr0 = suma_sqr0/muestras;
rms = sqrt(media_sqr0);
var_x0 = suma_des0/(muestras-1);
media_sqr1 = suma_sqr1/muestras;
rms1 = sqrt(media_sqr1);
var_x1 = suma_des1/(muestras-1);
```

This function finally returns four values that conform two vectors, each of them giving information about the contraction level of the muscles (power and strength).

$$v_{m1} = [rms_{m1} \quad var_{m1}] \qquad (3)$$

$$v_{m2} = [rms_{m2} \quad var_{m2}] \qquad (4)$$

2.4. Analysis of position

To determine the position of the servomotor that turns the wrist (pronation and supination) the vector resulting from the characteristics vectors was taken, this being the difference between the two vectors (from the two muscles).

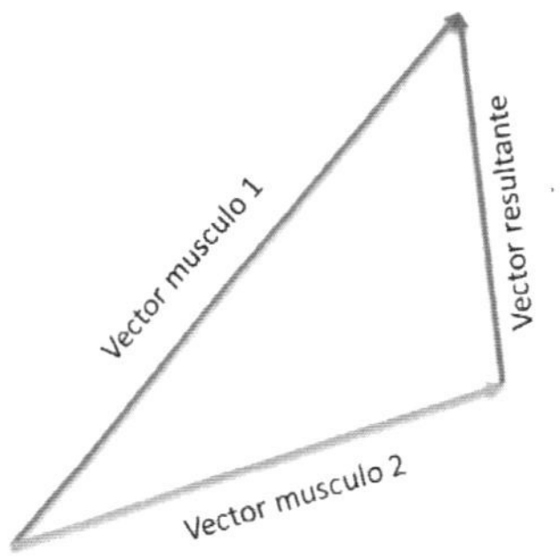

Figure 5. Vector sum between feature vectors

Being the resulting vector:

$$v_{res} = v_{m1} - v_{m2} \qquad (5)$$

This difference can be appreciated on figure 6, where there is the electric activity registry of the antagonistic muscles during forearm movement. First, the supination movement is present where the long supinator muscle contracts and the round pronator relaxes; then the pronation is made where there is a low level of contraction for both muscles. Finally, during the pressing movement of the hand it is appreciated a strong contraction on both muscles.

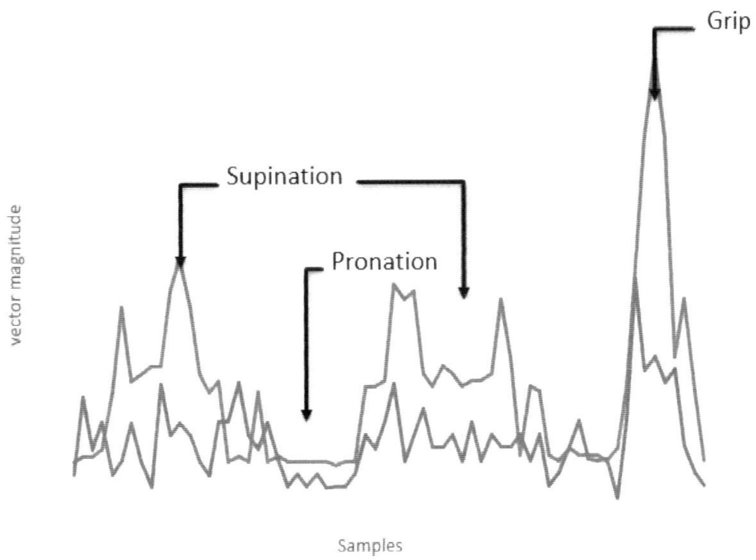

Figure 6. Forearm movement

Using the behavior on the magnitudes of the characteristics vectors a relationship can be obtained between the resulting vector and the PWM signal generated from the Arduino.

3. Results

Hereafter, it is presented the PWM signal generated from the EMG signal analysis. The PWM signal was not generated for any specific servomotor because the working range can be adjusted correlating the maximum amplitude value obtained on the resulting vector with the maximum width of the pulse.

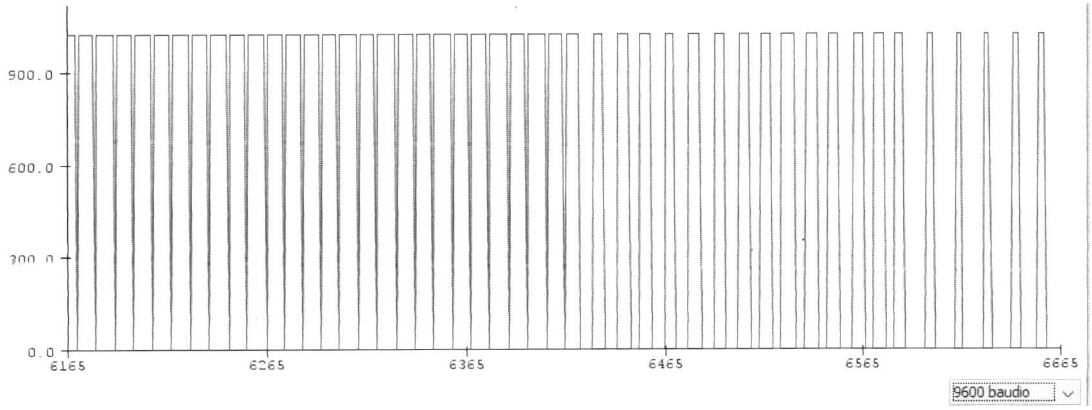

Figure 7. Pronation movement with an open hand

On figures 7 and 8 it is appreciated the PWM output signal according to the movements done by the forearm. On the first, it is seen the movement of pronation on the forearm and this is done with the hand open, so activation of the servomechanism that open and close the tweezers is not done. On figure 9, during the supination movement the hand is closed, so the tweezer servomotor is activated.

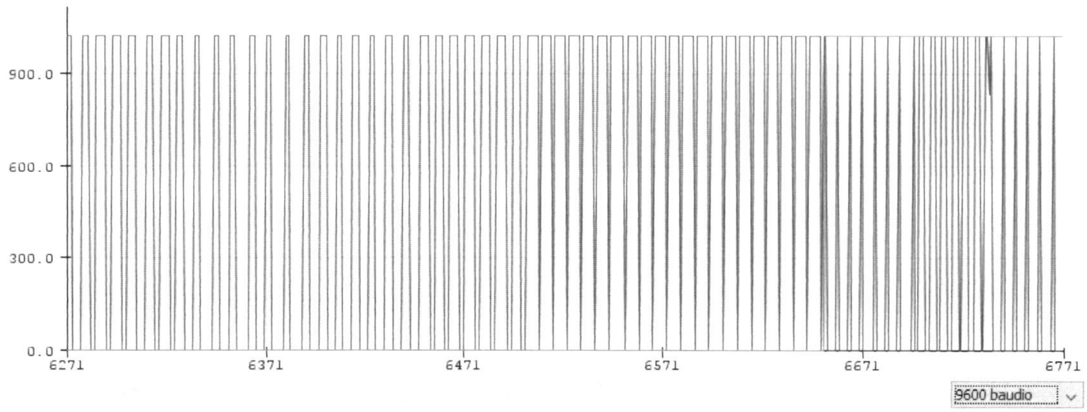

Figure 8. Supination movement and hand closing

4. Discussion

Based on the obtained results, a control system based on the time domain analysis is an economic alternative for the development of prosthesis that are accessible for people with low purchasing power. Even so, the control system must take in mind that the rapid variation of the signal produces errors on the servomotor position and this can be compensated with a bigger sample vector (number of obtained samples) being able to soften the oscillations produced on the signal or using working intervals for the resulting vector so the variation can be limited to these intervals.

A system that analyzes the signal on the frequency domain would make the system more expensive but can give a better control of the system, even though this can be used for control modules that are more robust directed to other type of population.

References

[1] DANE 2012 *Estadisticas demográficas de Colombia*

[2] C. De Luca 1997 *The use of surface electromyography in biomedichanics* (*Journal of applied biomechanics* vol. 13) pp. 135-163

[3] C. De Luca 2002 *Surface electromyography: Detection and recordeing* (*Delsys Inc.*)

[4] C. Quinayás and C. Gaviria 2015 *Sistema de identificación de intención de movimiento para el control mioeléctrico de una prótesis de mano robótica* (*Ing.. Univ.* vol. 19) pp. 27-50

IOP Conf. Series: Materials Science and Engineering **519** (2019) 012017 doi:10.1088/1757-899X/519/1/012017

Determination of Cross layer interference in Heterogeneous network using stochastic geometry

Y Tarriba-Lezama[1,2]
[1]Antonio de Arévalo technological foundation TECNAR,
[2]University of Cauca

yesid.tarriba@tecnar.edu.co

Abstract. This document describes the determination of the cross-layer interference for a heterogeneous network.HetNet Stands in the statistical parameters using the mathematical tool StoGeo stochastic geometry that is based on the poisson point process, PPP. A two-tiers network with differing power and distribution patterns is proposed, although defined as independent and identically distributed at each tiers in the low power nodes NBP, NHP respectively, and they are presented under the homogenous PPP. Power is considered at each constant tiers and fading and interference distributions are planned under approximation models for gamma distribution.

1. Introduction

The interference in a mobile network is the main limitation to maximize the use of the network, it is the tiers of the efficiency terms of the network with its spectral resources. For a heterogeneous network there are different sources, these are divided into two types which are described with the base stations of the same tiers and those that are at different tierss or external and different from the physical and electrical characteristics of the base stations.

In principle it is the description of a scenario that illustrates the growth of traffic and data usage in mobile networks and the solution that counteracts the inconveniences presented to cover the demand in traffic. For those who consider different scenarios [1], one of them based on the efficient use of the spectrum by aggregation of subcarriers, another based on the implementation and massification of the technologies in MIMO (Multiple In Multiple Out) antennas and finally a scenario where it is intended to optimize radio resources by implementing network topologies that increase the data traffic capacity, on which the research work is focused. To this, the available network elements present nonhomogeneous properties in terms of their electrical radio properties such as radiation patterns, frequency and power coverage. These proprietary characteristics define the network and are called heterogeneous networks of which for its acronym in English HetNet (Heteregeneous Network) present a solution to the demand for data traffic in mobile networks [1].

The operational technical foundation of heterogeneous networks is focused on the use and implementation of small cells. These are modeled and defined as network elements at different tiers.

Among the different effects caused in the implementation of heterogeneous networks is studied the interference effect of small cells at different tiers of the network. This interference called cross-layer interference or variation in literature also called cross-tier, which consists of degrading the power of the

Content from this work may be used under the terms of the Creative Commons Attribution 3.0 licence. Any further distribution of this work must maintain attribution to the author(s) and the title of the work, journal citation and DOI.
Published under licence by IOP Publishing Ltd

transmitted signals because of the superposition of other signals transmitted by other network elements at different tiers [2].

The solution to mitigate this type of interference is focused on algorithms that make a more efficient use of the available frequencies, among which we will focus on choosing the most suitable ones in terms of optimization and viability in the implementation, for this it is necessary to consolidate the studies and ongoing analysis of existing and implemented techniques. These techniques are based on stochastic processes or random variable calculations in which they are found [3]:

2. Modeling of the interference

One of the most important metrics of network performance and user experience is the SINR interference signal to noise ratio, defined as the total power of the received signal in relation to the power received from all sources other than the desired transmitter, plus the thermal noise power in the receiver (which is always present, even in the absence of interference) [4]. In a given link between a service BS and a user, the SINR determines the rate of bit errors in the link, if you want to optimize the operation of a heterogeneous network, you need to know the spatial distribution of the SINR and its dependence on the deployment parameters of the network. On the (That is, a link for transmissions from a BS to a user) [5].

In the heterogeneous network scenario as in figure 1, the origins of the interference are described, from which it can be said that the total interference.

$$I_T = \sum_{k}^{K} \zeta_k \qquad (1)$$

Where ζ_k is the interference at each tiers of the network, k = 1,2,3 ... indicates each tiers of the network Generally, the definition of interference is like all disturbances to a source in a base-station or user equipment.

Furthermore, the fixed transmission power conditions of the base-stations and the channel model (Rayleigh) are considered to be identical and independent distributed (iid) for all the links in the network. Under these conditions of interference raised.

If the nodes are distributed according to a stochastic poisson point process (PPP) Figure 1 the interference function can be stipulated as:

$$I(t) = \sum_{i=1}^{\infty} \ell(t - x_i) = \sum_{x \epsilon \Phi} \ell(t - x) \qquad (2)$$

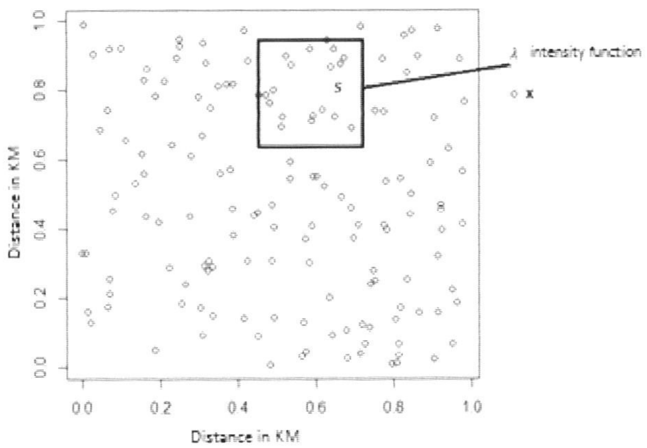

Figure 1. Distribution of the Base-stations under PPP

Where $\Phi = \{x_i\}_{i=1}^{\infty}$ is stationary poisson process point in $\mathbb{R}$, *and* ℓ is a loss function and is given by the impulse response which implies that the losses are given by the distance and without taking into account in the fading of the channel. Under this consideration interference estimation lacks realistic and practical meaning for it is defined.

$$I(y) = \sum_{x \epsilon \Phi} h_x \ell(y - x) \quad (3)$$

Where h_x represent the fading constant the model of the channel and is considered iid
Where in this sense it is posed to Φ as the PPP of the interference which must be measurable according to Slivnyak's theorem [6].

For access channels, ALOHA is a complement for PPP since it maintains the distribution properties of the PPP and the intensity λ for element (node) of the network. Therefore it retains the interference characteristics.

If we perform the derivation in brief instances of interference using Campbell's theorem, we obtain:

$$\mathbb{E}I = \sum_{x \epsilon \Phi} h_x \ell(\|x\|) = \lambda \mathbb{E}(\mathrm{h}) \int_{\mathbb{R}^d} \ell(\|x\|) dx \quad (4)$$

$$= \lambda c_d d \int_0^{\infty} \ell(r) r^{d-1} dr \quad (5)$$

Where $c_d = |b(0,1)|$is the volume of the d-dimension of the unit sphere.

In equation (5) the mean of the interference distribution is described. By using the function of the probability generator to characterize the statistical variables associated with the interference, we have

$$\mathcal{L}_I(s) \triangleq \mathbb{E}[e^{-sI}] = \mathbb{E}\left[\exp\left(-s\sum_{r\epsilon\Phi} h_r r^{-\alpha}\right)\right] \qquad (6)$$

This is equal to calculate the Laplace transform of the interference function and applying the mapping theorem [7].

$$\mathcal{L}_I(s) \triangleq \exp\left(-\lambda c_d \mathbb{E}\left[h^\delta\right]\Gamma(1-\delta)s^\delta\right) \qquad (7)$$

Where $\delta = \frac{d}{\alpha}$ is the symmetric difference between d (order of the dimension) and α exponent of the loss function and Γ is the distribution gamma function.

$$\Gamma \triangleq \int_0^\infty t^{p-1}e^t dt \qquad (8)$$

In the case of fading Rayleigh $\mathbb{E}\left[h^\delta\right] = \Gamma(1+\delta)$

$$\mathcal{L}_I(s) \triangleq \exp\left(-\lambda c_d \mathbb{E}h^\delta \frac{\pi\delta}{sin(\pi\delta)}\right) \qquad (9)$$

Considering a network under the distribution of the PPP with fading, one of the approaches that generalizes the characterization of the interference is the use of probability generating function. Where is given in equation 5 where the mementos to be generated are related to the mean, variance, symmetry and kurtosis in sequential order. That associated with its probability distribution function FDP and the accumulated probability function FPA. You determine the properties of the interference under the network with the PPP model [5] [7].

In a multi-tiers network, structure and topology are considered under different characteristics for the nodes. In essence, the low power nodes NBP small cell and the high power nodes NHP referring to the macro cellular network [8].

Among the characteristics to be taken into account for the analysis is the following:

Transmission power that is referring to the energy with which the signals are injected to the antennas, the radiation pattern that has a direct inference in the coverage of the cells, and frequency of operation of the carrier signals [9].

Under these premises the condition of a two-tiers network is established where the macro tiers refers to a homogeneous PPP distribution with intensity λ_m and the low power tiers refers to femto cells with a homogeneous PPP distribution with λ_f such that $\lambda_f > \lambda_m$.

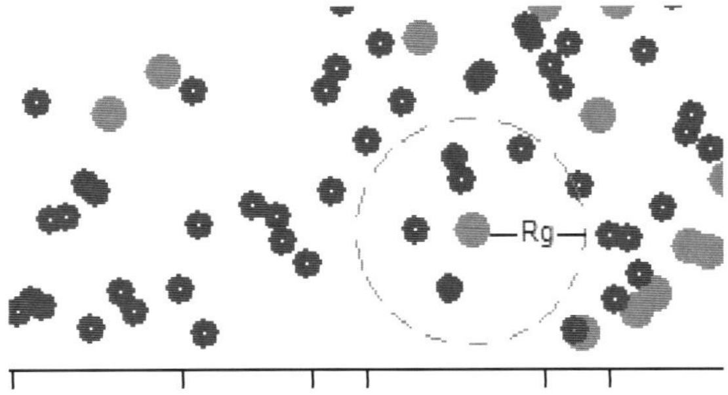

Figure 2. Effect of interference by power and coverage (blue) and NHP (red) between a NBP

For the proposed condition that each distribution of each tier is given by homogeneous PPP, the resulting interference will be parameterized by $I_k = (\alpha_k, \beta_k, P_k, \lambda_k)$, Where k = 1,2,3..M refers to the tier of the network α and β are the distribution parameters (exponential, gamma, chi square) P_k, λ_k are the transmission power parameter and the intensity of the PPP in each tier. Because the transmission power is different for each tier, the coverage region or Rg will be a function of the transmission power [6].

If we inscribe the center of the circle of radius Rg in the Voronoi cell condition as the principle of stochastic geometry we can analyze the performance of a particular cell is given. If you don't wish to use the Word template provided, please set the margins of your Word document as follows.

$$\frac{1}{2\sqrt{\sum_{k=1}^{M} a^2 \lambda_k}} \qquad (10)$$

Where $a = 1 + \dfrac{P_k}{P_l}$

In the case of the heterogeneous network, the total interference would be given by dominant interfering; from the base stations outside the guard region of tier k, which is the firing noise field. So the total interference would be given by

$$I_{tot} = \sum_{k=1}^{M} \sum_{x \in \Phi_k / B(Rc+Rg))} \frac{G_k^x H_k^x P_k^x}{\ell(R_k^{(x)})} \qquad (11)$$

Where Rc is the cell region smaller than a G_k H_k Rg It is the fading for the small scale network and for the large scale respectively.

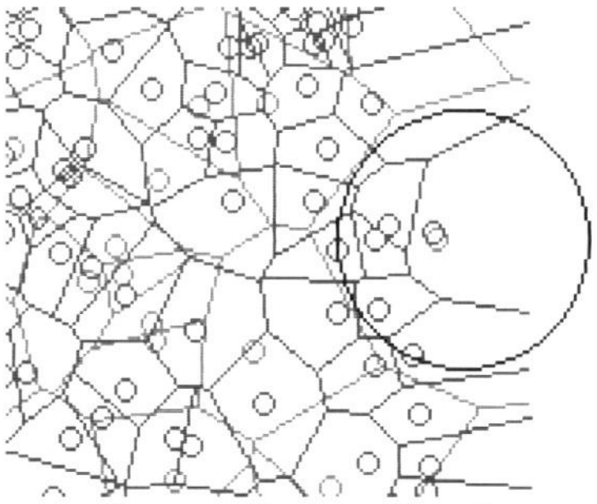

Figure 3. inscriptions of circumference Rg in cell with voronoi tessellation

Referring to figure 3 the interference caused by the nodes of different tiers is called cross layer and it is of special interest for this work in this situation is defined a region of excursion in the radio receiver R_{cross} and what we will call B_{cross}.

And the cross layer interference is defined as:

$$I_{cross} = \sum_{x \in \Phi_{cros}/B_{cross}} \frac{G_x H_x P_{cross}}{\ell(R_x)} \quad (12)$$

Where P_{cross} is the interference power.

3. Approximation Of Interference

The modeling of the cross layer interference has a strict and direct relationship with the distribution models of low power and high power base stations.

Among them, the distribution models under the specific processes that describe the location of the base stations or nodes define the behavior of the interference and in general the network.

For in the case of interference in heterogeneous networks is considered an approximation under the continuous distribution gamma [6] and the application of the first and second moments that coincide with the laplace transform as seen in [5]. For this we consider the parameterization exposed in section III and the substitution of equation 12 for $Z = G_x H_x P_{cross}$.

$$E[I(y)] = E[I] = E[p]\lambda \int_{\mathbb{R}^2} \frac{1}{l(|y|)} dy = \int_0^{\infty} \big(1 - G(s)\big) ds \, 2\pi\lambda \int_0^{\infty} \frac{r}{\ell(r)} dr \quad (13)$$

From which it is concluded that the expected value and the variance for the interference are given by [6]:

$$EI_0(\beta R_c) = \frac{1}{\lambda C R_c^4} \sum_{m=1}^{M} \frac{\lambda_m E Z^{(m)} a_m^2 (a_m^2 + \beta^2)}{(\beta^2 - a_m^2)^2} \tag{14}$$

$$EI_m(\beta R_c) = \frac{2\pi \lambda_m E Z^{(m)} a_m^2}{2CR_c(\beta^2 - a_m^2)^2} \tag{15}$$

$$VI_m(\beta R_c) = \frac{2\pi \lambda_m E\left(Z^{(m)}\right)^2 a_m^2 (a_m^4 + 6a_m^2\beta^2 + 3\beta^4)}{6C^2 R_c^6 (a_m^2 - \beta^2)^6} \tag{16}$$

What it indicates in equation 14 and 15 corresponds to the first and second moments of the gamma distribution function. The parameter $C = \prod_{i=1}^{n} (\theta_{\min} / \theta_i)$, θ is the argument of the gamma function.

4. Result

The interference at a point is given by two terms such that $I = I_o + I_1$ which must be in function of the position of a point r where I_0 is the dominant interference in the guard region Rg and I_1 is the intefrencia of the all stations base under the PPP outside the guard region [6].

We modelled a two-tier network by homogeneous PPP at each tier where λ_m for the high power tier or NHP is less than λ_f and the cell and guard region is a linear function of the transmission power [10].

As shown in figure 4, the heterogeneous network of two tiers with the indications of the guard region

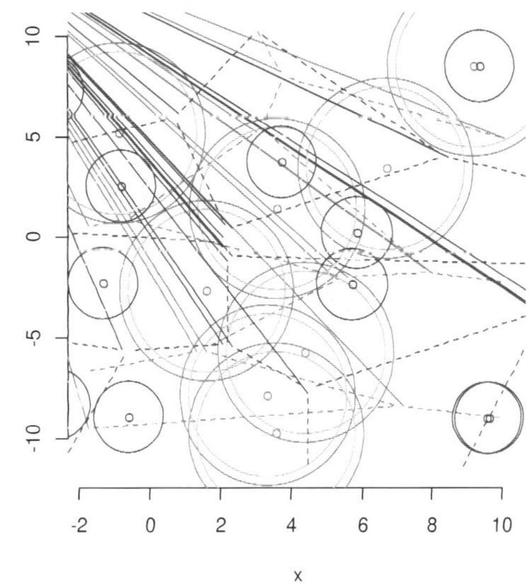

Figure 4. Two-tiered heterogeneous network with guard region

For this condition the variation of the interference in the guard region as a function of time is given by equation 14 from which it can be seen that Io is the first moment of the guard interference.

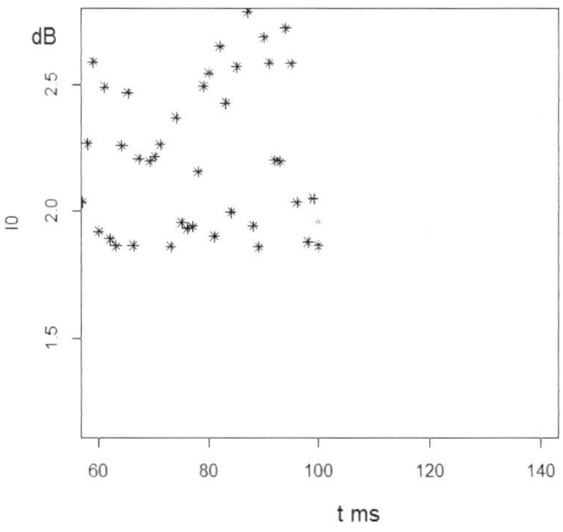

Figure 5. First moment of interference in the guard region.

Of which in equation 12 the fading of the large-scale and small-scale tier are approximated by the gamma distribution as defined above given in its FDP as that in figure 6.

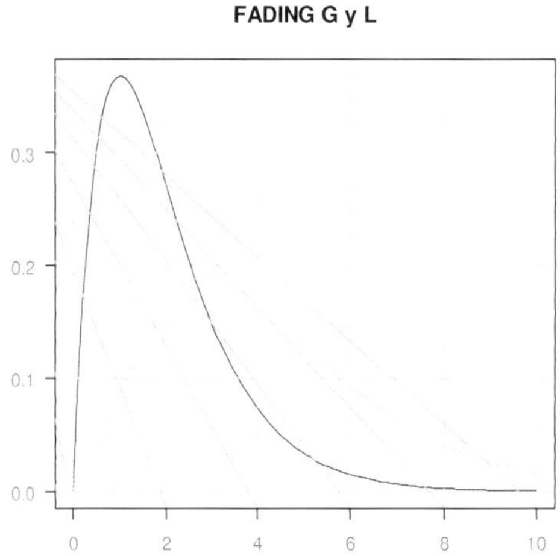

Figure 6. Distribution of fading at two tiers under the Gamma distribution

5. Conclusions

The models studied have maximum effectiveness depending on the problems faced, this problem fully chosen and identified limits the performance of the network in many areas or variable as coverage capacity number of users among others. The model of the generated interference is modeled by equation 12. Interference in the heterogeneous network of 2 tiers.

It is considered as the sum of the cross-references and is modeled under the parametrization of the intensity functions λ at each tier, the transmission power and under the approximation of the fading at each tier according to the parameters of the Gamma distribution.

The interference indicates the performance limit of a wireless network for this reason is vitally important to raise models of analysis and study to characterize or measure it.

IOP Conf. Series: Materials Science and Engineering **519** (2019) 012017 doi:10.1088/1757-899X/519/1/012017

The stochastic geometry as a mathematical tool for modeling random functions is very effective to roughly represent the behavior of a heterogeneous network. And in future statements the characterization of the SINR function that combines environment and noise variables defined by Equation 17.

$$SINR(y) = \frac{P_t(x_0)Ah_{x_0y}\ \|x_0 - y\|^{-\eta}}{W + \sum_{x\in\zeta} P_t(x)\,Ah_{xy\ \|x-y\|^{-\eta}}} \qquad (17)$$

Where A is the gain of the channel , h represents the fade factor $\|x_o - y\|^{-\eta}$, is the distance in terms of the model of slope-intercept channel [5] , Pt It refers to the transmitter power and W it's the noise in the channel.

References

[1] Huidobro J 2012 Redes heterogéneas. La solución para LTE, *revista antena de telecomunicación N* **189** p 61-64

[2] Nychis, G., Tsourakakis, C., Seshan, S., & Steenkiste, P. 2014, Centralized, measurement-based, spectrum management for environments with heterogeneous wireless networks. *In Dynamic Spectrum Access Networks (DYSPAN)*, IEEE International Symposium p 303-314.

[3] Lema Rosas M Á 2015 Contribution to the optimization of 4G mobile communications by means of advanced carrier aggregation strategies. *Doctoral dissertation*, universitad politecnica de catalunya.

[4] Al-Zahrani A Y 2015 Resource Allocation and Interference Management in Heterogeneous Wireless Networks (*Doctoral dissertation*, Carleton University Ottawa).

[5] Haenggi M, & Ganti R K 2009 Interference in large wireless networks. *Foundations and Trends® in Networking*, *3*(2), V**3** 127-248

[6] Heath R, W Kountouris M, & Bai T 2013 Modeling heterogeneous network interference using Poisson point processes. *IEEE Transactions on Signal Processing* **61**(16) p 4114-4126.

[7] Baccelli F, and Blaszczyszyn B 2009 Stochastic Geometry and Wireless Networks, *Volume I— Theory*. Hanover, MA USA:NOWPublishers **V 3**: N3-4 p 249-449.

[8] Maksymyuk T, Brych M, & Pelishok V 2015. Stochastic Geometry Models for 5G Heterogeneous Mobile Networks. *Smart CR* **5(2)** p 89-101.

[9] Hossain, E., Le, L. B., & Niyato, D. 2013. *Radio resource management in multi-tier cellular wireless networks*. John Wiley & Sons p 10-192

[10] Lin, X., Shroff, N. B., & Srikant, R. 2006. A tutorial on cross-layer optimization in wireless networks. *IEEE Journal on Selected areas in Communications* 24(8) p 1452-1463.

Low-cost automatic device for obtaining the coefficient of static friction

H A Rodriguez-Arias

Research group in Robotics, Automation, Telecommunications, Electricity and Electronics (GIRATEC), Faculty of Design and Engineering, Technological Foundation Antonio de Arévalo TECNAR, Cartagena, Colombia.

harold.rodriguez@tecnar.edu.co

Abstract. Considering that the coefficient of friction between materials is one of the important characteristics in the design, either by wear or by requiring friction for power transmission, and this value is usually achieved in tables and these values come out of experimentation or results with tests using devices. Furthermore, these devices are not often found in the laboratories of the universities. For this reason, it was decided to adopt one of the methods of calculation, in this case the inclined plane. This consists of finding the angle of inclination in which one material begins to slide on another, since calculating the tangent of this angle determines the coefficient of friction. To this end, a device was built, consisting of a platform that by means of a mechanism is tilted little by little, until the object under study begins to move. In addition, to exclude human error, the device has sensors and an adequate control, in such a way that the process is completely automatic. Finally, the device is based on open technology, such as Arduino, managing to build a device to obtain the coefficient of friction at low cost. Its development and the results obtained are shown in this paper.

1. Introducción

The following article describes the construction of a low-cost automatic device for obtaining the coefficient of static friction. Initially, the theory used is shown. In this case, the inclined plane method. Then, the elements used for the construction of the device are presented, taking into account even the way in which the angular position sensor is adjusted. In such a case, a linear potentiometer describes how the data is taken in order to convert the readings of the potentiometer into an equation that results in the angle of inclination, After this step, we proceed to the connection of the electronic and mechanical elements of the device, giving to the reader a clear idea of doing so in case of wanting to replicate this device. Finally, the program is attached to the Arduino IDE that controls the system.

This article seeks to supply the laboratories of the universities at low cost and with their own developments. In the case of this equipment, the cost is approximately US $ 50, and from here a lot of possibilities arise. The readers are left to discuss the functionality because at least for the Technological Foundation Antonio de Arévalo is contributing in the laboratory of applied physics, where the student does the experiment manually and then the results are contrasted with the automated equipment to get their own conclusions of the calculation of the coefficient of static friction.

Content from this work may be used under the terms of the Creative Commons Attribution 3.0 licence. Any further distribution of this work must maintain attribution to the author(s) and the title of the work, journal citation and DOI.
Published under licence by IOP Publishing Ltd

IOP Conf. Series: Materials Science and Engineering **519** (2019) 012018 doi:10.1088/1757-899X/519/1/012018

2. Calculation of the coefficient of static friction

The static friction force which is produced by the interaction between the irregularities of two surfaces (see Figure 1) is what prevents any relative movement between the two bodies to a limit where the movement already begins. That threshold of movement is characterized by the coefficient of static friction. The coefficient of static friction is typically greater than the coefficient of kinetic friction. [1].

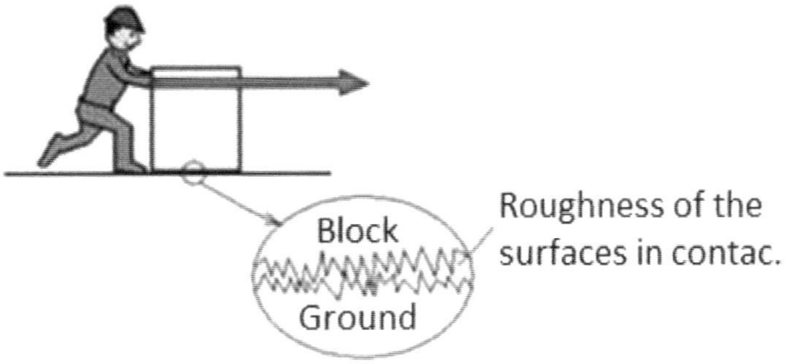

Figure 1. Irregularities between surfaces.

By making the free body diagram of the case of figure 1, we obtain the forces that intervene on the block (see figure 2), and from there we can attain the friction force.

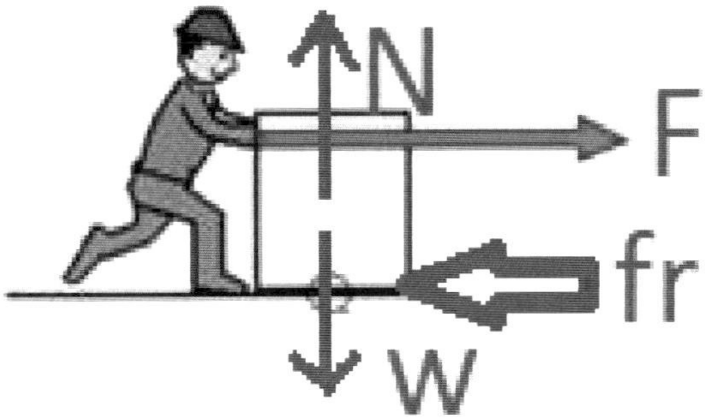

Figure 2. Forces exerted on the block.

Where:

F: External force or force applied by man on the block.
W: block weight.
N: normal strength of the block.
fr: Frictional force.

From here, it is obtained:

$$N = w \tag{1}$$

$$F = fr \tag{2}$$

The relation between fr y N is:

$$fr = \mu . N \tag{3}$$

In conclusion, to determine the friction force is required knowing the value of μ, and this depends on the materials that are in contact. The coefficient of friction (μ) is a dimensionless magnitude that expresses the proportionality existing between the force of friction (fr) and the normal force (N), which exerts a body that rests or moves on another. [2]

In this way, it is clear that it is necessary to have the value of μ, which is found in tables whose values have been found by experimentation. Since it cannot be said with certainty what the coefficient of friction is, taking into account that it plays an important role the polishing of the materials or if there is a lubricant between them. In table 1, it can be seen some and the most common data of friction's coefficients.

Table 1. Friction coefficients of some surfaces in contact. [3]

Surfaces in contact	μ. Static friction coeficiente
Wood on wood	0,5
Steal on ice	0,03
Teflon on teflon	0,04
Rubber on dry cement	1
Glass on glass	0,9
Ski (waxed) on snow (0ºC)	0,1
Wood on leather	0,5
Aluminum on steel	0,61
Human joints	0,02

In this way, whenever it is required to calculate the friction force, it is required to know the μ, and the question is, what if the materials in contact are not in a table? Or what if the finish of the surfaces in contact is special? Here, the need to find a method to calculate the coefficient of static friction arises. There are several methods but the one chosen for this case it is the inclined plane method (see figure 3), from which you can see the internal and external forces that appear on the block. For this case, any external force will be applied.

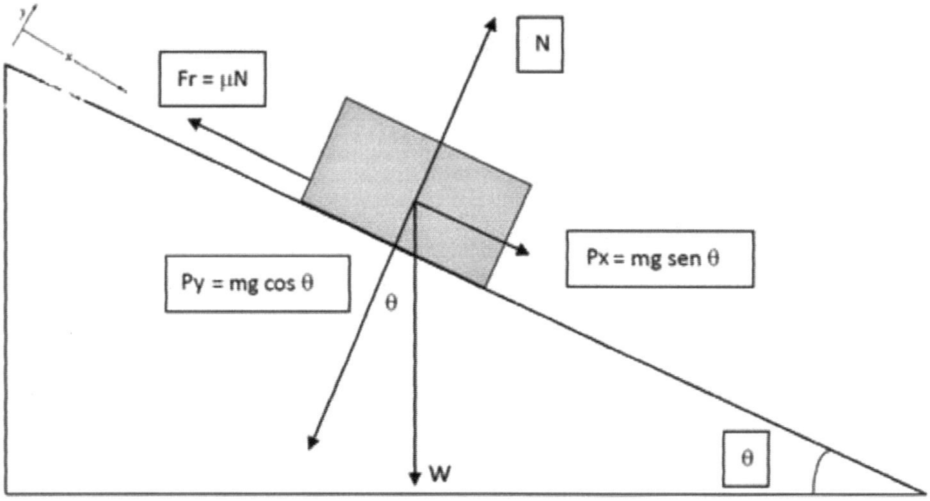

Figure 3. Block on inclined plane

Where:

N: normal
fr: it is the friction force
Px: component of block weight on the X axis

Py: component of the weight of the block on the Y axis
Θ: is the angle of inclination of the plane

It is determined from figure 3 that:

$$N = m.g.cos\Theta \quad (4)$$

$$fr = m.g.sen\Theta \quad (5)$$

Thus, by replacing the values of (4) and (5) in the equation (3), it is obtained:

$$\mu.m.g.cos\Theta = m.g.sen\Theta \quad (6)$$

From (6), it is determined that:

$$\mu = Tan\Theta \quad (7)$$

And this is how the coefficient of friction μ can be determined practically. In figure 3, the block is not moving due to the frictional force. However, if the angle continues to increase, it will reach a value in which the block will begin to slip, and with that value of Θ it is how the static μ is calculated, which it is the case of this study.

3. Development of the device to calculate the static μ

To calculate the static μ, the method of the inclined plane was taken into account.
It can be done manually (see figure 4) and to determine the angle at which the object begins to move and calculating the tangent of that angle, it is obtained friction coefficient.

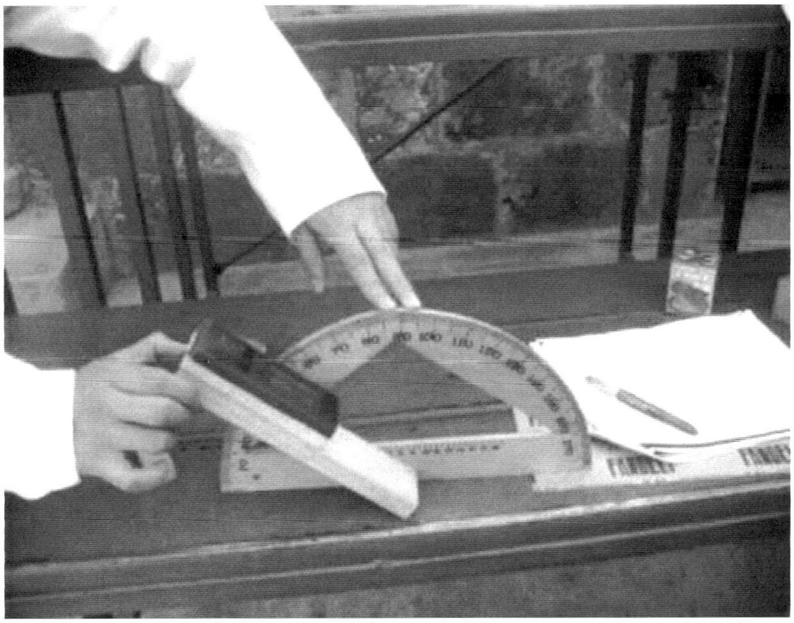

Figure 4. Calculation of the friction coefficient experimentally. [4]

This method is very easy to execute because it is only to have a goniometer, a plate that can be tilted and the two study surfaces. Nevertheless, the disadvantage is that the assignment of the value of the angle Θ is made by observation. Some errors can be presented by the human factor, and seeking to

IOP Conf. Series: Materials Science and Engineering **519** (2019) 012018 doi:10.1088/1757-899X/519/1/012018

eliminate this error factor, the low-cost automatic device for obtaining the coefficient of static friction arises.

3.1. Operation of the low-cost automatic device to obtain the static coefficient of friction

Before explaining the operation of the device, the main elements of the device and its function will be explained.

There are some devices in the market to determine the μ, and they cost a couple of thousand dollars, like the Friction Coefficient Meter Mod. LI-COF-P01 of Figure 5, which is automatic. Besides, it has the necessary sensors to angle calculation, motor with variable speed, RS-232 communication, network port and weighs up to 20 Kg. [5]

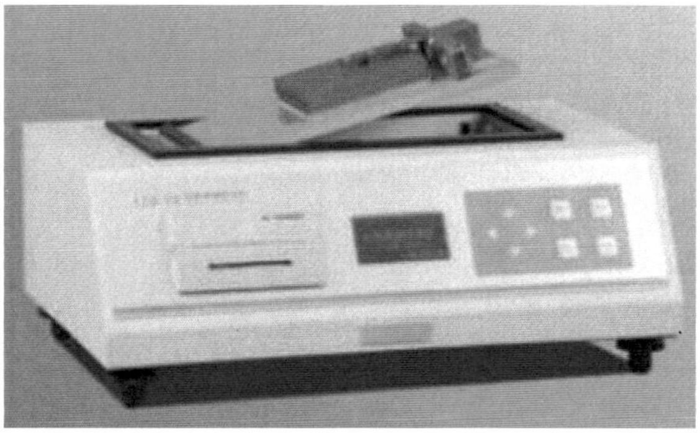

Figure 5. Friction Coefficient Meter Mod. LI-COF-P01 [5]

Based on the basic operation, the prototype of figure 6 was reached. It consists of a platform that pivots on an axis, and it is lifted by a rope that is wound on a reel that moves a gearmotor with variable speed. Moreover, a linear potentiometer is responsible for measuring the angle of inclination. There is a presence sensor, an Arduino UNO card, an LCD screen among others.

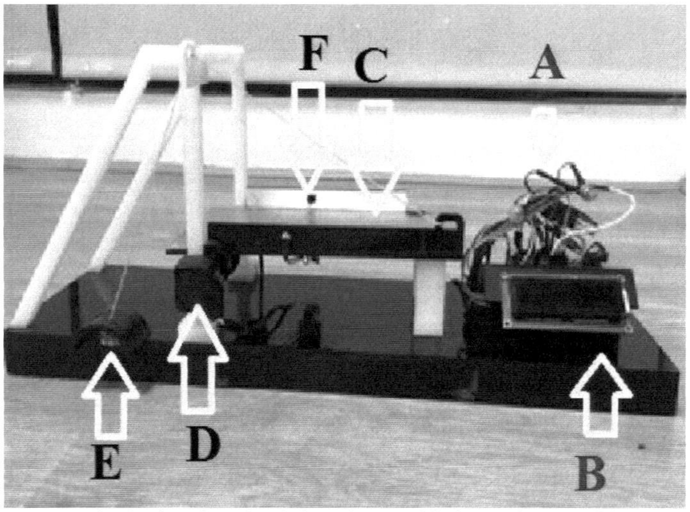

Figure 6. Low cost automatic device for obtaining the coefficient of static friction.
where:

- **A**: Control card Arduino UNO
- **B**: LCD screen of 16x2
- **C**: Platform
- **D**: Linear potentiometer
- **F**: Infrared motion sensor
- **E**: Variable speed gearmotor

The size of the device can vary according to the requirement. For this case, the base is 12 cm x 30 cm, and does not weigh more than 1.5 kg, which makes it very easy to move. The body of the device is made of acrylic which it is an economic material and easy to manufacture. Additionally, the connections of electronic devices are practically found in the network. However, the adjustment of the linear potentiometer will be shown in detail in the results stage of this article. In figure 7, the operating logic is clearly observed, even when the device is in operation.

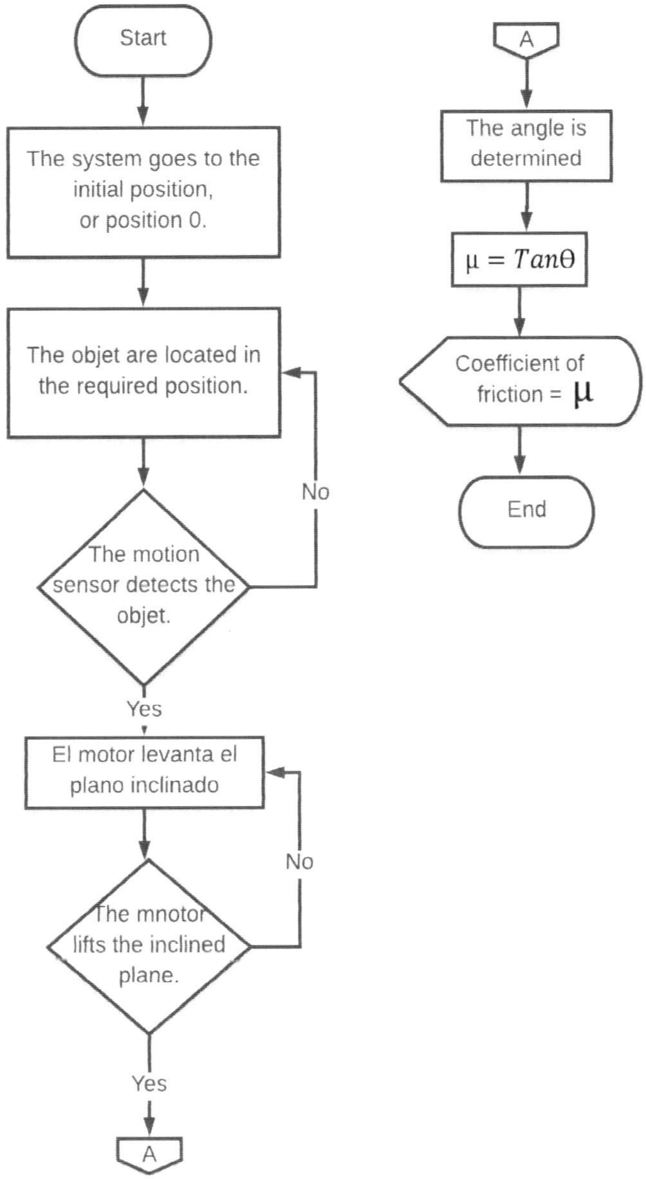

Figure 7. Operation of the low-cost automatic device to obtain the coefficient of static friction.

4. Calculation of the angle with a linear potentiometer.

To determine the angle of inclination in which the object begins to move, and which serves as input for the calculation of the coefficient of static friction μ, a linear potentiometer is used, which is coupled to the axis of the platform that is tilted. See figure 8.

This potentiometer gives as input to the Arduino card values between 0 and 1223, since the analog input of the Arduino is 10 bits, it must be determined the value of the angle of inclination with this input . For this case, measurements were made which can be seen in table 2, where in the first column is the value given by the linear potentiometer and in the second column the value of the angle of inclination measured with an external device. For this sense, the sensor of a Samsung Galaxy A7 cell phone was used.

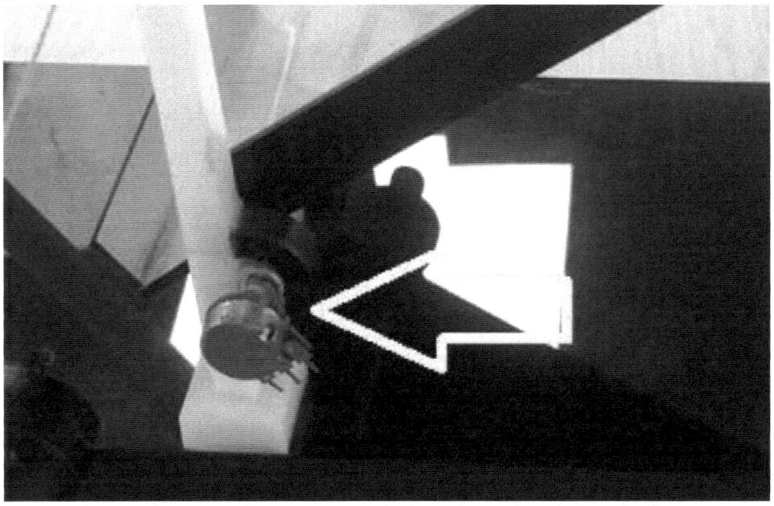

Figure 8. Potentiometer coupled to the axis of the platform.

Table 2. Relationship between the value given by the potentiometer and the Samsung Galaxy A7.

Analog input	Grades (A7)
370	0
395	8,4
413	14,5
439	21,9
465	30,6
479	36

Figure 9 shows the assembly to determine the data in Table 2, which lead to determine the equation that will govern the value of the angle of inclination.

Figure 9. Assembly to determine the relationship between sensor value Vs tilt angle.

With the data in table 2, the graph of figure 10 is made, and with the help of Excel, the equation that will govern the calculation of the angle of inclination can be determined.

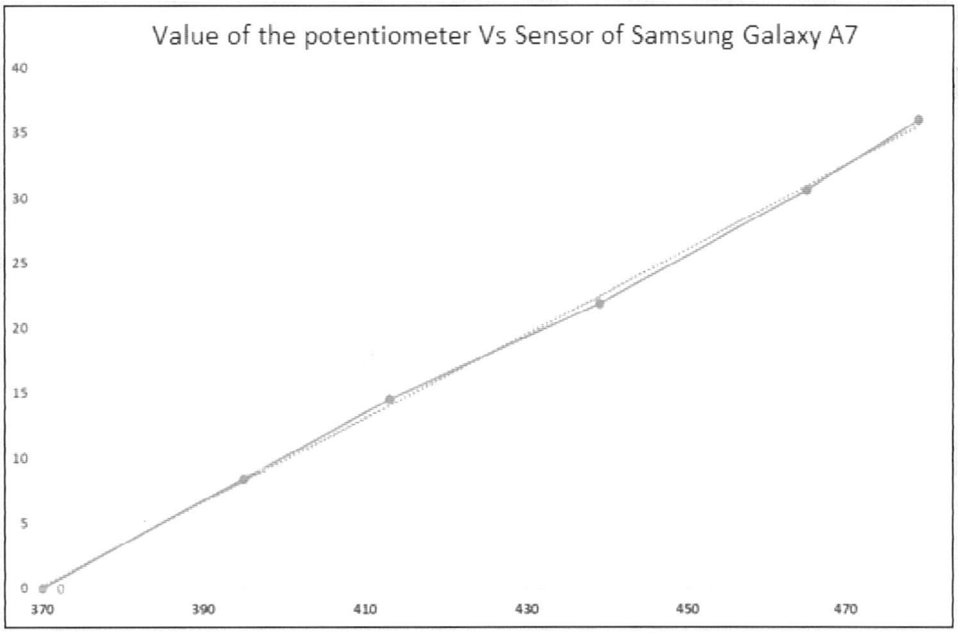

Figure 10. Value of the potentiometer Vs the angle of the Samsung Galaxy A7.

Using a linear adjustment in Excel, it can be found the function that will determine the angle of inclination. This equation is the one that is used for programming the device.

$$\theta = 0.3247.VP - 120.0 \quad (8)$$

Where

θ: Angle of inclination given by the device

VP: Value given by the Potentiometer

This adjustment gave a determination coefficient of:

$$R^2 = 0.9989 \quad (9)$$

IOP Conf. Series: Materials Science and Engineering **519** (2019) 012018 doi:10.1088/1757-899X/519/1/012018

The equation (9) is a good indicator that this linear adjustment has a very good approximation *[6]*. Furthermore, it was decided to leave the equation (8) as the function that will govern the system for calculating the coefficient of static friction.

5. Connections and Programming

5.1. Connections

The first test that is done, it is to connect the screen, since the I2C adapter will be entered to the Arduino UNO card and the 16x2 LCD screen. This adapter uses only 4 connectors which are, Vcc, GND and two more that go to the A4 and A5 pins, as it is shown in figure 11.

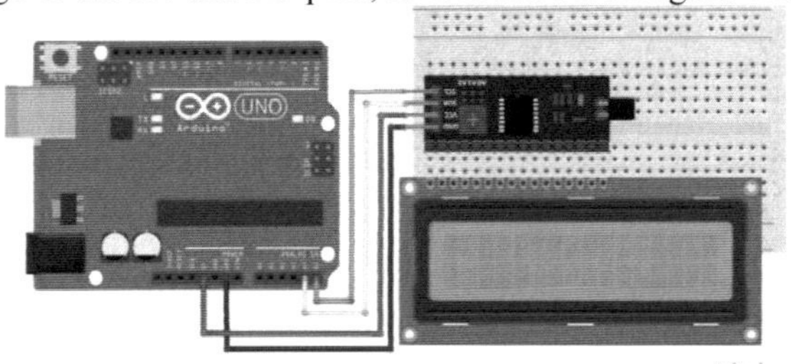

Figure 11. Diagram of connection between Arduino UNO, adapter for LCD I2C PCF8574 screen and LCD screen **[7]**

Having this connection working and bearing in mind that the A4 and A5 pins are reserved for the I2C LCD PCF8574 screen adapter, the rest of the elements of the device are connected. (See figure 12)

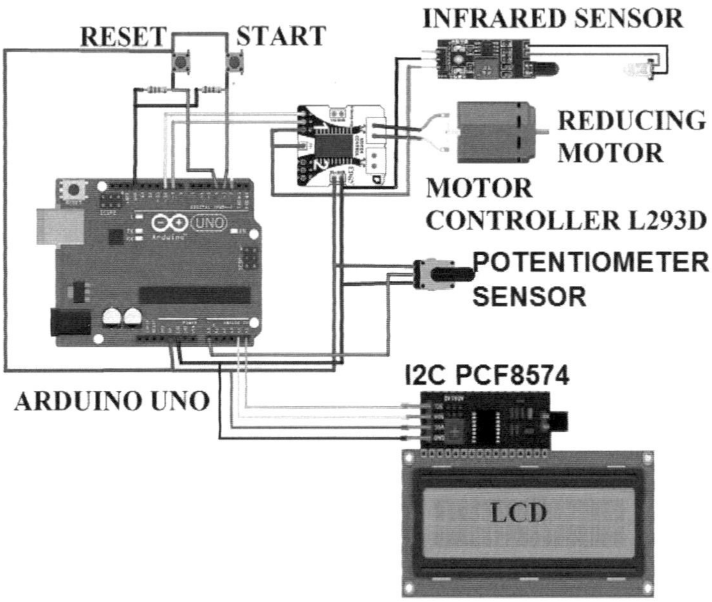

Figure 12. Low-cost automatic device connections for obtaining the coefficient of static friction

IOP Conf. Series: Materials Science and Engineering **519** (2019) 012018 doi:10.1088/1757-899X/519/1/012018

5.2. Programming

The programming of the low-cost automatic device for obtaining the coefficient of static friction will be presented below. This programming was done in the IDE of Arduino. However, the use of it, it is left to the consideration of the readers

```
/*
**********************************************************
*PROGRAM TO CALCULATE THE COEFFICIENT OF STATIC FRICTION*
*     BY: M.Sc. HAROLD ALBERTO RODRIGUEZ ARIAS     *
*              TECNAR              *
*             CARTAGENA               *
*             COLOMBIA              *
*              2018              *
**********************************************************
----REMEMBER TO DOWNLOAD THE LIBRARY LiquidCrystal_I2C---
--------------------AND INSTALL IT---------------------
*/
#include <Arduino.h>
#include <Wire.h>
#include <SoftwareSerial.h>
#include "LiquidCrystal_I2C.h"
double angle_rad = PI/180.0;
double angle_deg = 180.0/PI;
double pwm;
double ang;
void stop();
int conta;
int entrada;
int s;
int start;
double u;
int reset;
LiquidCrystal_I2C lcd_I2C_0x27(0x27, 16, 2);

void setup(){
  lcd_I2C_0x27.init();
   lcd_I2C_0x27.backlight();
  lcd_I2C_0x27.clear();
  lcd_I2C_0x27.setCursor( (1) - 1, (1) - 1 );
  lcd_I2C_0x27.print( "CALCULATION OF" );
  lcd_I2C_0x27.setCursor( (4) - 1, (2) - 1 );
  lcd_I2C_0x27.print( "THE u STATIC" );
  _delay(2);
  lcd_I2C_0x27.clear();
  lcd_I2C_0x27.setCursor( (1) - 1, (1) - 1 );
  lcd_I2C_0x27.print( "H. RODRIGUEZ" );
  lcd_I2C_0x27.setCursor( (2) - 1, (2) - 1 );
  lcd_I2C_0x27.print( "TECNAR-2018" );
  _delay(2);
  pwm = 30;
  baja();
  stopp();
  lcd_I2C_0x27.clear();
```

IOP Conf. Series: Materials Science and Engineering **519** (2019) 012018 doi:10.1088/1757-899X/519/1/012018

```
  lcd_I2C_0x27.setCursor( (2) - 1, (1) - 1 );
  lcd_I2C_0x27.print( "LOCATE OBJECT" );
  lcd_I2C_0x27.setCursor( (4) - 1, (2) - 1 );
  lcd_I2C_0x27.print( "WAITING..." );

  pinMode(8,OUTPUT);
  pinMode(9,OUTPUT);
  pinMode(A0+0,INPUT);
  pinMode(2,INPUT);
  pinMode(3,INPUT);
  pinMode(4,INPUT);
}

void loop(){

  s = digitalRead(2);
  if(((s)==(1))){
    lcd_I2C_0x27.clear();
    lcd_I2C_0x27.setCursor( (4) - 1, (1) - 1 );
    lcd_I2C_0x27.print( "PRESS THE" );
    lcd_I2C_0x27.setCursor( (4) - 1, (2) - 1 );
    lcd_I2C_0x27.print( "START BUTTON" );
    start = digitalRead(3);
    if(((start)==(1))){
      while(!(((s)==(0))))
      {
        _loop();
        s = digitalRead(2);
        sube();
        medir();
        lcd_I2C_0x27.clear();
        lcd_I2C_0x27.setCursor( (1) - 1, (1) - 1 );
        lcd_I2C_0x27.print( "TETA =" );
        lcd_I2C_0x27.setCursor( (4) - 1, (2) - 1 );
        lcd_I2C_0x27.print( ang );
      }
      stopp();
      u = tan(angle_rad*ang);
      lcd_I2C_0x27.clear();
      lcd_I2C_0x27.setCursor( (1) - 1, (1) - 1 );
      lcd_I2C_0x27.print( "u ESTATIC =" );
      lcd_I2C_0x27.setCursor( (4) - 1, (2) - 1 );
      lcd_I2C_0x27.print( u );
      reset = digitalRead(4);
    }
  }
  if(((reset)==(1))){
    baja();
    lcd_I2C_0x27.clear();
    lcd_I2C_0x27.setCursor( (1) - 1, (1) - 1 );
    lcd_I2C_0x27.print( "CALCULATION OF" );
    lcd_I2C_0x27.setCursor( (4) - 1, (2) - 1 );
    lcd_I2C_0x27.print( "THE u STATIC" );
```

```
  }

  _loop();
}


void sube()
{
  digitalWrite(8,1);
  analogWrite(9,pwm);
}

void baja()
{
  while(!(((ang)==(0))))
  {
    _loop();
    analogWrite(8,pwm);
    digitalWrite(9,1);
    medir();
  }
}

void stopp()
{
  digitalWrite(8,0);
  analogWrite(9,0);
}

void medir()
{
  conta = 0;
  for(int __i__=0;__i__<10000;++__i__)
  {
    entrada = analogRead(A0+0);
    conta = (conta) + (entrada);
  }
  entrada = (conta) / (10000);
  ang = ((entrada) * (0.3247)) - (120.01);
}


void _delay(float seconds){
  long endTime = millis() + seconds * 1000;
  while(millis() < endTime)_loop();
}

void _loop(){

}
```

6. Conclusions

It can be conclude that low-cost laboratory devices can be developed by using free technology such as Arduino and applying basic theoretical bases. It is important to note that complex calculations are not required for the construction of equipment. Although, this device automatically achieves a simple exercise if this calculation is done manually, it can involve human error when taking the data.

An equipment of less than 50 dollars was achieved, and in this way the goal of building low-cost equipment is fulfilled. Furthermore, it is recommended to use a multi-turn potentiometer to improve the accuracy. Also, it must be connected to a pinion box in such a way that by 90 ° of movement of the mobile platform, the potentiometer rotates for about 8 full turns.

Finally, at the moment the results can only be compared with the experiment done by hand. However, a certified device is being searching for the comparison of results. Currently, this accomplishes with the practical calculation applying the theory of the inclined plane.

Acknowledgement

The author expresses his gratitude to the students of the SIRATEC research center for their work and support in this project, especially to the student José David Betin Payares, for his commitment to research.
Special thanks to my beloved wife Karen Julio Cárdenas for her contribution translating this document.

Referencias

[1] M. Olmo y R. Nave, «Fricción estática,» 2018. [En línea]. Available: http://hyperphysics.phy-astr.gsu.edu/hbase/index.html. [Último acceso: 1 Agosto 2018].

[2] J. Villasuso, «Fuerza de rozamiento,» [En línea]. Available: http://newton.cnice.mec.es/materiales_didacticos/rozamiento/index.htm. [Último acceso: 7 agosto 2018].

[3] Universo Formulas, «www.universoformulas.com,» 2018. [En línea]. Available: https://www.universoformulas.com/fisica/dinamica/fuerza-rozamiento/. [Último acceso: 20 Agosto 2018].

[4] Huertocha, «Pratica de física "fricción",» [En línea]. Available: https://www.youtube.com/watch?v=pG3U2b70cyM. [Último acceso: 20 Julio 2018].

[5] Representaciones Adendorf, «Representaciones Adendorf, Equipos y suministros para la industria.,» [En línea]. Available: http://www.adendorf.net/medidor-de-coeficiente-de-friccion-mod-li-cof-p01-p-169.html?osCsid=50da92d0a917106b2712f92f59a2ce05. [Último acceso: 25 Julio 2018].

[6] Addlink Software Científico, «Cómo interpretar un modelo de regresión con bajo coeficiente de determinación R-cuadrado y bajos valores P,» 26 Febrero 2018. [En línea]. Available: https://www.addlink.es/noticias/minitab/2736-como-interpretar-un-modelo-de-regresion-con-bajo-coeficiente-de-determinacion-r-cuadrado-y-bajos-valores-p. [Último acceso: 19 Septiembre 2018].

[7] Geekfactory, «LCD 16×2 por I2C con Arduino usando solo dos pines,» 29 Mayo 2017. [En línea]. Available: https://www.geekfactory.mx/tutoriales/tutoriales-arduino/lcd-16x2-por-i2c-con-arduino/. [Último acceso: 15 Septiembre 2018].

IOP Conf. Series: Materials Science and Engineering **519** (2019) 012019 doi:10.1088/1757-899X/519/1/012019

SafeWalk App, mobile application in health for people living in motor disabilities

E A Osorio[1,2], J D Osorio[1,2] and D A García[1,2]

[1] Institución Universitaria de Envigado, Envigado 055420, Colombia
[2] Member of the research group GIOSTIC and member of the working line Humans & Apps of the Latin American Network of Mobile Technologies in Mobile Health (Red CYTED-RHYTHMS), Colombia.

eaosorioy@gmail.com

Abstract. The research had the objective of develop a mobile application in health for people with motor disabilities oriented to be multiplatform and will allow integrate different accessibility functionalities. The exploratory method was used to aproach reality and as instrument, information search was carried out in government sources, main mass media of the country, observation in public places and semi-structured interviews to health area professionals. With that information was possible identify the need of this population in term of mobility issues, either because of its near or far enviroment. The application will gather information of the places available to serve this population and will be used to accessibility to their interest places. Users will be able to part of a community of people with similar conditions, interact with them and share information. Also they will have the possibility of communicate with tutors that can provide various kinds of help and contact them through a button available across the interfaces of the applications. It will generate information that will can be rated by users to estimate if the registered places are really accessibles and have good rating from those who have already been there.

1. Introduction

In Colombia there is a social inclusion policy, which consist in guarantee the rights and duties of persons with disabilities (PCD) [1], this population daily find barriers that limit the full enjoyment of life. Government has developed projects that allow work with different organizations in the country to strengthen the quality of life of these people, but still find problems in their development and is not evidenced a contribution to their welfare.

The main feature of this population is a high poverty incidence, many because of their condition, do not attend educational establishments generating high illiteracy rates, in the best case, they only have primary basic education and very few PCD have high school studies or higher [1], limiting their employment opportunities. All this caused by obstacles and few opportunities that they have in the middle.

To analyze the problematic it is necessary to mention that, according to figures from DANE, people included in the Location Registration and Characterization of People with Disabilities (RLCPD by its

Content from this work may be used under the terms of the Creative Commons Attribution 3.0 licence. Any further distribution of this work must maintain attribution to the author(s) and the title of the work, journal citation and DOI.
Published under licence by IOP Publishing Ltd

acronym in Spanish) the majory of alterations present are: the movement of the body, hands, arms and legs with a figure of (34.05%) [2] being the highest number of different disabilities currently reported and continues rising. It is understood that these people have physical problems that limit their movement and many of them by their own way, adapt themselves so that their life is in equal conditions.

The investigation of this social problematic is performed in the interest of providing a solution to this population. The development of a mobile application that generates information for people with a motor disability [3], and because of their condition, require information from their environment about accessible places, attention in these places, points of interest that will be reported according to preferences that the user has previously set, ability to assign tutors (People without disabilities) which together with users generate information from the environment and the ability to interact with a community within the application, that is exclusively for people with disabilities and tutors.

In order to determine the functionality of the applications, continuously are proposed new utilities that can be integrated and form a final product because it seeks to obtain reliability, ease of use for the user and the community, contribute a possitive and differentiating impact, pertinent to the solution and informative support for the mobility of the population with motor disability.

2. Methodology

The cualitative research finds stablish a serie of generalizers of analytic and theoric types about topic; that allows the interpretation of the different human expressions and the understanding of the subjective reality (Yin, 2009) [4], originated by the sense that things adquire according the social actors in their natural envirnoment from their own values (Taylor & Bogdan, 2000) [5]. The exploratory method was used that approximate to the understanding of a reality, their characteristics and dynamics, also because combine techniques that allow describe situations without exhibit problems of results validity. As sources of information it was consulted in governmental bodies, information media, observation of some access spaces of this kind of population and semi-structurated interviews carried out to health and engineering specialists of Medellín city.

Through the literature review it evidenced that a big part of studies about the topic are a cuantitative cut that suggest that throws indicators, but not subjective interpretations of this reality, what gives relevancy to perform studies of cualitative order. Observation in public and work places was carried out. Semi-structured interview was used, every time that possibilities the change according interview dynamic and ease that other questions flow to deepen or clarify interpretations (Vargar J., 2012) [6].

Once adquired the information, it proceeds to identify and structurate the main needs, that have the people in motor disabilities and taking into account the barrers that must be faced on a daily basis. Such as access to different means of transport, access to commercial stablishments, education centers, heatlh centers, among others [7]. The development of an application oriented to the mobility of people with motor disability was proposed. This is proposed in discussion with trained professionals that contribute to the development of each utility. Their contributions provide a panorama that generalizes the lack of social inclusion of these people and of adequate facilities in public places of the city, this solution is proposed from ICT knowledge (Information and Communication Technologies), guite users and be a support for government social inclusion policies.

2.1. Design considerations for application

SafeWalk App is a cross-platform application with graphic individualities by operating system provided by the development environment (Ionic Framework) [8], which has its previously designed components for free use, which are used for the development of the graphic environment of the application.

IOP Conf. Series: Materials Science and Engineering **519** (2019) 012019 doi:10.1088/1757-899X/519/1/012019

Some modifications are added to the components to adapt it to its own criteria that facilitate navigation and highlight the functionalities of the application to provide accessibility and maintain a more fluid user-application interactivity.

2.2. Application Development

The application is developed in a web-based structure that provides the Ionic Framework (Framework A framework, work environment or framework is a standardized set of concepts, practices and criteria to focus on a particular type of problem that serves as a reference, to face and solve new problems of a similar nature) development framework with customized components in its graphic line and functionalities development. The programming is done in the Typescript language; a web language that controls the communication between the graphic interface and the logic of the application.

The programming language used implements other development frameworks such as Angular JS, which makes the system more robust and allows to perform more complex tasks such as view loading, navigation and communication with servers. Additionally it has the Apache Cordova framework that allows to communicate with the hardware of the device granting access to components such as the camera, the geo locator, among others. This framework acts as an interpreter between the web language (Javascript / Typescript) and the native language of the device (Objective-C / Swift for iOS and Android / Java for Android) enabling the use of these components.

For the storage of data, authentication and management of users of the application, it makes use of a development platform owned by Google called Firebase. This platform allows the creation and authentication of users using different authentication providers such as Github, Google, Facebook and Twitter, as well as authentication using email and password or with the cell phone number via SMS. In the case of the application, we will use the authentication by email and password, as can be seen in 'figure 1'. Firebase also provides a database to store user information such as your profile and registered entities.

Figure 1. Starting screen

3. Result, application SafeWalk App

According to information compiled in the main communication media and governmental entities, the need to create a tool to support the population in disability status in mobility issues was identified [9,10]. SafeWalk App is an application that brings users closer to establishments with accessibility for them.

The picture of disability that is not addressed in a relevant way, only in Colombia corresponds to a quite significant number. The policies that currently exist, although they cover the priority needs of people with disabilities, do not represent systems that at all times address problems that they are in many occasions very diverse, more when talking about issues of mobility and accessibility, because the systems that have been implemented as they are the respective signaling, buses with preferential seats and services of first necessity in shopping centers, hospitals and educational centers they are known by all the population for which they are destined, most of the times, for the ignorance or little access to this type of information.

With the type of research carried out, information was obtained that gave rise to the guidelines that defined each functionality and each utility, thus, the main parameters were established in favor of the development, evaluation and automation of the application in order to generate information in mobility. for people with disabilities is a fact that does not depend only on the policies implemented by the government, but is a system that owes its operation to the joint work of different entities that provide information.

The application is developed in a way that is transversal for smart mobile devices with iOS and Android operating systems, which allows a percentage of 96% of the global market [11,12]. It has two interconnected developments, the first, the development of the mobile application for people with motor disabilities; the second, for future development, a registration website for commercial, governmental, health and educational entities in order to collect the information that will be available to the target public.

For the development of the mobile application you will have a side menu to access the functionalities 'figure 2' and you will make use of the geolocation provided by components of the mobile device, this in order to obtain in real time the establishments of preference of the user 'figure 3'. In order to support the mobility system of the application, it is also proposed to have a user tutor and if you need additional help, the application will provide information such as your location, suggest places of interest and have the possibility to contact you through the community 'figure 4' or by call.

The interaction with the tutor can also be evidenced when the user is in trouble and use the panic button [13] 'figure 5' that will serve to contact him or a local emergency entity and be able to transmit audio or video to know the real situation of the user [14].

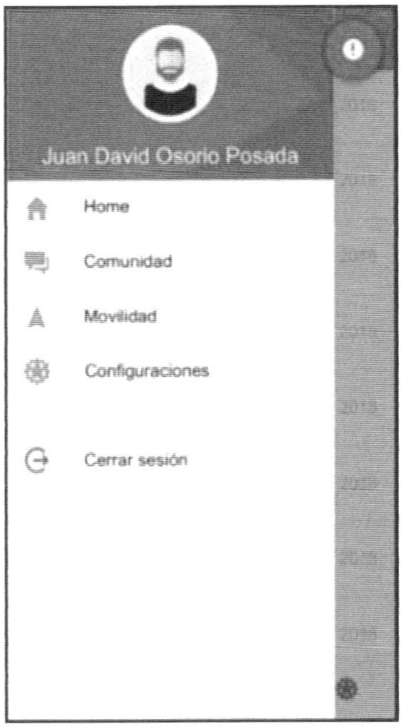

Figure 2. Initial Utilities

Figure 3. Navigation map

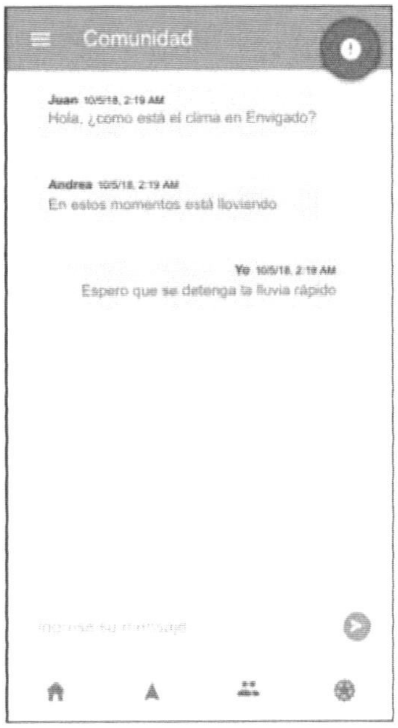

Figure 4. App comunity

Figure 5. Configuration panic button

4. Conclusions

Mobile health technologies in recent years have shown an improvement in the quality of life of users. Improvement in the accompaniment and generation of information that solve doubts and inconveniences, but due to the wide range of applications available in stores. They are considered developments that are far from being considered a complete system of sanitary support of first necessity, that is why it is necessary to create support plans with broad functionality that are inclusive and at all times promote the safety and quality of life of the users.

The needs faced by the population with some motor disability due to social or cultural factors that are often discriminatory and relegate disability as an impossibility, implants in a negative way, a vision of an environment to which one does not belong if one has disability. Factors such as the low possibility of accessing employment, traveling and freely accessing any health establishment and service, among others, make the person move away from the environment and not dare to meet and interact with him.

The way in which a mobile application is really useful, for the accessibility of people with disabilities is determined by the quality of its development, its functionality and ease of interaction, if the facility is spoken, mobile devices such as cell phones and tablets have advantages over other equipment, size, integration of functionality and the possibility of being connected to the internet; the use of mobile devices constantly increases, which is why it is to be expected that they are suitable devices that allow the generation of information regarding mobility issues.

Acknowledgments

Engineer Javier Darío Cadavid Restrepo, Electronic teacher at Institución Universitaria de Envigado; Jorge Iván Jaramillo doctor and magister in community protection and safety promotion and Diana María Rendón Correa magister in administrative sciences for the advise and technical accompaniment during the development of the Project.

References

[1] Ministerio de Salud y Protección Social 2014 Política pública nacional de discapacidad e inclusión social 2013 – 2022 https://www.minsalud.gov.co/sites/rid/Lists/BibliotecaDigital/RIDE/DE/PS/politica-publica-discapacidad-2013-2022.pdf
[2] Ministerio de Salud y Protección Social 2017 Sala situacional de la discapacidad (PCD) https://www.minsalud.gov.co/sites/rid/Lists/BibliotecaDigital/RIDE/DE/PES/presentacion-sala-situacional-discapacidad-2017.pdf
[3] Arévalo J A 2016 Aplicaciones móviles en medicina y salud XIII Jornadas APDIS España, Universidad de Salamanca)
[4] Yin R K 2009 Investigación de estudios de caso: diseño y métodos . Thousand Oaks, CA: Sage
[5] Taylor S and Bogdan, R 2000 Introducción a los métodos cualitativos de investigación. La búsqueda de significados. Barcelona: Paidos. doi:10.1017/CBO9781107415324.004
[6] Vargas J 2012 La entrevista en la investigación cualitativa: nuevas tendencias y retos. Revista CAES **3** 119-139
[7] Col prensa 2015 Ni el Gobierno sabe cuántos discapacitados hay en Colombia http://www.vanguardia.com/actualidad/colombia/293839-ni-el-gobierno-sabe-cuantos-discapacitados-hay-en-colombia
[8] Ionic framework 2018 Obtenido de Ionic: https://ionicframework.com/
[9] Nación con información de corresponsales 2017 La deuda del país con su población en condición de discapacidad El tiempo

[10] Saigí F, Novillo D an Piette J D 2017 Red CYTED-RITMOS: hacia la búsqueda de soluciones para fomentar la salud móvil en América Latina. *Rev Panamá Salud Publica.*
[11] Valero N 2018 Consumo móvil en Colombia Siempre conectados: ¿Bendición o maldición? *Deloitte.*
[12] Arévalo J A and Canelo J A 2017 Aplicaciones móviles en salud: potencial, normativa de seguridad y regulación *Rev Cubana información en ciencias de la salud,*
[13] Atlas España 2018 Aplicaciones del botón de pánico *ABC España*
[14] Pont-Sorribes C, Codina L and Pedraza-Jiménez R 2009 Comunicación de riesgo y sistemas de información en la web: cinco modelos. *El profesional de la información* **18** 389-397

IOP Conf. Series: Materials Science and Engineering **519** (2019) 012020 doi:10.1088/1757-899X/519/1/012020

Architecture for the management of a remote practical learning platform for engineering education

D A Segura-Torres[1], E F Forero-García[1]

[1]Facultad de Ingeniería Electrónica, División de Ingenierías, Universidad Santo Tomás, Carrera 9 No. 51-11, Bogotá, Colombia.

dariosegura@usantotomas.edu.co

Abstract. In the Scientific, Technological and Academic environment in which the world of today is developed, it is important to permanently update the instruments, platforms and elements with which educational requirements are met. In this sense, engineering is one of the disciplines with the greatest changes, emergencies and innovations per unit of time. Learning tools therefore require constant evolution to facilitate the appropriation of knowledge in an easy and continuously available way for students and teachers. So far it is clear that the intervention of laboratory practices as learning strategies, guarantee a solid education for students in engineering. These practices are the scenarios where the student verifies the theoretical knowledge and validates with real results the application of mathematical and statistical models, which are set around a specific learning in engineering. Continue the development of these activities not only in a traditional way, but including innovations that the same engineering allows from the use of different technologies is a challenge, which can be assumed with the application of the same discipline and be projected to other fields of the education, to facilitate personalized and asynchronous learning. This paper presents the design and architecture for the implementation and management of a new educational platform for the development of Remote Practical Learning experiences for electrical machine laboratories. This platform integrates the possibility of remote interaction with the machine and its connection, as well as obtaining test parameters through measurement interfaces and real-time visualization. In this way engineering and technology at the service of Engineering Education is used, which will increase access to higher education to a greater number of people and will make the educational experience much more pleasant

1. Introductio

Advances in communications allow proposing structures in man-machine interfaces that are used in machining and manufacturing centers that reduce operator contact with electrically energized or moving parts. However, these systems are implemented locally at very high costs and are not designed to carry out effective and efficient educational work. Establish an electrical and mechanical arrangement of engines, generators and transformers, which together with the use of information and communication technologies as facilitators of active learning and the safe development of educational practices is the challenge of the future of remote laboratories in this area, consequently increasing flexibility in learning times for each student, considering a permanent availability of the systems to be used.

Content from this work may be used under the terms of the Creative Commons Attribution 3.0 licence. Any further distribution of this work must maintain attribution to the author(s) and the title of the work, journal citation and DOI.
Published under licence by IOP Publishing Ltd

The use of automatic systems for characterizing electrical machines then involves the use of electronic devices and control platforms that manage the connections desired by the student for each specific practice remotely. This requires the design of electronic switches governed by the appropriate logic from the implemented control. Added to this is a work network that manages the online operation of each practice carried out remotely [1].

2. Remote Laboratories in Engineering Education

In the conventional laboratories of electrical machines, the practices carried out by the students require a greater amount of time to process and generate graphs and values from which an analysis of information may arise and much more feedback from the teacher. However, it is of greater concern when there is no mechanical and electrical structure previously configured for the connection, at least, manual of the machines. In this case the students must transport the machines, make the electrical connections without terminals arranged and identified with the risks that this practice carries with it [2]. On the other hand, the results obtained will be influenced by the variation of connection impedances and the corresponding time involved in making the arrangement. Consequently, with these characteristics the practical verifications of the theory of existing electrical machines are limiting in the active learning of the students [3].

The remote operation laboratories in this area include the possibility of automatically acquiring the experimental data for different configurations of electrical machines (in AC and DC, motor-generator groups, transformers) remotely in computer equipment with analysis software of data for the interpretation of results and the identification of possible sources of error, which fed back the students' learning [4]. When designing and implementing a laboratory of remote operation electric machines for execution of practices in time, the following added values would be available:

- Innovation in terms of active student learning.
- Reduction in the execution times of student practices and substantial reduction in classroom attendance, as proposed by educational trends.
- Innovation in data acquisition, storage and analysis system [5].
- Automatic construction of equivalent electric models of electric machines.
- Automatic feedback of obtained results.

3. Management Philosophies for remote practical learning platform

Among the developments carried out so far in terms of online laboratories have been different management philosophies depending on the architecture implemented. The architecture of a common remote laboratory includes a virtual working interface, which animates the laboratory, usually in HTML code for the management of the end user. This interface is generally a website executed in a browser that connected to a server interprets the data that the user includes in the virtual access platform. The server sends the respective information to the real actuating elements arranged, so that the latter executes the corresponding action according to the nature of the remote practice implemented. In the same way, the instrumentation records the corresponding measurements and returns said information to the end user through the server provided. The communication systems used operate with the different protocols available by standardized standards [6]. What has been described above is the basic management architecture and philosophy for the operation of a single laboratory, however, with the technological tools available today it is possible to include isolated remote laboratories by modifying their protocols or intervening their coding if these allow [7] [5].

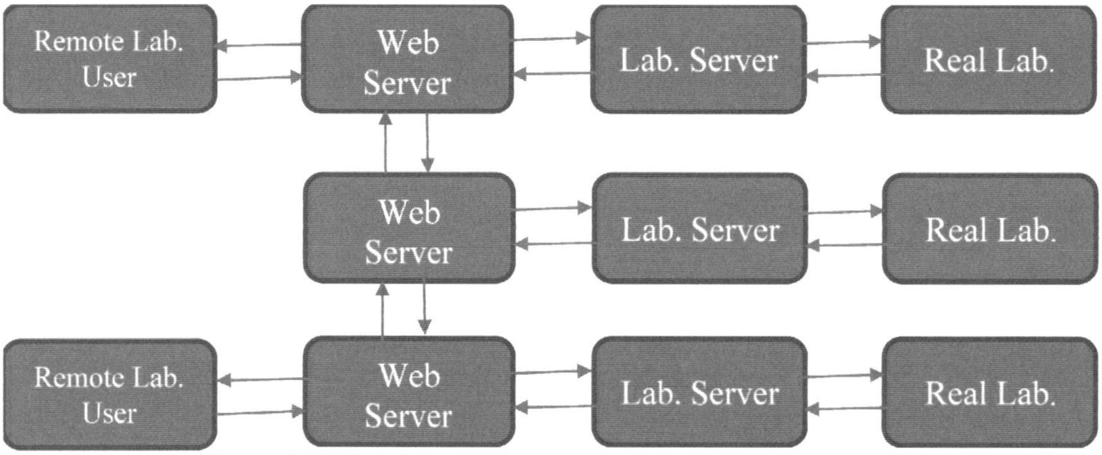

Figure 1. Ideal architecture for a remote laboratories management

A summary of the management architecture of a remote operating laboratory network for different practices and users is presented in Figure 1 [8]. The ideal would then be access to different real practice execution platforms, from different terminals and user interfaces, from any location, taking advantage of the universal access network, internet [9] [10] [11] [1] .

4. Architectural Proposal for a remote practical learning platform

This proposed architecture attempt to collect the most important requirements from others projects to get a generic architecture that allows four aims: security, integration, scalability and decoupling in a learning remote platform. Understanding security like a secure access to the information and resources, extending this definition to any resource schedule access, understanding integration like the ability to provide and consume services between others remote laboratories and the own one, scalability like the ability to increase or decrease the informatics resources according to the new needs of the local and remote users (systems and people) [12], and decoupling like production of the independent modules for the system [13]. The following sections describes the construction of the proposed architecture adding components steep by steep.

4.1. Requirements from previous projects

From [4], is taken the strong need to implement any standard authentication system like SAML, OpenID y OAuth2, in the "StandardAuthentication" component merged with some institution's Identity Provider (IdP) [4], to manage the access to the provided services by the Learning Manager System (LMS). These authentications methods are in the security layer which will protect and control the access to the LMS (Figure 2), even more, the "StandardAuthentication" module can implement others authentication methods like in [5] using social network authentication methods according to the policies for the institutions. The "StandardAuthentication" component also must work like functionality provider of the "IdP" component, which should take the final decision about the access over resources or services, like information about the experiments, results, data raw captured in an experiment, student information and results, etc. This security layer must control the access to information, resource, and experiment for a user according to a given role in the system, example an administrator can see all information, but a student only can use a remote experiment. The component Laboratory Management System (LMS) is the learning and experiment process in itself.

IOP Conf. Series: Materials Science and Engineering **519** (2019) 012020 doi:10.1088/1757-899X/519/1/012020

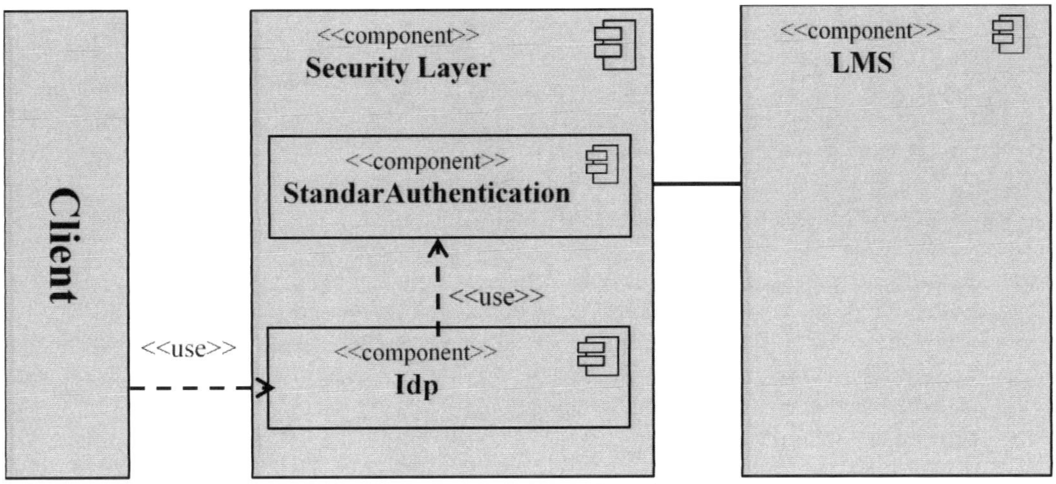

Figure 2. LMS with Security layer.

The scheduling is also an important part for a LMS, because give to the system the ability to optimize the use of the resources, managing the user access to the system avoiding unexpected users concurrence, that may will cause a failed access to the experiment representing a bad user experience [14]. Then appear the component called "Scheduling" to provide information to "IdP" about the user access rights over a resource or information in a specified time. This component have to provide all needed interfaces to: check and consult for resources availability, schedule available resources, release reserved resources, and manage the resource use log. In this way the "Scheduling" component interact with the "IdP" component and any system or user that want to use the system, like shows the Figure 3.

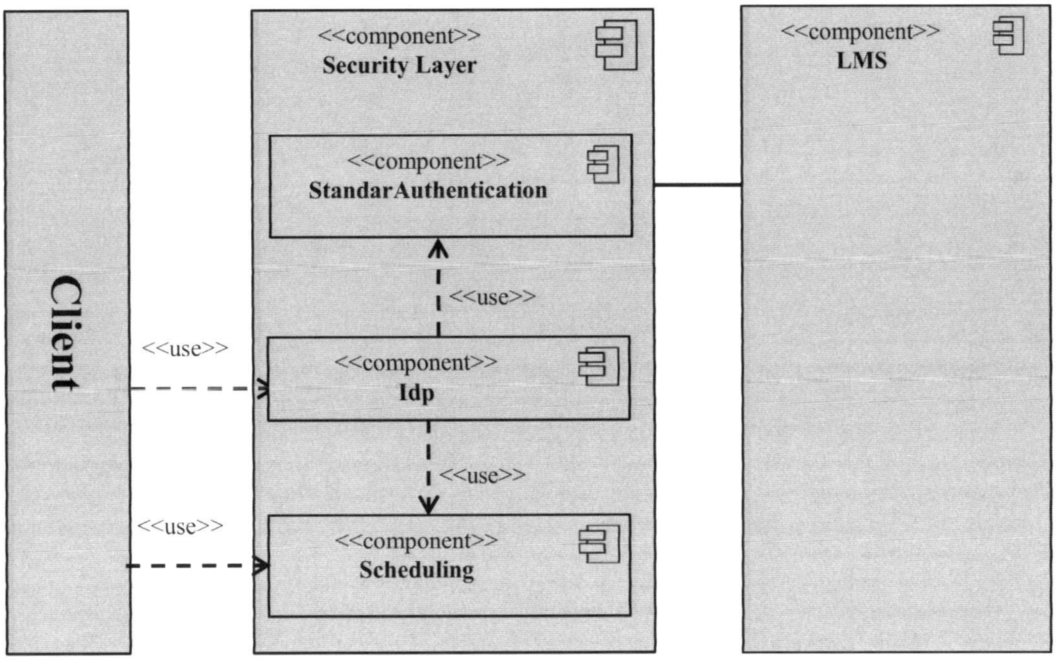

Figure 3. LMS with Security layer and scheduling.

IOP Conf. Series: Materials Science and Engineering **519** (2019) 012020 doi:10.1088/1757-899X/519/1/012020

4.2. The functional requirements from the learning business

The LMS have to contain some components that help to reach its main goal, learning and practice, then is necessary some basic services like, the first one, access to remote lab and practices, evaluation sub systems, laboratory guides, and others, each one create a new component into the "*LMS*" component, evolving the architecture to the showed in the Figure 4, into the "*LMS*" component.

The proposed components for a basic system are:

- RemoteLaboratory: Allow interactions between the real laboratory and the remote user, for example will get the measurement from a network analyzer and send it to a client, receive a command from a client and close an automatic switch, or read the image from an oscilloscope and send it to the client, or the oscilloscope data. Also can be associated with an experiment guide, or an evaluation items.
- ExperimentGuides: Store the procedure that the user should follow to accomplish an experiment successfully.
- Evaluation System: Manage the items to evaluate an experiment, the expected results, the results obtained by a user, may be a student, and the evaluation for this results.

The use of the modules by some client are independent, which means a client can use just the guides or the evaluation system, or a full experiment, always according to the rights in the "Scheduling" and "IdP" modules.

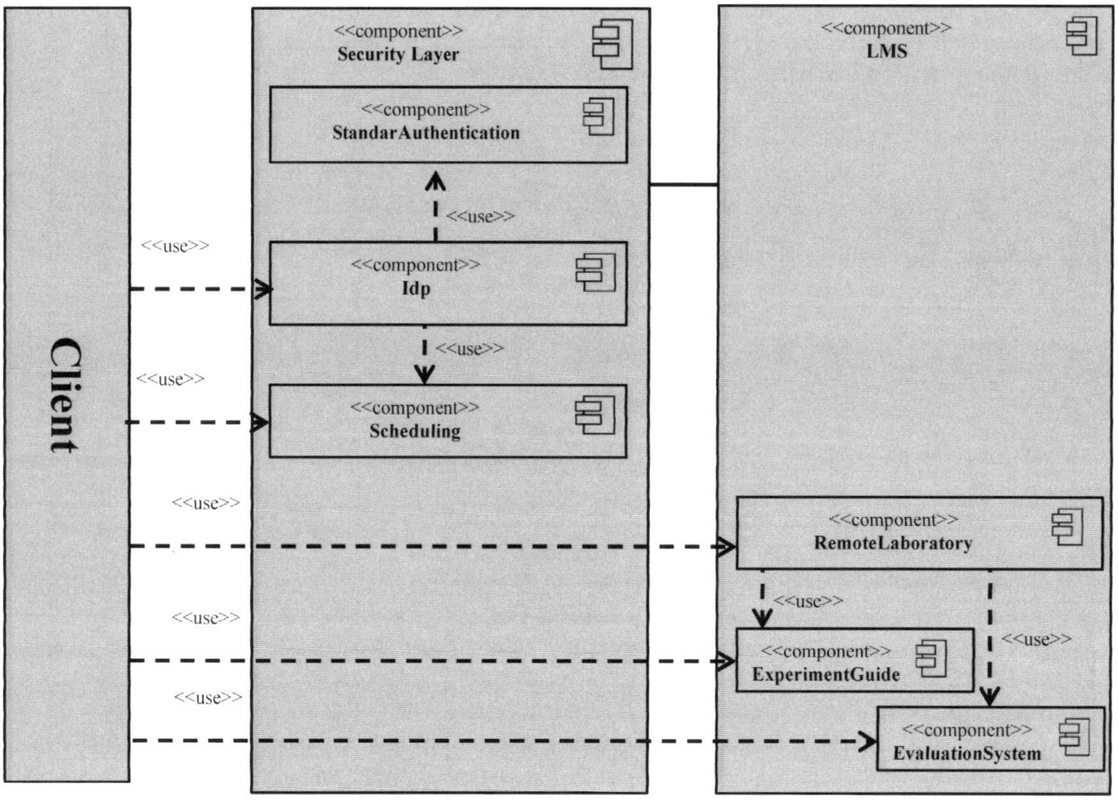

Figure 4. LMS architecture with business requirements.

4.3. Applying non-functional requirements

Some non-functional requirements detected for any actual system can be included in the model:

- Persistent storage: Is usually the database system to storage the system information, like configurations, user's roles access, scheduling by user, evaluation criteria and the results, data from the experiments and so on.
- Log user or client actions: This component store any action made by any user or client, should be classified according to the specific needs, but is very important log any activity. Some calcifications can be: access granted, access denied, access to a resource by user o client, etc.

From above, the architecture now looks like in the Figure 5.

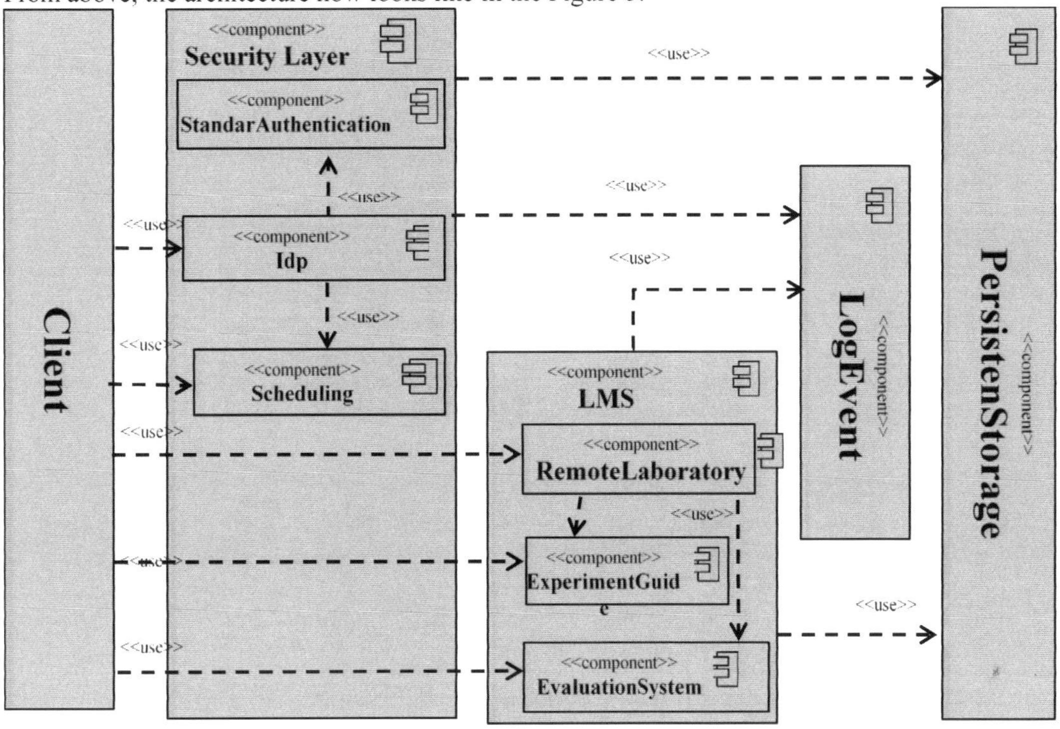

Figure 5. LMS architecture with non-functional requirements.

4.4. Appling new technology tendencies

To create an architecture focused on provide an API through services enabling an easy development of system clients in standard programing languages like JavaScript, Java or C# [15], but this architecture cannot take in account use components from other systems and not specify how to connect to the experiment in self. The current systems try to serve its remote laboratories but are not oriented to use other laboratories services or other system services, creating frameworks instead real services.

Also in most documented cases, the presented architectures are oriented to single, and may be small, development teams, and the development efficiency is not evaluated, that will generate an unsupported systems and sort time systems, especially because the current LMS are development by educational institution teams, these teams are very volatile, change the developers frequently.

Appling the growing concept of Microservices to the above architecture can solve some of this problems. The Microservices architecture is the evolution of Services Oriented Architecture (SOA), and can be defined like "… an autonomous software service which is built to perform a single, specific, and granular task …" [16]. The Microservices architecture is showed in the Figure 6. The API Gateway is the entry point for a client, when a client request for a service the API Gateway receive the request and call the appropriated Microservice, decoupling the client and the services used in fact. The Identity

IOP Conf. Series: Materials Science and Engineering **519** (2019) 012020 doi:10.1088/1757-899X/519/1/012020

Provider is the manager for authentication and access to the system. The Microservices are small and very simple services that can be development in different technologies, by different person, by different development team, that mean they are loosely coupled in many levels. Management and Service Discovery are responsible for managing nodes and update entry points for Microservices. Remote services are services consumed by the system. Finally the architecture provide options for some classical web content through the Content Delivery Network and static content modules.

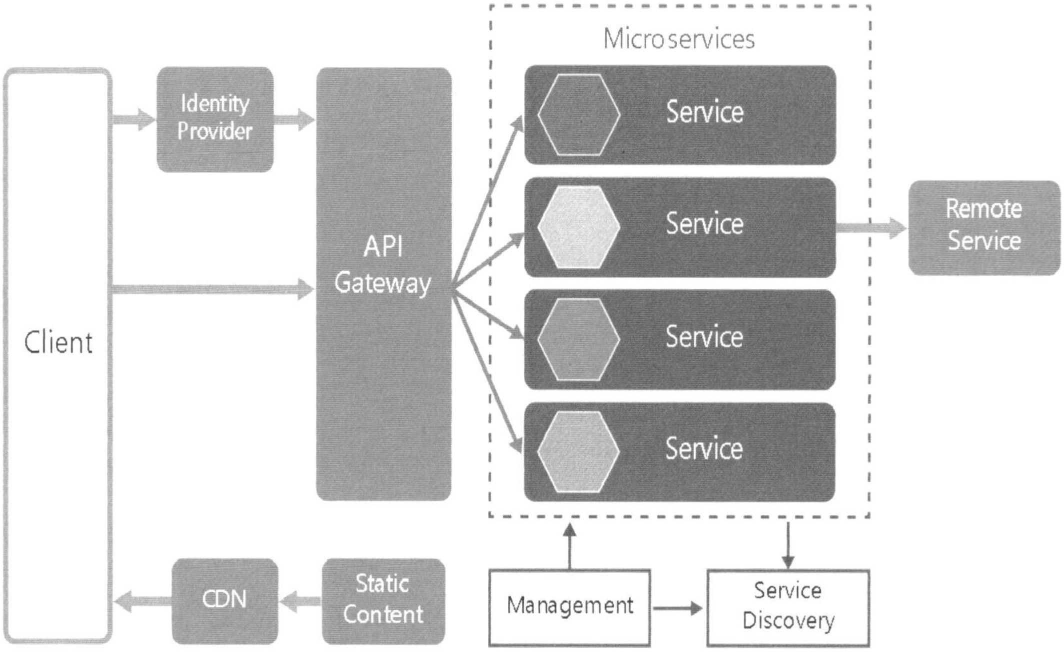

Figure 6. Microservices architecture. Image taken from: https://docs.microsoft.com/en-us/azure/architecture/guide/architecture-styles/microservices

A Microservices architecture warranty scalable systems, and like SOA, provide loosely coupled applications and high systems integrations and shared functionalities [17] [18] [19][20][21]. To apply the Microservices architecture philosophy into the proposed architecture, is necessary include the API Gateway services layer, the identity provider role is taken by the "SecurutyLayer", and all components into "LMS" component turn into Microservices. The Management and Services Discovery will not implement at this moment, but can be implemented. Is important define in this architecture every sensor and every actuator like a service that require a little high effort to develop interfaces between hardware and the LMS, but give to the LMS the capabilities to use, and reuse sensors components, the new proposed architecture is showed in the Figure 7.

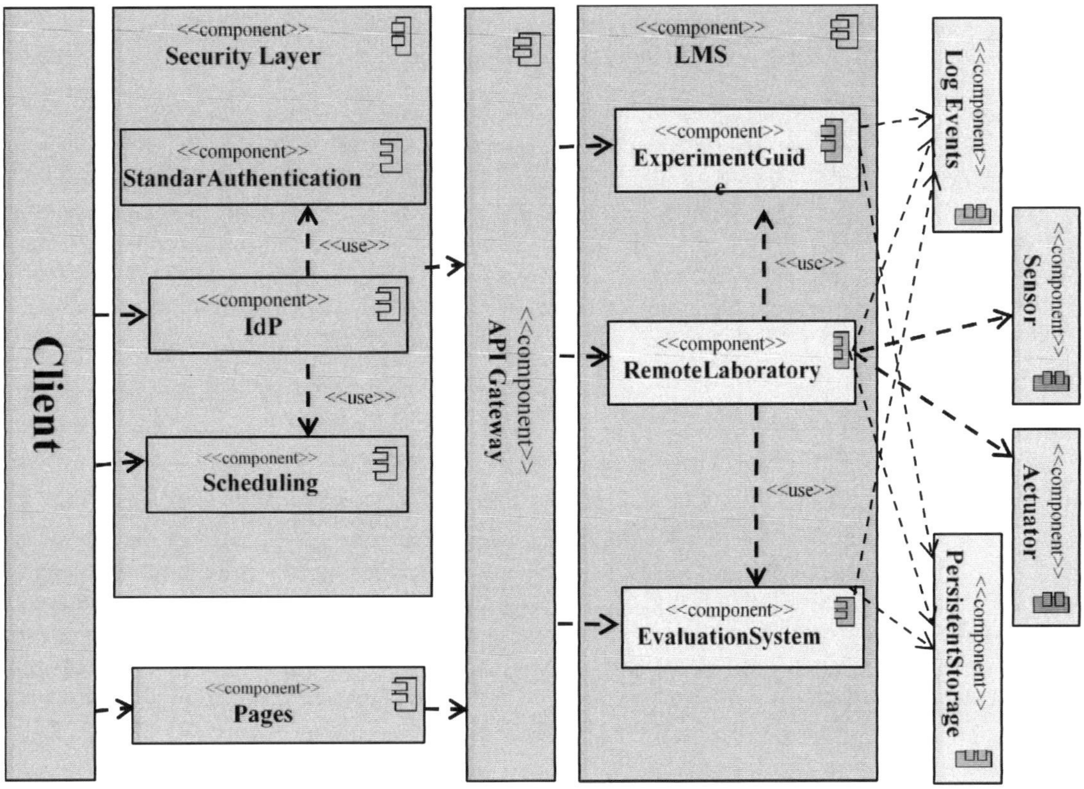

Figure 7. Final proposed architecture applying Microservices architecture.

In a real implementation some Microservice component will be splitted into many Microservices and some not. For example "ExperimentGuide" Microservice can serve all guide documents and process about all experiments, but the "RemoteLaboratory" component should create one Microservice for each experiment, anyway, the granularity for these Microservices is up to the solution's architect.

5. Analysis of Proposal Operation

The *"StandardAuthentication"* component will be so strong like the implemented methods, and will give an integration limitation according to how many standard methods implement, that mean, you can only give services or access services with the same authentication protocols implemented in "*StandardAuthentication*" module, more protocols implemented in this component open the system to use other systems and provide services to more other ones.

The implementation of this architecture, will allow access any resource independently and consume resources from other providers, and give the flexibility to use just one physical or virtual server, or implement different servers for each component. Even more, allow the possibility to have the same experiment twice, for example, use one like main experiment, and use the other like backup, if the first one is busy or not available, the user just access the same page and access the second experiment that is the same like the first one but implemented in other place, of course, is important implement Manager and Service Discovery layers.

In [9] the author concludes that on line laboratories are not a trivial issue, this way, many groups are growing and working in get a standard, like Global Online Laboratory Consortium, or adopting a common platform o framework. The above concept, can create a segmentations of groups, not a standard, because everyone try to push to their initiative, this is the reason because the authors of this paper thinks that is necessary a created standard by an international and recognized institution. Anyway, the proposed architecture, because its implementation of micro-services architecture style, is ready to the integration with any service oriented standard, in the future, because is decoupled throw its service layer, and inclusive can serve other protocols to interchange services.

6. Conclusion

In this paper was presented the architecture for implementation and management of an educational platform for development Remote Practical Learning experiences in electrical machine laboratories. This platform integrates the possibility of remote interaction with the machine and its connection, as well as obtaining test parameters through measurement interfaces and real-time visualization. In this way engineering and technology are uses as service of Engineering Education.

The article includes a proposed architecture attempt that collected the most important requirements from others proposals to get a generic architecture that allows four aims: security, integration, scalability and decoupling in a learning remote platform.

References

[1] B.-A. DEAKY 2013, "Contribution to Online Laboratory Implementation and Standardization," in *2013 IEEE Global Engineering Education Conference (EDUCON)*, Berlin, Germany.

[2] j. s. rojas-aguilar, s. contreras-castillo, j. s. gomez and e. f. forero-garcía, "diseño e implementación de un curso teórico sobre la energía fotovoltaica apoyado por una planta física conectada a una plataforma web para prácticas," in *latin american and caribbean consortium of engineering institutions: xvi laccei international multi-conference for engineering, education and technology*, lima, peru, 2018.

[3] E. Lindsay, S. Murray and B. D. Stumpers, "A Toolkit for Remote Laboratory Design & Development," in *2011 First Global Online Laboratory Consortium Remote Laboratories Workshop*, Rapid City, SD, 2011.

[4] T. J. Mateo Sanguino, I. F. Viana Gonzalez, J. Espejo Fernandez and A. Garcia Dominguez, "Using Identity Provider and Automatic Resource Management to Improve a Remote Networking Lab," *IEEE Latin America Transactions,* vol. 16, no. 5, pp. 1547 - 1556, 2018.

[5] F. Lerro, P. Orduña, S. Marchisio and J. García-Zubía, "Development of a Remote Laboratory Management System and Integration with Social Networks," in *International Conference on Interactive Computer aided Blended Learning*, Viena, Austria, 2013.

[6] L. F. Zapata-Rivera and M. M. Larrondo-Petrie, "Design of a Latin American and Caribbean Remote Laboratories Network," in *2016 IEEE Frontiers in Education Conference (FIE)*, Erie, PA, USA,.

[7] L. Rodriguez-Gil, J. Garc´ıa-Zubia and P. Ordu˜na, "An architecture for new models of Online Laboratories: educative multi-user gamified hybrid laboratories based on Virtual Environments," in *2016 13th International Conference on Remote Engineering and Virtual Instrumentation (REV)*, Madrid, Spain, 2016.

[8] H. Saliah-Hassane and A. Reuzeau, "Mobile open online laboratories A way towards connectionist massive online laboratories with x-API (c-MOOLs)," in *2014 IEEE Frontiers in Education Conference (FIE)*, Madrid, Spain, 2014.

[9] M. Auer, A. K. Azad and A. E. Ton de Jong, "Online Laboratory Architectures and Technical Considerations," in *Cyber-Physical Laboratories in Engineering and Science Education*, Cham, Switzerland, Springer, 2018, p. 135.

[10] H. Benmohamed, A. Leleve and P. Prevot, "Remote Laboratories: New Technology and Standard Based Architecture," in *Proceedings. 2004 International Conference on Information and Communication Technologies: From Theory to Applications, 2004.*, Damascus, Syria, 2004.

[11] M. Tawfik, E. S. Cristóbal, A. Pesquera, R. Gil, S. Martin, G. Diaz and J. Peire, "Shareable Educational Architectures for Remote Laboratories," in *2012 Technologies Applied to Electronics Teaching (TAEE)*, Vigo, Spain, 2012.

[12] A. B. Bondi, "Scalability and performance," in *Foundations of Software and System Performance Engineering*, Crawfordsville,Indiana, United States, Addison-Wesley, 2015, p. 273.

[13] F. Alonso Amo and L. Martínez Normand, "Modelos de desarrollos de programas," in *Introducción a la ingeniería del software*, Madrid, España, Delta Publicaciones Universitarias, 2005, p. 209.

[14] R. Zamora, "Laboratorios Remotos: Actualidad y Tendencias Futuras," *Scientia et Technica,* vol. 51, pp. 113-119, 2012.

[15] J. Zornig, S. Chen and H. Dinh, "RESTlabs: A Prototype Web 2.0 Architecture for Remote Labs," in *2012 9th International Conference on Remote Engineering and Virtual Instrumentation (REV)*, Bilbao, España, 2012.

[16] N. Tanasseri and R. Rai, "What are Microservices?," in *Microservices with Azure*, Packt Publishing, 2017, pp. 7-17.

[17] W. Hasselbring and G. Steinacker, "Microservice Architectures for Scalability, Agility and Reliability in E-Commerce," in *IEEE International Conference on Software Architecture Workshops*, 2017.

[18] S. Stallin Kapembe and J. Quenum, "Lihonga — a Microservice-based Virtual Learning Environment," in *IEEE 18th International Conference on Advanced Learning Technologies*, 2018.

[19] C. Fetzer, "Building Critical Applications Using Microservices," *IEEE Security & Privacy,* vol. 14, no. 6, pp. 86-89, 2016.

[20] D. Sprott and L. Wilkes, "Understanding Service-Oriented Architecture - Microsoft," 01 2004. [Online]. Available: https://msdn.microsoft.com/en-us/library/aa480021.aspx. [Accessed 09 2018].

[21] I. Titov and E. Smirnova, "Network Architectures of Remote Laboratories," *International Journal of Online Engineering (iJOE),* vol. 9, no. 6, pp. 41-44, 2013.

Use of Augmented Reality for the Simulation of Basic Mechanical Physics Phenomena

A D Morales[1], S A Sanchez[1], C M Pineda[1] and H J Romero[1]

[1]Corporación Universitaria Antonio José de Sucre, Sincelejo, Colombia.

alex_morales@corposucre.edu.co

Abstract. The augmented reality (AR), has been cataloged like the technology that will have greater penetration in the education sector in the second decade of the XXI century. According to the report horizon 2016, the first research of AR in education point out a positive impact, in collaborative learning and autonomous learning; because it allows potentiating reality with virtual information that can be supported in curricular contents. The present article exposes the development of a prototype with AR technology that simulates the uniform rectilinear movement and free fall of the subject of mechanical physics. The research is of technological development type. Included in three phases: Initially the functional and non-functional requirements were determined, then the architecture was designed using UML diagrams and finally the prototype was developed. As a result, we obtained a mobile prototype with AR with two functions, the first one allows to simulate the movement of a vehicle at a constant speed, the second allows to simulate the fall of objects that are being attracted by gravity. A technical feasibility study was carried out in which the prototype was installed in different mobile devices, without errors. It is concluded that AR has several characteristics that can be applied to educational environments to increase students' motivation and make them live experiences that are limited access in classrooms.

1. Introduction

The integration of information and communication technologies (ICT) in the field of education has allowed the evolution of the educational model. In the beginning, the model was centered on the faculty who were responsible for finding and providing the necessary information while the students appropriated knowledge through technology as a resource for transmitting information; in this model the theories of Behaviorism and Objectivism were the protagonists. Recently the model has undergone changes focusing on the students, who have gone from being receptors to being knowledge providers and the teacher has to do the guiding tutor role, looking for the student to generate knowledge using technology in an interactive way; this new model the theories that star it are Cognitivism, Constructivism and Constructionism [1].

One of the technologies that is penetrating the field of education in recent years is the so-called Augmented Reality (AR) [2]. This technology allows digital information to be enriched with real physical scenarios using a digital camera; the applications with this technology are generally developed for operating systems of mobile devices (Tablets or Smartphones) that, due to their portability, can contribute to delocalized learning, where any scenario can be transformed to be used in a formative way [3].

Content from this work may be used under the terms of the Creative Commons Attribution 3.0 licence. Any further distribution of this work must maintain attribution to the author(s) and the title of the work, journal citation and DOI.
Published under licence by IOP Publishing Ltd

Locating the use of AR in regard to higher education can be applied in various areas of knowledge. In social science subjects recreating heritage, allowing knowing historical and cultural events of a region. In areas of mathematics they facilitate the explanation of abstract models, illustrating mathematical objects and facilitating the understanding of their location in space; in subjects such as physics they allow the simulation of laboratories that recreate physical phenomena; those and more applications can be given to AR, and be used by students both inside and outside the classroom.

Finally, it can be said that technology is fundamental in the life of man. Allowing to revolutionize the behavior of this in the different sectors (economic, social, educational); in what comprises education, day by day the use of technologies contributes to the teaching and learning processes, however it must be emphasized that the technological factor must be accompanied by human talent to generate transformative processes.

2. Augmented Reality (AR)

According to Horizon Report: 2016 Higher Education Edition, augmented reality increases access to information, generating new uses of technology in learning processes [2]. Currently the use of AR is more accessible by different users due to the boom that mobile devices have had. The characteristics of this technology allow students to build knowledge, which makes the student an active subject in their training process [4].

The following authors Buitrago-Pulido, Cabero Almenara & Díaz Marín, Gutiérrez, Duque, Chaparro, & Rojas, Cabero Almenara, Vázquez Cano, & López Meneses, Martínez-Pérez & Morales Segura, Diaz, Fava, Banchoff, Schiavoni, & Martin, with their publications that AR is a technology that allows to combine in real time digital information and physical information through different technological supports, to create This forms a new enriched reality. The elements that must interact to allow the use of reality increases are: An electronic device with camera, a software that has the properties of reflecting the information on the screen when activating the camera of the device and a trigger known also as "trigger" or activator of the information (2D images, marker, objects, and 2D codes). The combination of these three elements must comply with three basic properties to make up an AR system: a) combine real and virtual objects in a real environment, b) alignment of real and virtual objects with each other, and c) execute them interactively and in time real [5].

This technology can be cataloged in different levels depending on the type of components used when implementing it (see table 1) [6].

Table 1. Augmented Reality Levels

Level	Description
Level 0: Linked with physical world.	By using bar codes or QR codes, the RA application creates a link to the real physical world by generating content associated with that code.
Level 1: RA with markers.	At this level the application makes use of the so-called marked which can be in 2D or 3D, for the first we have black and white images and color images, for 3D we use recognition of real objects.
Level 2: RA without markers.	At this level, the applications use the GPS, the compass and the accelerometer of the device to determine the location and orientation of the user and add points of interest in the real world.
Level 3: Increased Vision.	This last level involves devices such as Google Glass, high-tech contact lenses or others that, in the future, will be able to enrich the user's context with additional information.

3. Methodology

This study is of technological development nature. According to Colciencias "this type of research generates as a result the development of software that produces advances in the generic approaches for the capture, transmission, storage, recovery, treatment or presentation of information" [7]. The purpose of this research is to generate a software prototype that simulates the physical phenomena of uniform rectilinear motion and free fall as the first stage.

To carry out the investigation, three phases were developed (see figure 1):

In the first phase, the requirement survey of the system was carried out using different techniques. With the documentation analysis technique, an analysis of the mechanical physics course plan was made to obtain information on the functional requirements and non-functional requirements of the prototype. Through interviews conducted with teachers in the area of physics, it was possible to have specific aspects of the dynamics of the activities to be developed in the prototype. And finally with the work tables (Workshops) it was possible to have relevant information about the different technologies to be used for the development of the prototype.

In the second phase after having the requirements of the program, we proceed to design the architecture of the augmented reality prototype using UML language (unified modeling language) creating the different diagrams (use case, interaction, state and classes) that allow defining the structure of the system. To be able to precede with the codification stage and creation of the graphic interface of the prototype.

Finally, in the third phase, the prototype of the different interfaces of the prototype is developed and the creation of the C # programming language script that allows the program's functionality.

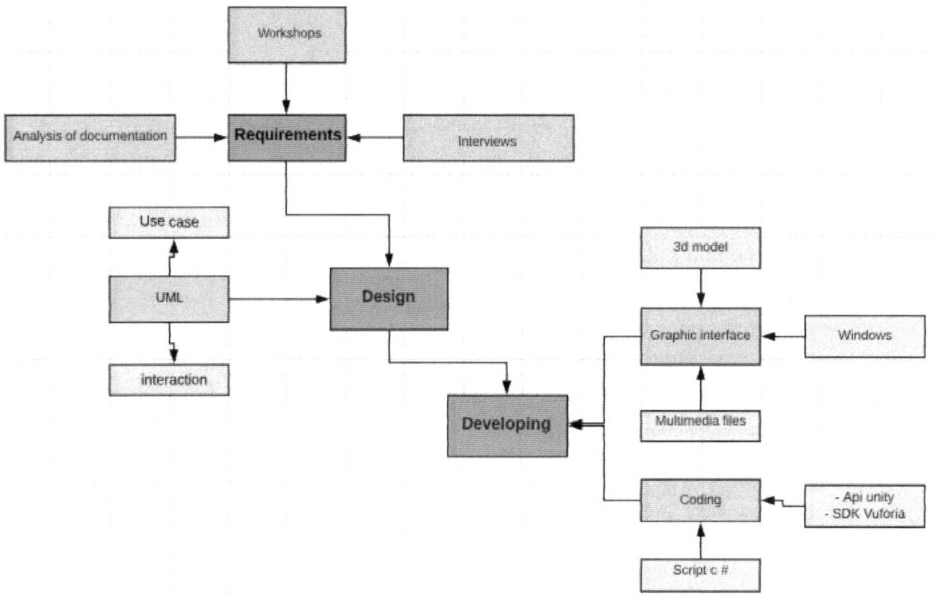

Figure 1. Prototype diagram

4. Results

In the present investigation, the following results were found:

In the phase of raising requirements, a search was started in different academic databases, where it is evident that AR is being studied by different authors of the XXI century, being implemented in the different levels of student training, starting from the basic primary to higher education.

IOP Conf. Series: Materials Science and Engineering **519** (2019) 012021 doi:10.1088/1757-899X/519/1/012021

Emphasizing the advantages that this technology offers to the educational sector. the following are listed: it creates a friendly, playful and educational environment; fosters a positive and creative attitude in the student; strengthens the learning of curricular contents through its association of real and virtual information; it is a constructivist resource; promotes ubiquitous and mobile learning; contributes to the increase of attention and concentration; allows the simulation of remote laboratories and phenomena difficult to experience; It can be applied to existing educational physical resources.

In addition, in this phase the Mechanical Physics course plan of the Antonio José de Sucre University Corporation, Corposucre is analyzed, where it is observed that the subject is subdivided into 4 academic units (see table 2).

Table 2. Units of Subjects Plans for Mechanical Physics – Corposucre

Academic Unit	Unit Name
Academic Unit nº 1	Movement in one Dimension
Academic Unit nº 2	Vectors
Academic Unit nº 3	Movement in two Dimensions
Academic Unit nº 4	Laws of Newton
Academic Unit nº 5	Work, Power and Energy Transfer

Unit 1 focuses on the explanation Movements with Constant Acceleration and Free Fall of the Bodies and Launching Up. it is also specified in the subject plan that the learning outcome for this unit is to allow the student to recognize the basic concepts of mechanical physics, as well as the different approaches methods of physics to solve problems of specific situations in the environment. And the resources that are used to teach the classes of this unit are ICT tools (slides and activities of documents written in moodle platform), texts and documents, work guides.

In the design phase, the IDE Netbeans platform was used to create the UML diagrams.

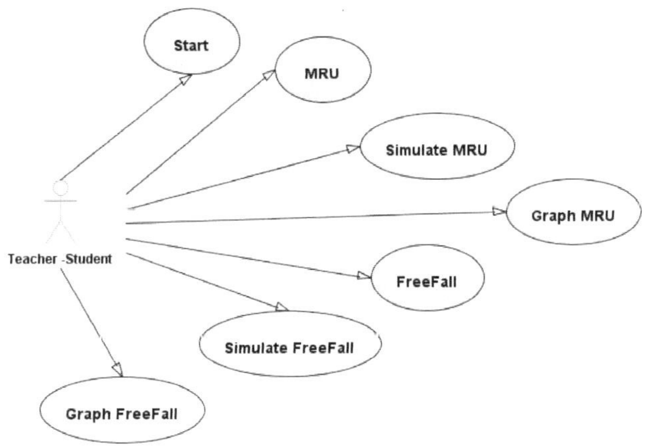

Figure 2. Prototype use, Case Diagram. The following diagram illustrates in a general way the relationship of the actor (teacher - student) with each of the characteristics of the system.

In the development phase Unity 3D platforms were used, which provides us with free of charge the ease of manipulation of virtual objects and the Vuforia SDK that allows using augmented reality technology.

4.1 Prototype AR (Graphic interface and operations)

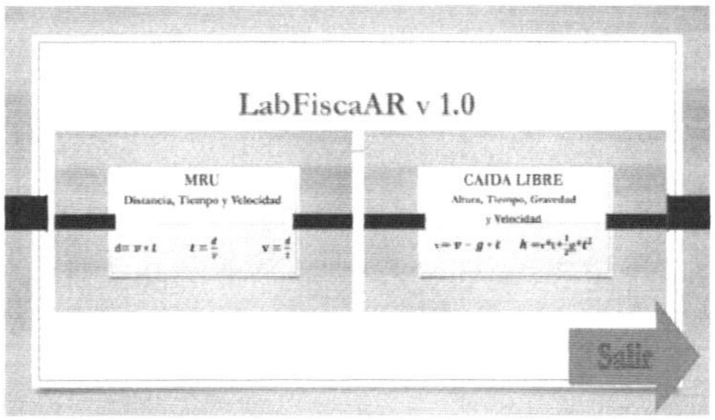

Figure 3. Initial screen of the prototype. Illustrates the name of the application, the exit button, plus the options of uniform rectilinear movements and free fall.

Figure 4. MRU interface. Illustrates a definition and the equations that are involved in the uniform rectilinear movement, allows to perform calculations knowing two of the variables that interfere in the equations.

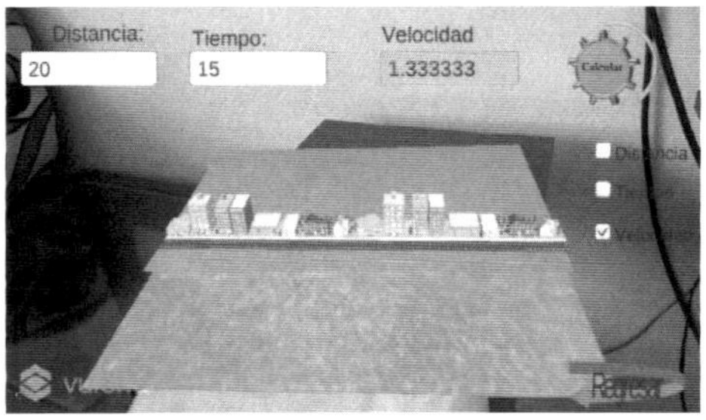

Figure 5. MRU simulator. Reflects a scenario of a vehicle that travels a track at the speed that indicates the equations that are involved in the phenomenon.

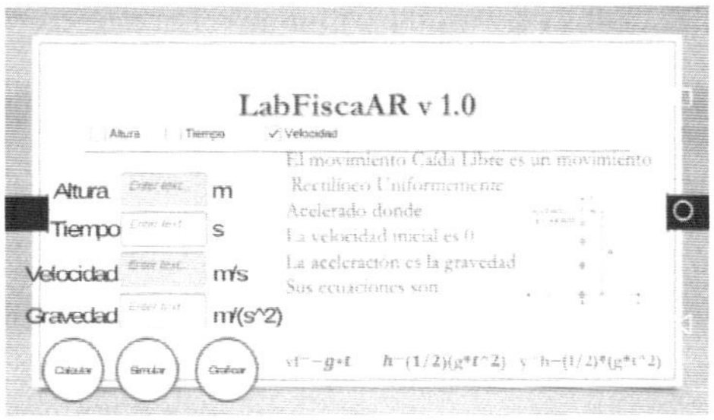

Figure 6. Free Fall Interface. Illustrates a definition and the equations that are involved in the free fall movement, allows to perform calculations knowing two of the variables that intervene in the equations

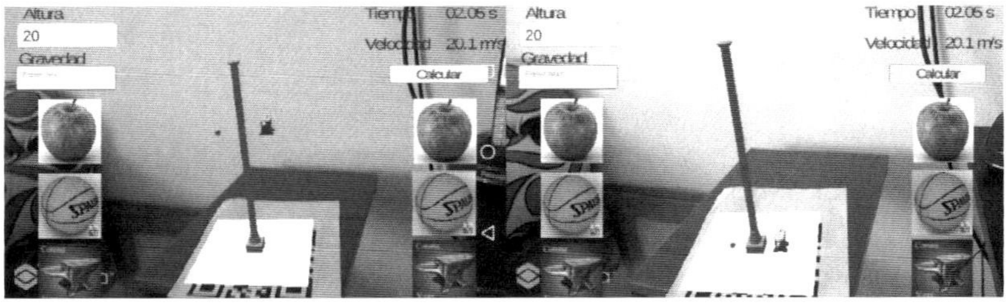

Figure 7. Free Fall Simulator. It reflects a scenario of two objects which can modify their height and the gravity that intervenes to know the time and the final speed.

4.2 Feasibility study

Once the software prototype has been developed, the feasibility of both the technical and the economic part is reviewed.

4.2.1Technical feasibility.

Checking the technical aspects of the prototype to be able to carry out its implementation in Corposucre, It is observed that the minimum hardware and software requirements that a device must have for its execution are (see table 3):

- Mobile device with integrated video camera.
- 1.5 Ghz processor.
- RAM capacity 1.5 GB.
- Android operating system version 5.0.

This allows the prototype to be at the reach of Corposucre's community of the programs of the faculty of engineering sciences, due to the great part of the students and teachers of its community has mobile devices of mid-range.

Table 3. Tests on mobile devices.

Device	Operating system	Version O.S	Processor	RAM	Performance	Resolution
Samsung Note Galaxy 3	Android	5.0	Qualcomm quad-core , 2.15 GHz	3 GB	Excellent	Excellent
Moto 4G	Android	7.0	1.5GHz Qualcomm Snapdragon 617 eight-core	2 GB	Very good	Very good
Samsung Galaxy S5	Android	6.0	quad-core a 2.5GHz	2 GB	Very good	Very good
Samsung Galaxy j5	Android	6.0	Qualcomm Snapdragon 410 a core 1,2 GHz	1.5 GB	Very good	Very good
Moto E XT1021	Android	4.4.2	1.2Ghz Dual-Core ARM Cortex-A7	1 GB	Acceptable	Acceptable
Huawei p20	Android	8.1	Kirin 970, eight-core, 2.36GHz.	6 GB	Excellent	Excellent

4.2.2. Economic feasibility.
Reviewing the cost-benefit relationship once the prototype was developed for the corporation, it was determined that it has the human and technical resource for the development of software prototypes with augmented reality technologies. On the other hand, there are different multimedia materials with free license for their use, which can be used for the development of this type of applications. It is recommended to acquire the license of some tools used in the development of the project to avoid watermarks.

5. Conclusions
It is concluded that thanks to the fact that mid-range mobile devices have a medium-low cost, the students community can have access to application with technology AR. AR allows the simulation of different scenarios or events that are limited access for teachers and/or students due to economic, abstract or risk factors, ensuring that a large part of the academic community accesses these resources. The study of AR in the education sector should be further deepened and more educational experiences should be generated, in order to validate the real impact it can have on the teaching and learning processes. The construction of applications with AR is available to all, because there are several free platforms for the development of these, as well as a large community that provides information on how to implement it in software development.

References

[1] J C Marino, Dodge y I, Stefanell De León 2012 Uso de Realidad Aumentada para Enseñanza de Conceptos Básicos de Física Mecanica *Ingeniare*, Universidad Libre-Barranquilla, pp 11 - 26.

[2] N. M. C. NMC y E. L. I. ELI 2016 Horizon Report: 2016 Higher Education Edición *Educause* pp 40-41.

[3] T Brasero Castellano, P L Santacruz Valencia 2018 EnseñazaAPP: Aplicacion Educativa de Realidad Aumentada para el Primer Ciclo de la Educacion Primaria *Revista Iberoamericana de Tecnología en Educación y Educación en Tecnología* N°21 pp 7-14.

[4] Johnson L, Adams Becker, S Cummins, M Estrada, V, Freeman 2016 Horizon Report Edición Educación Superior *The New Media Consortium* Texas.

[5] Gabinete de Tele-Educación 2017 Realidad Aumentada en Educacion Madrid.

[6] E Garnica Estrada y J A Franco Calderon 2018 REALIDAD AUMENTADA Y EDUCACIÓN *Rev. Ingeniería, Matemáticas y Ciencias de la Información* pp 59-65.

[7] Colciencias 2016 Tipología de proyectos calificados como de carácter científico, tecnológico e innovación (vol 4) Colombia.

[8] G Veletsianos 2010 A Definition of Emerging Technologies for Education pp 3-4.

[9] C Belloch 2017 Las Tecnologías de la Información y Comunicación en el aprendizaje *Noticias DCyTIC - BUAP*.

[10] A Mejía Jálabe, C P Villarreal Mora, C A Silva Giraldo, D A Suarez Suarez y C F Villamizar Niño 2018 *E*studio de los factores de resistencia al cambio y actitud hacia el uso educativo de las TIC por parte del personal docente *Revista Bolentin Redipe:* Vol. 7 Núm. 2 Ética, educación e investigación pp 53 - 63.

[11] R L Chaparro Aranguren y J N Barbosa Sánchez 2018 Incidencia del Aprendizaje Basado en Proyectos, implementado con Tecnologías de Información y Comunicación, en la motivación académica de estudiantes de secundaria *Revista LOGOS CIENCIA & TECNOLOGÍA* pp 167 - 180.

[12] R Gutiérrez, E Duque, R Chaparro y N. Rojas 2018 Aprendizaje de los Conceptos Básicos de Realidad Aumentada por Medio del Juego Pokemon Go y sus Posibilidades como Herramienta de Mediación Educativa en Latinoamérica *Información Tecnológica* – Vol. 29 N° 1 pp 49 - 58.

[13] J Cabero Almenara, E Vázquez Cano y E López Meneses 2018 Uso de la Realidad Aumentada como Recurso Didáctico en la Enseñanza Universitaria *Uso de la Realidad Aumentada* Vol 11 pp 25 - 34.

[14] I Martínez-Pérez y M Morales Segura 2018 Recorridos interactivos de realidad aumentada como nueva herramienta multimedia en la enseñanza de asignaturas de Construcción Arquitectónica *Advances in Building Education / Innovación Educativa en Edificación* pp 31 - 50.

[15] J Cabero Almenara y V Díaz Marín, 2018 Blended learning y realidad aumentada: experiencias de diseño docente RIED. *Revista Iberoamericana de Educación a Distancia* pp 57 - 74.

[16] M A Mendoza Pérez, R G Cruz Flores, A A Villalba Hernández, J A Calderón Rodríguez y E. Arreola Patiño 2017 *A*plicacion de realidad aumentada para la enseñaza de la robotica *Pistas Educativas* Vol. 39 pp 340 - 353.

[17] T Castellano Brasero y L P Santacruz Valencia, 2018 EnseñazAPP: aplicación educativa de realidad aumentada para el primer ciclo de educación primaria *Revista Iberoamericana de*

Tecnología en Educación y Educación en Tecnología N°21, pp 7-14.

[18] R D Buitrago-Pulido 2015 Incidencia de la realidad aumentada sobre el estilo cognitivo: caso para el estudio de las matemáticas *Educ. Educ.* Vol. 18 No. 1 pp 27-41.

[19] J Diaz, L Fava, C Banchoff, A Schiavoni y S Martin 2018 Juegos serios y aplicaciones interactivas usando realidad aumentada, *XX Workshop de Investigadores en Ciencias de la Computación* pp 829-834.

[20] H Martínez HungI, A García López y J C Escalona ArranzI 2017 Modelos de Realidad Aumentada aplicados a la enseñanza de la Química en el nivel universitario Rev. *Cubana Quím* Vol. 29 no.1 pp 13 - 25.

[21] M Vidal Ledo, B Lío Alonso, A Santiago Garrido, A Muñoz Hernández, I d R Morales Suárez y A M Toledo Fernández 2017 Realidad aumentada Educación Médica Superior.

[22] J Cabero Almenara 2017 Aplicaciones de realidad aumentada en la educación *Revista de educacion mediatica y TIC* pp 4-8.

[23] C A Solano Villanueva, J F Casas Díaz y J C Guevara Bolaños 2015 Aplicación móvil de realidad aumentada para la enseñanza de la clasificación de los seres vivos a niños de tercer grado.

[24] J Cabero Almenara, J Barroso Osuna y M d C Llorente Cejudo 2016 Technology acceptance model & realidad aumentada: estudio en desarrollo *18 REVISTA LASALLISTA DE INVESTIGACIÓN* - Vol. 13 No. 2 pp 18-26.

[25] E Garnica Estrada y J A Franco Calderon 2015 Realidad aumentada y educación *Rev. Ingeniería, Matemáticas y Ciencias de la Información* Vol. 2 / Núm. 3 pp 59-65.

[26] X Basogain, M Olabe, K Espinosa, C Rouèche y J Olabe 2016 Realidad Aumentada en la Educación: una tecnología emergente *Information and Communications Technology*, ESIGELEC – Rouen, France.

IOP Conf. Series: Materials Science and Engineering **519** (2019) 012022 doi:10.1088/1757-899X/519/1/012022

Stimulation program for selective and sustained attention in preschool based on ICT

T L Paternina[1], D A Rodríguez[1], D Sibaja[1] and A D Morales[1]

[1]University Corporation Antonio José de Sucre – CORPOSUCRE, Sincelejo, Colombia.

alex_morales@corposucre.edu.co

Abstract. Attention is the key cognitive system of mental activity, which influences the development of higher processes and executive functions. Indeed, the need to design an innovative training program for preschoolers arose. Because here the most important evolutionary development of individuals is determined. Because of this, record potentiates the program and improves attentional levels in children, emphasizing two respective types of attention which are sustained and selective, and focusing in two modes, which are visual and auditory, but based on software of ICT implementation. The research type is quasi-experimental and descriptive, through which a documentary review of psychological theories and apps were made with relevance for this purpose. As a result, a specialized program to stimulate attention, with a complete file containing the theoretical directives of the program and its respective protocol program was obtained. It is concluded that the use of ICT to stimulate selective and sustained attention in children of this age, can generate many benefits for being interactive and innovative, contributing to the motivation of preschool, while it enhances mental abilities and improves their attentional levels.

1. Introduction

The XXI century is known as the digital era, the major role of new information and communications technology in several daily activities that people do; which implies that at an early age, children be encouraged to use these tools, guided by professionals for proper development in computer literacy in the infant.

According to Jean Piaget, children pass through specific stages as it matures his intellect and ability to perceive relationships of all kinds, developing cognitive processes; attention is a fundamental cognitive process of human beings because it is the access mechanism for the implementation of any mental activity [1]. According to Portellano and Garcia attention is a functional complex information filtering system, which focuses on hierarchy and organization, addressed to the selection and coding of stimuli to develop a response process in the human body. Similarly, basic cognitive processes are foregoing higher processes and executive functions of the preschool.

For Rothbar and Posner, during childhood, fixing attention on an object is in competition with multiple distractors; the development of that attentional process in this period is characterized by a significant increase in the time span in which are sustained attentional resources on a stimulus, it means that it increases the number of elements that can be retained in a time specific and also improves the skills to ignore and inhibit irrelevant stimuli to a task [2].

Attention turns to modify the structure of psychological processes determining these to appear as activities that are aimed at certain objects, which occurs as the content of the activities proposed and guide the development of all the psychic processes passing attention to be a fundamental facet of

Content from this work may be used under the terms of the Creative Commons Attribution 3.0 licence. Any further distribution of this work must maintain attribution to the author(s) and the title of the work, journal citation and DOI.
Published under licence by IOP Publishing Ltd

psychological processes. That is why, it is necessary to stimulate this basic cognitive process through interactive, fun and innovative tools, features containing ICT tools.

2. Attention

Attention is the access mechanism to perform any mental activity, the anteroom of cognition, and functions as a filter system able to select, prioritize, process and monitor information.

Attention is an inclusive process when learning, that is why preschoolers need stimulation during the learning process because they are in a developmental stage where they begin to develop their cognitive, physical and social skills. Because of this, and maturational changes in the central nervous system begin to act under a sphere of pre-operational thoughts, not leaving out the need to socialize and interact with their peers.

In this sense, attention is a process that allows input of information with the aim of facilitating integrated learning, which helps the individual to recognize his surroundings; Apart from that, Reategui proposes that attention is responsible for the leak of information for internal adaptation of the organism in relation to external demands process [3].

Attention is considered an independent function, however there are factors that influence it, such as perception, emotion, intelligence, memory and motivation, primarily talking about the existing fusion between motivation and voluntariness including the alert level that the individual possesses, which happens to contribute to the performance of tasks, because the more you are motivated the more will be the effort needs to be applied, in the same way, the more will a subject is to concentrate the better, because with this, he can eliminate all distractions and the more this heightened his alert levels are, his attention levels will be better [4].

2.1 Sub-categories of attention

2.1.1 Selective attention. Selective attention is the one that includes the ability to select and integrate specific stimuli and mental images, also targeting skills and alternation of these stimuli as adequate information processing [5]. In addition, it is considered as the type of care that allows people to perceive the important stimuli, put aside that which is irrelevant, that means it focuses the individual to attend a specific stimulus even though this contains simultaneously, additional and irrelevant information, generating the capacity of processing a section of information as whole.

2.1.2 Sustained Attention. Sustained attention or vigilance does not refer to cognitive functions but motivational aspects, including complex interactions of functions, where the attentional focus is maintained with effort and resistance to increased fatigue and conditions of distractibility that can occur. That is how sustained attention happens to include warning and component contributions in terms of motivation. It is important to highlight that sustained attention acts on an attentional focus, and therefore within the hierarchy of the system it happens to be linked directly with selective attention.

Posner, defines sustained attention as the ability to retain the attentional focus, either on an activity or on a stimulation which occurs during a period which allows attention in an activity to concentrate for the required time so that it can happen. Also it is divided into surveillance and concentration despite the distractions that can occur. So all the activities that require sustained attention, are influenced by various factors that may come to favour their performance, among which we highlight physical characteristics of stimuli and in the same way the number of stimuli presented, the rhythm required by the presentation of these stimuli [6].

2.2 The physiology of attention.

Physiology of attention is explained from two specific models, highlighting areas of the brain involved, the functions of each and intersections made:

Meshullam model, it states that attention consists of two brain subsystems; attentional matrix (state function) which regulates the overall processing capacity information, the detection efficiency of stimuli, the potential targeting capacity, the level of vigilance, the resistance to interference and signal to noise ratio; these processes are directly related to what is known as the alert level. While attentional channel (vector function), regulates the management of attention in all dimensions: Personal extra, mnesica, semantics, visceral, among other dimensions.

Furthermore, the Mesulam Model understands selective or targeted attention as a neural network distributed attention. In the network, cortical regions participate: the dorsolateral posterior parietal cortex, prefrontal cortex, and the cingulate gyrus, which are associated with perceptual components, motor and limbic; each of these components form a local network [7].

Moreover, Posner and Petersen Model, proposed the existence of certain anatomical and functionally independent neural networks responsible for attentional processes. Despite its anatomical and functional independence, these networks may also interact during responses to stimuli, so it is considered to maintain a functional relationship; two of these networks could be found anatomically distributed by the anterior cerebral cortex. The third of these networks, which is the orientation of attention network, would be distributed by the posterior regions of the cerebral cortex. In first instance, the network would be compromised in the guidance of the selection of sensory information and it would justify the visuospatial attention. Tasks that are used for the functional evaluation of this network involve preferentially either finding a particular stimulus in a scene with distractors (visual search tasks), brain areas of this network would be; parietal cortex, the precentral gyrus, the front oculomotor cortex, and thalamus upper calículos [7].

2.3 Evolutionary development of attention up to 5 years.

According to Lopez and Garcia, the evolutionary process of care is as follows:

At the age of 0 to 18 months attention is determined by the characteristics possessed by the stimulus, that is, the striking color, larger or more novel depending directly from the position where it was or intensity that is desired from the stimulus. Approximately at the age of two, no differences in the processes of eye fixation, the complexity of the stimulus or task increases, other differences among children aged 3 to 11 years were observed, the younger ones tend to give up, because in the new stimulus early age becomes a very considerable force and can easily and efficiently serve for a period of seven minutes. In contrast, children less than 5 years serving the salient aspects of objects, then they begin to attend to visual cues that determine them. Within this novelty is no longer the central and attractive part of the stimulus. It should be noted that since the early ages they maintain the ability to focus more intensely on stimuli that are irrelevant provided by the environment, but which are demanding an interest and are thus able to maintain a strong attention up to fourteen minutes so they recognize the existence of variables which determine the attention, and there are some objects that hinder the process. Also at this age they are fully capable of recognizing certain situations or stimuli that are easier to deal with and that in other circumstances can educate strategies to draw attention to relevant stimuli.

3. Information technology and communication (ICT)

The information and communications technology are technological devices (hardware and software) that manage digital information through several processing techniques and data transmission.

3.1 ICT in education

ICT in the education sector has had positive results being used to create major changes in intra- and intermental processes involved in teaching and learning; taking into account the characteristics and properties of symbolic environments that allow to create and design some particular process.

These technologies have features that contribute to and influence the teaching and learning processes described below:

- *Immateriality*. Creating ICT is generated in some cases without real references, but through simulation processes emulating a behaviour or specific process.

- *Interactivity.* In Education is a fundamental characteristic, because it obtains an exchange of information between the user and the computer, to adapt the resources used to the needs and characteristics of users, depending on the specific interaction of the person with the device.
- *Interconnection.* It refers to the creation of new technological possibilities from the connection between two technologies, creating new opportunities strategies in teaching and learning.
- *Instantaneity.* A feature that allows interaction between different users quickly without limitation of distance.
- *Digitization.* Transforming information into a single format that allows the transmission and processing of it by the different electronic devices.
- *Innovation.* ICT are subject to constant changes allowing inroads into new forms of applications of these tools to achieve the desired results in the field of education goals.

Therefore the use of ICT in ages 4 to 5 years old is motivating; because children are fascinated with interactive elements that a computer teaches them. It is not the same teaching English with a blackboard or headphones, to create an English class through specialized software, including games, music, and colors, children thus learn more and in the same way are entertained, without doubt that the motivation and fun play a conclusive role in the desire to study and learn.

4. Methodology

The research is documentary because the study is based on the search, retrieval, analysis, criticism, and interpretation of secondary data, that is to say, obtained and recorded by other researchers in documentary sources: print, audio-visual or electronic, as in any research, the purpose of this design is the contribution of new knowledge. But at the same time, it is not experimental because it is not seeking to manipulate the variables of this research, but it is empirically directed to observe and study the original context and/or the evolution of the subject matter until the present time. The research focuses on exploring the basis of relevant data for the study, for the construction of such a program; and also the results of this research according to Arias, are set in intermediate terms for the acquisition of new knowledge.

In turn, it is descriptive and, according to Hernandez, Ferandez & Baptista, this type of study seeks to characterize phenomena, people, things or communities that may be subject to a process of analysis, without manipulating variables. This research aims to design a program to stimulate two sub-categories of attention, sustained attention and selective attention through ICT tools, thus a characterization of psychological models and existing theories that support the development of research is done, ICT tools that serve as support in the process of encouraging attention are also characterized.

Literature and sources were selected based on a search criterion, where the main subject is the stimulation of attention in preschool, also keywords such as ICT, selective, sustained attention were used. The research was conducted in the databases of academic google, redalyc, scielo, dialnety, pshicothema, among others; a review of different university articles, undergraduate and graduate research focused on the subject under study was also made; where keywords such as TIC, selective, sustained attention were used; meanwhile, the subtopics as ICT in education, attention physiology, evolutionary development of attention, information technology and communication (ICT), types of attention more influence in the results and concept of attention, with an antiquity of 5 years old or less.

5. Result

After an arduous review of documents in several scientific journals Redalyc, scielo, dialnety, pshicothema, among others, and undergraduate and graduate research related to the topic of study. Thus, a filter was performed to select the literature and application software relevant to the design of the training program that acts as a stimulus that reinforces sustained and selective attention specifically on preschool.

For his part, finding applications that simulate the behaviour of these activities took place, so a complete design record of the program was made, which contains and classifies the type of attention that needs to be worked on initially, the protocol of functionality, time determined for each session, duration of overall program, also activities for each type were specified. What is evident in Table 1.

Table 1. Tab stimulation program execution for attentional stimulation.

App name	*Type of activity*	*Type of Attention*	*N° session*	*Time*	*Application mode*
child coloring book	color	sustained	6	20 minutes	Individual
Coloring pages for kids	color	sustained	6	20 minutes	Individual
Kid Mazes	labyrinth	sustained	6	20 minutes	Individual
Finger Paint	drawing	selective	6	20 minutes	Individual
Color for kids	Find the color	selective	6	20 minutes	Individual
Intruder	Find the intruder	selective	6	20 minutes	Individual

In addition, a record format for monitoring was generated to write down the behaviour of each pre-schooler during the session (see table 2).

Table 2. Format for observations by session. Facilitate the monitoring process during and after implementation of the intervention program, since it contains the specific punctualities to be recorded to observer the progress and benefits.

Date		Observations
Start Time:	*Ending time:*	
Name of patient		
Age		
Sex		
Session number		
Type of care that works		

Finally, it generated a record of characterization of ICT resources that will be selected for program implementation, as shown in the following example. (See Table 3).

IOP Conf. Series: Materials Science and Engineering **519** (2019) 012022 doi:10.1088/1757-899X/519/1/012022

Table 3. Tab characterization of ICT application software. It describes the functionality of the app and pertinence of this to be used to train sustained attention.

Software name	Coloring Pages for Kids
Company that produces	Sunny Kid Games
Version	1.45
link	https://play.google.com/store/apps/details?id=com.sunny.coloring.book.kids.paint
Attention	Sustained
Recommended age	2 or more years.
Topic	Educational
Language	Spanish
Platform	Android
Format	APK
Equipment requirements	60M space It requires 4.0.3 and later
Navigation and how to use	- For the use of this educational software, it is important to have a counselor to strengthen and secure the child in the use of video games and technology. - Navigation is aimed at children in a simple way, where interaction is important, since the sounds are present and visibly indicated the steps. - Because the software is designed for children, there is no pressure and all errors are represented as learning.
A brief description of the content	Over 130 fun coloring, doodling and painting with educational content, simulating the use of paint, tempera and oil paintings with your own fingers drawings.
Educational aspects: *-Goals* *-Materials* *-Activities* *-Evaluation*	The application allows children to display their creativity by coloring, drawing, painting and decorating their works with multiple seals. It is an educational game that encourages and helps the child to learn, develop and grow intelligently and creatively. In addition to amuse, enhance their creativity and entertain them, it contributes positively to growth and development.
Pedagogical projections	Coloring will grant the child a great satisfaction to realize the beautiful things he can do alone. It develops his ability to detect, investigate and memorize. He will create drawings that will be valued by their loved ones encouraging their self-esteem. This entertainment gives them the opportunity to express their inner world and to channel their mood and feelings and giving us, parents another way to understand them.
Personal opinion	

5.1 Pilotage

As a pilot test, a sample of twelve children was chosen, which was divided into two groups, one of control and the other to intervene.

Initially it applies to both groups the preschool neuropsychology battery - BANPE that allows to measure the level of sustained and selective attention in children from 3 to 5 years old through the items digits in progression, cubes in progression, visual cancellation, detection of digits and visual search.

Subsequently, the program was applied for a period of 3 months with a periodicity of twice a week to the group to intervene.

Finally the BANPE instrument was applied again to both groups to perform a correlation and to know the real benefit of the program, the results obtained were.

When making a comparison between the previous test and the subsequent test of the control group, it is observed that there are minimal changes in the oscillation of the values, from which it can be argued that the daily activities of children aged 3 to 5 years in a lapse of time at 3 months varies little in the parameters referring to sustained attention and selective attention (see graphic 1 and 2).

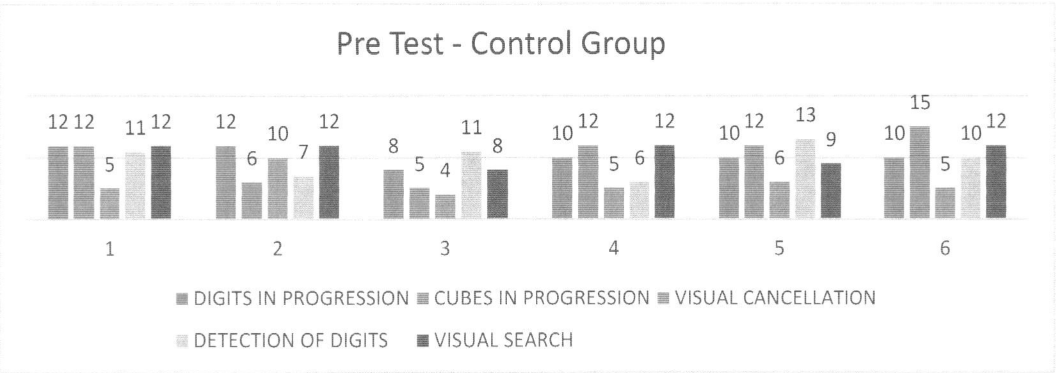

Graphic 1. Pre-test – Control Group

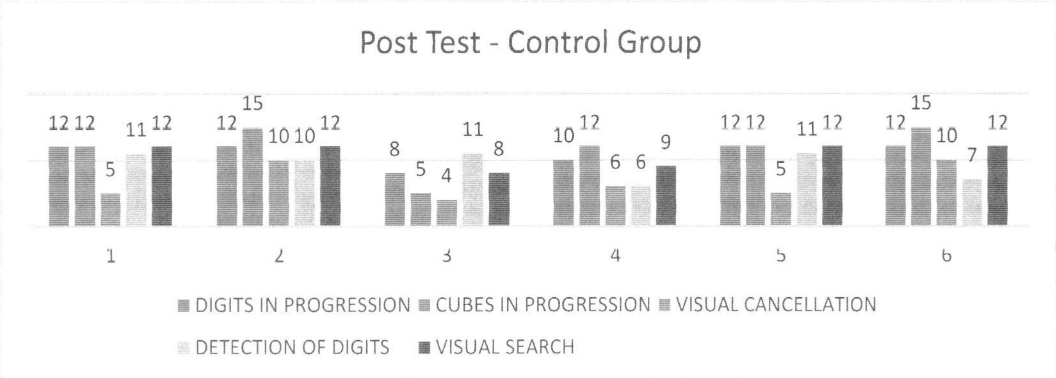

Graphic 2. Post-test – Control Group

When reviewing the comparison between the previous test and the subsequent test of the intervention group, it is observed that there are significant changes in the aspects of digits in progression, cubes in progression, visual cancellation, detection of digits and visual search, of the previous ones (see graphic 3 and 4). The program proposed in this research can contribute positively to the stimulation of sustained and selective attention, so in a subsequent investigation it is recommended to perform the intervention with a larger group of children between the ages of 3 to 5 years.

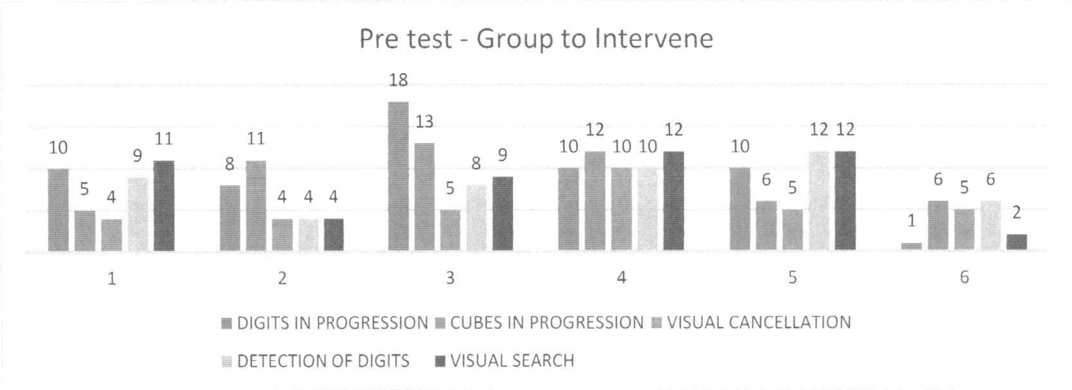

Graphic 3. Pre-test – Group to Intervene

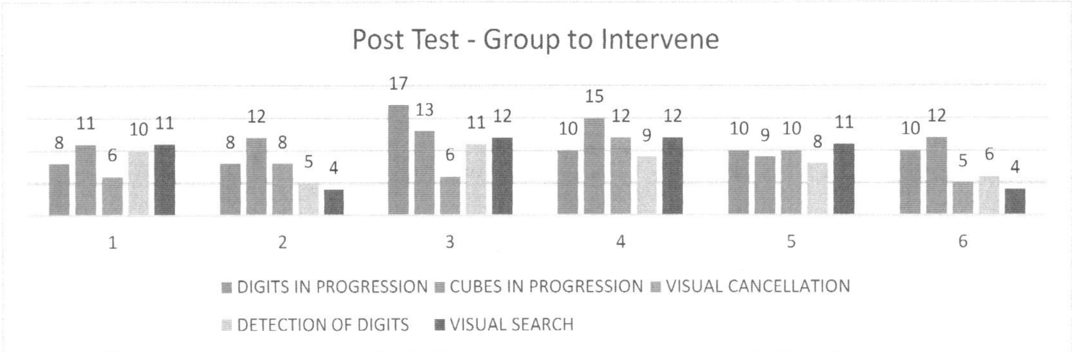

Graphic 4. Post-test – Group to Intervene

6. Conclusion

With the results presented in this research, it can be seen that stimulation of attention through a specialized innovative program in the use of these transversal tools such as ICT specifically at early ages, with proper guidance and support, promote improvements in attentional levels, mental ability and promote social relations. Also, in contrast with the results obtained by July and Pimentel, it was evidenced that attentional factor interventions enhance skills and competencies in subjects. Therefore, to make use of ICT for preschoolers, gives positive results in cognition, because these tools currently act as positive stimuli, fostering in children that intrinsic motivation that leads them to engage in academic or daily activities, which are difficult to perform or are boring for them if traditional methods are used. This is due to the characteristics and properties of ICTs that transform something boring and mechanic into something dynamic and playful.

For its part, during the course of the research, it appears that there are several free applications on the market that simulate the behavior of intervention activities created by experts for attentional stimulation. Most of them are available for the Android platform. In addition, when selecting the application it is recommended that they be offline applications because Internet connectivity can be a distracting agent.

Finally it should be mentioned that the stimulation program focused on selective and sustained attention through the use of ICT, according to other authors as Martinez, Olivo, and Herrera, Deck, and Callejas and Romero, is a viable study result to help preschoolers to increase and improve their attentional levels and influence the rapid acquisition of higher cognitive processes and in turn, facilitate the arrival of executive functions they should have for life.

Referencias

[1] M. SERIGOS, 2014, Focus on girls and children in the preschool level educational institution beef Cartagena District, Cartagena: Universidad del Tolima.

[2] F and Lapiáñez, 2014 The attentional theory Posner: a task to measure attentional functions Guidance, alert and cognitive control and interaction between them, Psicothema, pp. 260-266.

[3] P and GARCÍA 2014 Neuropsychology of attention, executive function and memory.

[4] T and DELGADILLO 2016 The playful as a strategy to improve care processes in children of first grade American College of Ibague, Ideals Magazine.

[5] C Gonzalez 2016 The use of ICT to improve attention deficit disorder in children from the first year of Basic Education of the Education Unit Lauro Damerval Ayora Loja city.

[6] R and RODRIGUEZ 2014 Journal ResearchGate.

[7] Fernández 2014 Neuropsychology of attention. Concepts, alterations, and Evaluation. *Argentina journal Neuropsychology* pp. 6-14.

[8] A A 2013, Evaluation of a program to improve attentional deficits in children 4 and 5 years (Introduction to educational research) Aspe, España, National University of Rioja.

[9] N. &. H. BELMAR, 2013 attentional processes involved in the disorder attention deficit hyperactivity disorder (ADHD), Convergence Education, pp. 9-19.

[10] B and COHN 2013 Strengthening selective attention through an educational software for children 4 to 5 years garden beautiful child gymnasium years.

[11] J Briceno 2015 uses of ICT in preschool: towards curriculum integration *Virtual Magazine of the National University*.

[12] A F G 2012 The research project: Introduction to scientific methodology.

[13] F and LAPIÁNEZ 2014 Posner's attentional theory: a task to measure attentional orientation functions, alertness and cognitive control and the interaction between them, Pshicotema, p. 260-266.

[14] H and BAPTISTA 2014 Research Methodology.

[15] O and H MARTÍNEZ 2015 Treatment and prevention of scattered attention through digital tools in transition students of School Jose de la Vega city of Cartagena, Cartagena.

[16] A MUÑOZ 2014 Inventory of ICT resources for children with special educational needs.

[17] R and CALLEJAS 2016 Program "attentive learn" as a strategy to improve selective attention, Magazine educational horizons.

[18] S and MONTOYA 2016 Improving care through popular games children transition from school grade based Tabora c, *Journal of Educational Research.*

[19] M Albeza 2013 Evaluation of a program to improve attentional deficits in children 4 and 5 years.

[20] M Decks 2018 ICT favoring selective and sustained attention: ICT as a mediator in teaching element to favor processes of selective and sustained attention, Chia, Colombia: Universidad de la Sabana.

[21] I B and S S 2009 Evaluation of sustained attention of children with attention deficit hyperactivity disorder, Psicodidáctica p. 13-27.

IOP Conf. Series: Materials Science and Engineering **519** (2019) 012023 doi:10.1088/1757-899X/519/1/012023

Overview of Data Governance in Business Contexts

L M Jimenez[1], J A Polo[1], N A Duarte[1]

[1]Remington University Corporation

luisa.jimenez@uniremington.edu.co

Abstract. Taking into account the amount of information that is currently handled in business contexts, it is mandatory to develop a good data governance program. It guarantees that all the information is available to the company when it needs it, as it needs it and where it needs it. Through this data governance, the main goal is to increase profits, decrease costs, improve knowledge management and interact with customers, suppliers, employees and processes. Besides, this type of programs are not only part of a section in the company, but also they must have a transversal collaboration to all the technology and business teams. During the first stage of this research, it seeks to perform a review of the state of the art on the data governance. It allows to unify criteria through a reference item later, to provide a tool that helps companies to support the information necessities through the development of data governance projects that respond to business needs and strategic decision making.

1. Introduction

Nowadays, organizations are changing all the time. Effective decisions play an important and decisive role in front of competition. It implies a series of attributes of the information quality. It has to be precise, coherent, consistent, available, easy and immediate accessible, but the variety of formats and the volume of data in the business contexts have grown considerably in the last years. It is due to current companies work online worldwide and accumulate a huge amount of information, mainly about their products, goods, services, customers and their behavior. On the other hand, companies add unconventional information from social networks, multimedia, heterogeneous data sources and many of these are unstructured ones.

Consequently, the complexity of the information is meaningful and companies need to find strategies to organize, manage and rule information as an income of the company and can be used as an input that supports business strategies. Data governance is an important variable in strategic management within a corporate government, and it must follow its principles [1]. An efficient data governance can help people of an organization to create a clear mission, increase confidence in the use of the organization's data, establish roles and responsibilities, quantifiable principles and indicators [2]. More and more organizations have realized the growing importance of controlling their data. However, there is a growing focus on data governance aspects to ensure security, high quality and better data management [3]. Nevertheless, there is no single standard or approach for the implementation of data governance programs in organizations nowadays [4][5].

During the initial stage, this research goal is focused on showing a review of the state of the art, which allows an analysis to unify criteria by means of a reference model in order to guide companies in the implementation of data government projects.

Content from this work may be used under the terms of the Creative Commons Attribution 3.0 licence. Any further distribution of this work must maintain attribution to the author(s) and the title of the work, journal citation and DOI.
Published under licence by IOP Publishing Ltd

2. Main concepts of data governance

The Data Government is the field in charge of treating the data as a company asset. It is about the decision policies so as to improve, protect and take advantage of the data as a business asset. It involves the orchestration of people, processes, and technology and organization policies, to obtain the optimal value of the company's data. The Data Government plays the main role in the alinement of disparate and often conflicting policies that generate anomalies in the master data. It defines who has the right and responsibility for taking decisions about the company data [6][7][8]. There are several meanings of data governance, as follows in Table 1.

Table 1. Conceptualization of Data Governance

Concept	Source
It is the set of decisions, processes, standards, policies and technologies required to manage, maintain and exploit some information as a business resource.	[9]
A data governance plan hast to be part of an IT program. It involves processes and controls to ensure that information at the data level is true, accurate, and unique.	[10]
Data governance can be defined as an organizational approach to data management that establishes a set of policies and procedures to include the data entire life cycle.	[11]
Data governance can be seen as a formal orchestration of people, processes and technology to allow an organization to take advantage of data as a business asset.	[12][13]

Data governance and data management are different, the second one refers to the implementation of the decisions made about architecture level, policies and procedures in order to improve and manage the life cycle of the data in a specific context.

3. Information governance concepts

The information governance is considered as the value driving and it moderates the risk of data exploitation in organizations. An institution can take advantage of information as a competitive asset [14]. The governance of information, mainly, is the set of principles, policies and processes through which an organization guarantees that the information is protected and according to its needs and business goals [15][16][17][18][19][20][21]. Information is an asset of the organization that includes both the data and the context in which data has meaning [14]. Architects, system administrators and data center administrators are key performers in the governance of information; this incorporates processing, infrastructure management and security requirements [17].

4. What it works and what it doesn´t in data governance

The Data Government is not a new concept. Professionals of knowledge and information management have faced the quality and confidentiality of data for most of the computing history. Data doesn´t have not only some value within the application and the users who captures it, but also in the organization. These functions demanding, and cross-architectures of the same data is forcing companies and IT leaders to recognize dependencies and perceptions of data value conflict. Many information managers are betting on data governance to decrease these conflicts [22]. This proposal does not define neither decision dimensions nor a methodological guide for data governance, but it addresses important issues that must be considered when designing and implementing data governance.

It is important for a data governance project success to take into account some topics like catalysts that help to launch a data governance project joined to another: how can a BI project be. An organizational structure for the project has to be led by a CXO of the company, an exclusive project manager, accompanied by a responsible for the business and another IT. It is mandatory to identify the main risks that can cause the data governance initiative failures and decide how to decrease them: the risks can be aspects such as lack of explicitness about business processes, lack of participation of business representatives, among others. Defining the team in charge of the master data and who are involved in the project is necessary. The government of data should be designed according to the company policies: it means that, if a company decides in a decentralized manner for each field of business, it is suitable to implement a data governance in the same way. Although the government is decentralized, decisions should be taken centrally, but the business data can be taken in a decentralized manner.

5. Data domains

The information architecture, according to [23], describes the principles and guidelines that allow a coherent implementation of information technology solutions: how data and information are ruled and shared in the company, as well as, what it needs to guarantee information confidence. There are several types of data that can be used in each business lines. The data can be structured or unstructured and it can also be viewed from a perspective of how it is stored. In [23] it explains, how to divide business information into 5 domains, according to the use and format of the information within a company. The domains are: metadata, master, operational, unstructured and analytical data.

Metadata is defined as "data on data". It is the information to describe the characteristics of each piece of corporate data and other entities. The master data refers to the data instances that explain the main commercial entities, such as customer and product data. Operational data is also known as transactional data. They are derived from day-to-day business transactions. Unstructured data, also known as content is usually managed by a business content management application. Analytical data is usually derived from the transformation of operational systems to address the specific requirements of decision support applications [23].

6. Reference models

Data governance expresses how the organization manages all the data and information that is transversal to all its processes [24]. It is not a one-time project, but a constant program. The reference models provide a set of tools and guides that help organizations through the implementation of data governance projects to get their business goals.

6.1. Federated enterprise architecture

The information architecture, according to [23], describes the principles and guidelines that allow a coherent implementation of information technology solutions: how data and information are ruled and shared in the company, as well as, what it needs to guarantee information confidence. There are several types of data that can be used in each business lines. The data can be structured or unstructured and it can also be viewed from a perspective of how it is stored. In [23] it explains, how to divide business information into 5 domains, according to the use and format of the information within a company. The domains are: metadata, master, operational, unstructured and analytical data.

6.2. The unified process

Many companies apply a manual procedure to accomplish the instructions for a data governance program. Obviously, all of them will run it differently, especially due to the different goals of the companies. Some of them could focus on data quality, others on customers, and the others on ensuring the privacy of sensitive customer data. Some organizations will adopt a formal Data Governance program, meanwhile others use something lighter and more tactical.

The unified process of data governance shows 14 steps to set up an effective program, from step No. 1 to 9 and 14 are required, the other ones are optional. The company selects one or more of the four optional steps. It means: Government of Master Data, Analytical Government, Security and Privacy Management and the management of the Information Life Cycle. Finally, the unified process of data governance has to be checked, and the results showed periodically to the executive promoters [6].

6.3. ISO 38500

It proposed an IT governance model [19][20][25], to offer a guidance on behavior in decision-making and responsibility measurements, strategy, acquisitions performance, compliance and behavior of personnel. It explained that organization's leaders should focus on testing the use of current and future IT, guiding the preparation and implementation of the government and establishing a model cycle of evaluation, direction and monitoring. At the same time, ISO believed that IT was also available to data governance.

6.4. IBM Data Governance Council (IBM DG Council)

It explained a data governance element model and it was divided into four levels: support and central domains, contributing factors and results. The results are seen as the key to data governance in this model. The organization could obtain the results and realize the value of the data by means of taking into account the function of the support domain, the central one and the contributing factors. IBM DG Council combined the characteristics of the data with its practices and proposed the data governance maturity model, which divided this model into five levels, including the initial stage, basic, active, quantitative management and permanent optimization [26].

6.5. Information Systems Audit and Control Association (ISACA)

ISACA [26][27] believes that data governance is an important element for IT. He considers data governance as a principles-based methodology for process control. COBIT 5 provides an IT and management framework based on the Top-Down principle and gives the strict distinction between government and management. Besides, COBIT 5 explains key areas of data governance, such as stakeholders, contributing factors, governance and management objects, among others.

6.6. DAMA-DMBOK2 Framework

DAMA International defines 11 areas of knowledge for most of data management areas: governance, architecture, modeling and design data, Operations and data storage, security, integration and interoperability data, Documents and content, Master Data and Reference, Data Warehouse and Business Intelligence, Metadata and Data Quality. Each area has topics per section and group activities. There is also an additional section of Data Management that contains topics to describe the knowledge requirements for data management professionals. The new area of is integration and interoperability data [28].

6.7. Decision domains

The data governance design, according to [7], defines 5 related decision making domains. These are as follows: data principles, data quality, metadata, security and data life cycle. For each of them, it defines the type of decisions that must be made. It also proposes a set of possible roles in the company that may be responsible for each one.

Data quality. Data quality impacts the company at an operational and strategic level. Poor quality in the data generates operational costs and, strategically, the information is not reliable for taking strategic decisions. The quality of the data refers to its capacity to satisfy the requirements of use within the company. The quality of the data has multiple dimensions such as: precision, timeliness, integrity and credibility, among others, these are relative and should be defined according to the context of data use [7]. Data quality problems are showed in [29]. Despite all the efforts and investments in technology, the departments of the companies identified a lot of problems related to data quality and reliability, at the time of the implementation of analytical projects, which compromises the established deadlines and the investments they made.

Data principles. Data principles establish the link with the business. To align it with the use of the data, its principles set up its measurement as an asset of the company and define which are the specific policies, standards and appropriate guidelines. Data principles also establish and foster opportunities to share and reuse it. Each one is supported by a fundamental reason and a set of implications [7].

Metadata. It is defined as "data on data". It provides a mechanism to obtain a concise and coherent description of the data. It helps to understand the meaning or semantics of the data [7].

Data access. Accessing to data is based on the ability of participants to assign a value to different categories. An effective risk analysis by security agents identifies the company necessities and defines policies to ensure confidentiality, integrity and data availability [7].

Data life cycle. Being conscious about that all data moves through the stages of the life cycle is mandatory to the design of data governance. By understanding how data is used, and how long it should be retained, organizations can develop methods to allocate usage patterns for optimal storage media in order to decrease the total cost of data storage during their cycle of life [7].

7. Mature Models of Data Governance

There is a series of maturity models to improve the planning and implementation of data governance. Each one has strengths and can provide valuable perspectives, current characteristics and form the basis for the subsequent planning of a data governance implementation process. Its review and evaluation must occur at the beginning of the process to establish an understanding of the final state. It is necessary to plan a suitable data governance program. The implementation process is a continuous operational discipline to improve the job as business architecture. The successful implementation of data governance requires a business perspective [30][31][32].

7.1. The Data Warehousing Institute (TDWI)

TDWI explains that many of the goals and tasks related to data governance can be reduced to four main imperatives or requirements. They are organized into two organizational imperatives and two technical requirements [32][33].

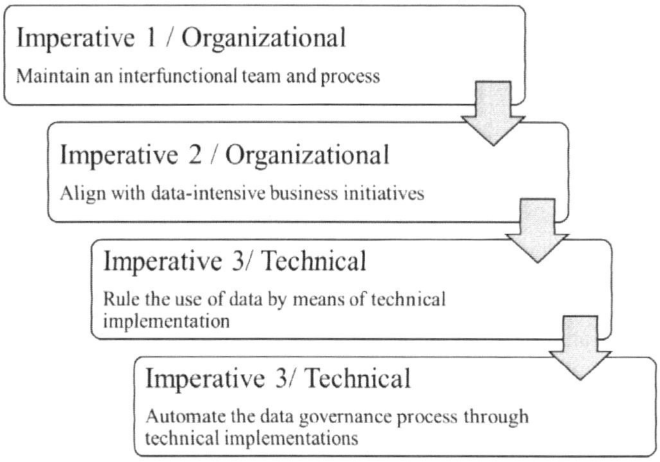

Figure 1. Imperatives of data governance. Adapted from [32]

There are two main characteristics in the data governance imperatives classification. First, everyone has stages which develop over time. This advance means development. Second, imperatives as a group imply a sequence of time. For example, it is clear that imperative 1 must create an interdisciplinary team before imperative 2 coordinates the team's objectives with the business initiatives. Figure. 1.

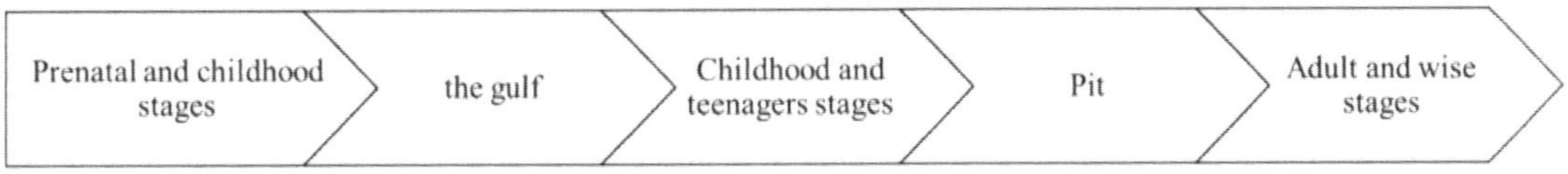

Figure 2. TWDI maturity model. Adapted from [32]

According to imperatives 3 and 4, [32] it explains: The imperative 3 must rule IT systems before imperative 4 begins to use IT systems to automate governance processes. Although these dependencies determine an order to begin the imperatives, they must coexist and interact. The TDWI maturity model is showed in the Figure. 2 and it is described in Chart 2.

Table 2. Description of stages in the TWDI maturity model

Maturity stages	Description
Prenatal and childhood stages	In this stage, organizations depend on manual means applied ad hoc as a low-level solution for their business or technological concerns. As we move towards the infant stage and over time, a study of the requirements has led the organization to a specific technology or practice.
The gulf	Leaving this obstacle depends on the organization regularizes the concepts of solution verified in the previous step.
Childhood and teenagers stages	In this stage, the organization extends its technological initiatives or practices with it has recently committed. However, growth could arise in a limited context, such as divisions. As a result, growth slows down in the adolescent stage
The pit	Crossing the abyss successfully requires important global changes, such as business adoption or re-architecture of solutions
Adult and childhood stages	In the adult stage, the best practices and the details of technology implementation established and developed in the adolescent stage continue to grow. Silo implementations of the previous stage have been replaced by the wise stage of centralized organizational control and technology integration

7.2. Analytical Business Maturity Model

[34] It takes a different approach according to the TDWI. It shows the analytical business maturity model, with five levels of maturity and explains that it allows organizations to specify current and objective maturity levels and what it needs to be achieved in each phase. The stages are the following: ad hoc, foundational, competitive, differentiating and breaking, see Figure. 3.

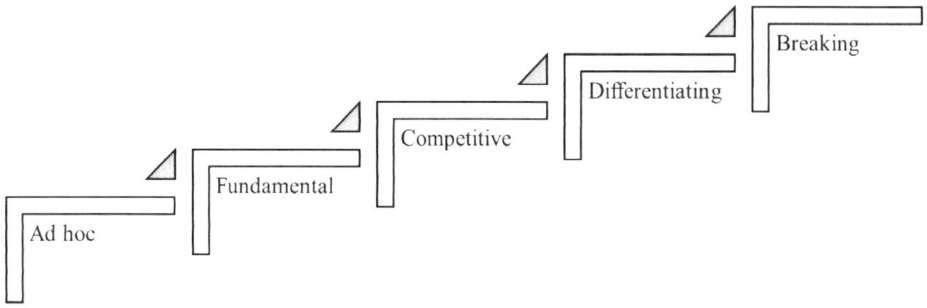

Figure 3. Maturity model of analytical business. Adapted from [34]

In the ad hoc stage, the company is generally behind other companies in the corresponding information area. It is an initial phase, the company begins to develop the capacity to gather information in its functional areas. Different from basic reports, information is not available, or it spends time. Human efforts are usually required to gather the necessary information in order to take daily business decisions.

In the fundamental stage, the company cannot gather the key information yet and is still behind most of its competitors. Information is constantly not available to take business decisions and some human effort is still required to get it. On the contrary, the competitive scenario represents companies whose capacities in gathering information is online, like most of similar companies. This level of maturity is also the starting point to establish a certain consistency in the main business metrics in the company.

In the differentiating stage, a company whose commercial strategies are carried out through the use of information is considered better than most other companies in its sector. Management has the ability to be adapted to commercial changes. Business leaders and users have the opportunity to get the main information and the respective metrics so as to take effective decisions.

Finally, when a company is in the breaking stage, it is generally considered as the best in executing key business strategies. The information is used in the company for multidimensional decision and key indicators of predictive performance are used to model the results.

7.3. DataFlux Maturity Model

This is another model of data governance maturity. According to this one, the organization progresses in the sequence from stage one to stage four, the value increases and simultaneously the risk associated with "incorrect data" decreases. Table 3. Tony Fisher, president and general manager DataFlux explains that the term "data governance" can be lost in different conversations such as big data governance, data and knowledge management, data assets and information assets. DataFlux has improved its maturity model to highlight a business to manage data as a business asset and achieve the necessary levels of data quality [35][36][37][38][39].

7.4. Gartner Maturity Model

Gartner's maturity model, see Table 3, explains that business information management is not a project itself. It is a long-term program that evolves over time. Gartner developed the maturity model to provide information and guidance to organizations that take information management seriously as a strategic asset [30][16].

Table 3. DataFlux data governance maturity model

Level	Characteristics
Level 1: Undisciplined	There are redundant and duplicate data, sources and divergent records at this stage. The main problem is that poor data and information will lead to wrong decisions and missed opportunities.
Level 2: Reactive	This is the beginning of data governance. The improvements are experienced at the departmental level.
Level 3: Proactive	The company recognizes the value of a unified vision of information. It thinks of Master Data Management (MDM). The IT and business groups collaborate in this stage. In this phase, culture is also preparing to change.
Level 4: Ruled	The information is unified in the company. Business needs to manage IT projects. It has a sophisticated data strategy and framework. Employees have realized that information is a key asset of the company.

Table 4. Gartner maturity model

Level	Characteristics
Inconsistent	• Strategic decisions are made without adequate information. • Lack of information architecture, principles, rules and procedures to share information. • Lack of control of information, privacy, security and responsibility.
Consistent	• Identifying the value of information. • Realizing the risks associated with the lack of adequate management of information and the exchange of information.
Reactive	• Companies realize the value of information as an organizational asset. • Information is shared in multifunctional projects. • Measurements are taken to exchange data among departments.
Proactive	• The roles and structure of the government are structured. • Data governance is incorporated to the systems development methodology.
Managed	• Policies are developed to achieve coherence and are accepted throughout the organization. • A government organization has been established to address concerns related to the management and exchange of multifunctional information
Effectiveness	• Senior management considers that a competitive advantage is obtained by adequately improvement information assets. • The strategies of the management of business information are linked to the objectives of productivity and risk management.

7.5. IBM Data Governance Maturity Model

Data governance has reached such importance that IBM has created a data governance Council. One of the proposals of this council is a data governance maturity model based on the Capacity Maturity Model (CMM) of the Software Engineering Institute [26][40]. The Maturity Model of the Data Governance Council defines a set of domains that comprise data governance. The revision of these domains is a first step to understand the IBM Maturity Model. The 11 domains reside within four main groups: Outcomes, Enablers, Central Disciplines and Support Disciplines. The interactions among these groups are showed in the Figure No.4.

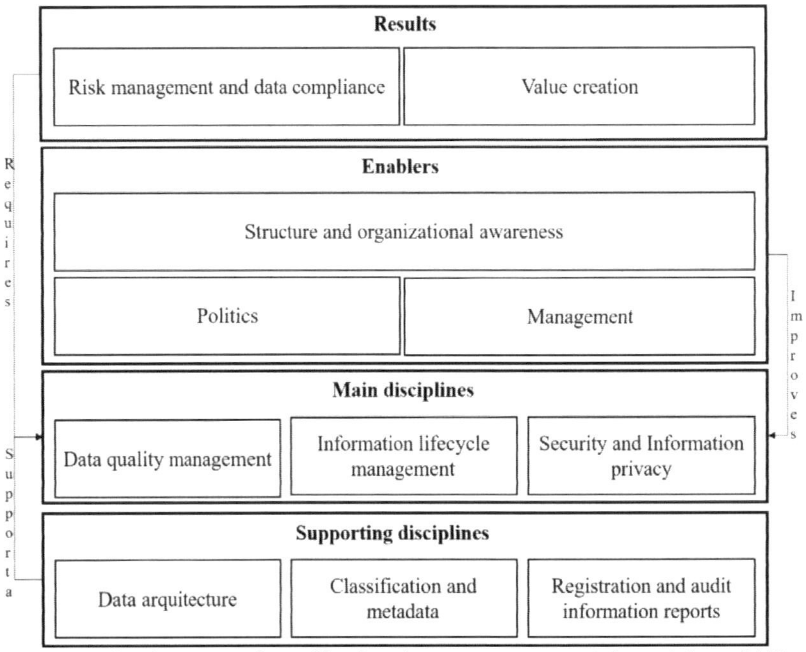

Figure 4. Elements of an effective data governance. Adapted from [41]

IBM developed the maturity model (Figure 5). The maturity model is the "criteria" for evaluating and measuring progress within each of the 11 domains [30].

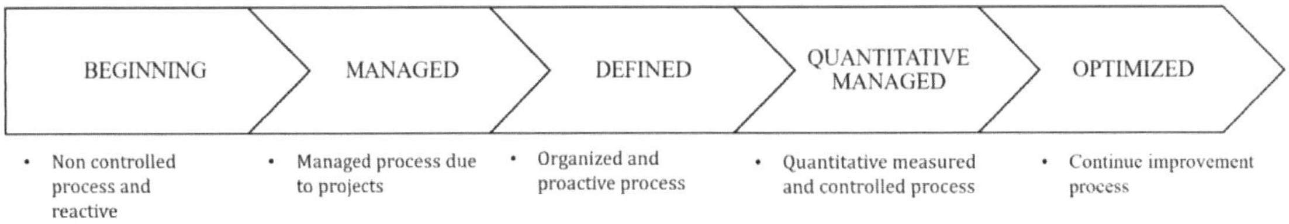

Figure 5. Maturity model of the IBM Data Governance Council. Adapted from [42]

7.6. Oracle Data Governance Maturity Model

Oracle is well known for its emphasis on a well-designed underlying data architecture. Oracle has data management experience with the government policies about data. It is showed in [37]. Effective data governance must correctly align people, processes and technology to convert data into strategic information and knowledge assets [30].

It is necessary to rule the data applied in a wide variety of heterogeneous applications and business intelligence systems. Most data quality problems start in these segmented applications. The origin of this data makes it difficult to manage and creates challenges for data governance [30].

Oracle considers that a data governance maturity model help the company determine where it is according to the evolution of its data governance discipline and identify the short-term steps required to move to the next level, see Table 5. Each step has key performance indicators, measurable on real investment to assume the cost [30].

Table 5. Oracle data governance maturity model

Stage	Level	Characteristics
1	Marginal	This stage describes an organization that has understood the data governance needs. They will expand the scope of ongoing data quality initiatives and add data management capabilities.
2	Balanced	This stage is featured by data governance initiatives at the division or department level with functioning data governance teams. It socializes the successes achieved at this level and helps to increase the progress requirement. Teams should be organized in the company and interdisciplinary conflicts have to be solved. Master Data Management solutions must be implemented.
3	Best practices	Companies are applying the best data governance practices. The data management policies are automatically executed by the Master Data Management engines and the feedback loops that report the results to the committees.
4	Transformational	This stage integrates the quality data tested in the applications and business intelligence tools in every business process to achieve the transformational state for the company

8. Big Data Government

Most of large organizations have multiple application and format data. The mobile revolution is experiencing today. It has completely changed the way they capture data and build intelligent systems. Large data tends to be operational in nature and is featured by "3 Vs", large volumes, high speed and a variety of formats such as structured, unstructured and semi-structured [34][43][44][45]. The figure. 6 shows the three dimensions of Big Data. These three dimensions and related characteristics are a useful lens to understand the nature of the data and how they can be understood.

Therefore, challenges of Big Data are different and are featured by a combination of the "3 Vs" that make the information and knowledge of the business difficult to obtain. In addition, when large-scale data is processed and stored, other dimensions such as government, ethics, security and policies take part of it [12][45][46][47][48][49][50].

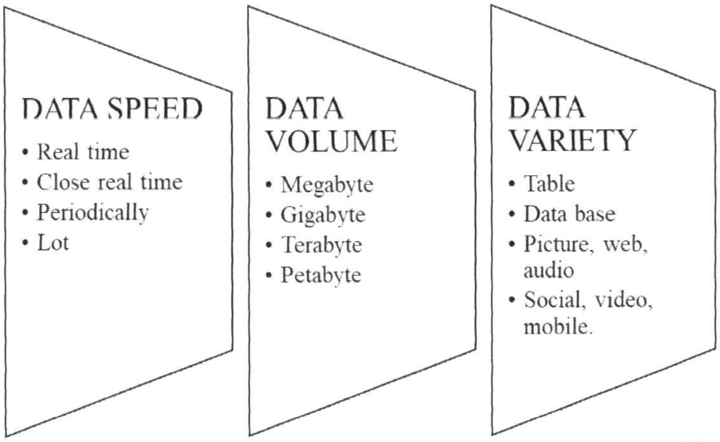

Figure 6. Big data growth dimensions and related characteristics. Adapted from [42]

9. Conclusions

Reference models, data domains, what it works and what it does not, should be taken into account as a necessary tool when planning and executing a business data governance strategy or program. These reference models are useful to organize concepts and establish a perspective about a data governance initiative in order to fit and respond to business necessities and strategies.

A comprehensive data governance strategy should not only include a reference model, but also maturity models, business processes and, the organizational context and culture. Maturity models should be used as a measure of the evolution in data governance by means of the data and information recognition with one more asset of the company or organization.

The quality of the data depends directly on a comprehensive strategy of data governance and without it, the objectives, strategies, business indicators cannot be defined, understood, communicated and measured properly.

There are several reference and maturity models as proposals for planning and executing data governance programs. They are different based on their characteristics but complementary at the same time. Each proposal addresses the issue from different perspectives and equally important issues. Consequently, if someone wants to be part of this type of program, it is necessary to review each of the proposals, analyze and determine which of the proposals is the most appropriate or choose another one according to the company needs.

According to the previous work, the opportunity to propose a data governance model for a specific context is identified in order to facilitate the planning, execution and evolution of a data governance program in high-level educational institutions.

References

[1] Wende K 2007 A Model for Data Governance – Organising Accountabilities for Data Quality Management *18th Australas. Conf. Informatin Syst.* 417–25
[2] Otto B 2011 A Morphology of the Organisation of Data Governance *ECIS 2011 Proc.* 272
[3] Nwabude C, Begg C and Mcrobbie G 2014 11.CE *4: Data Governance in Small Businesses – Why Small Business Framework should be Different *Int. Proc. Econ. Dev. Res.* **82** 101–7
[4] Weber K, Otto B and Osterle H 2009 One Size Does Not Fit All — A Contingency Approach to Data Governance *ACM J. Data Inf. Qual.* **1** 4:1-4:27
[5] Begg C and Caira T 2012 Exploring the SME Quandary : Data Governance in Practise in the Small to Medium-Sized Enterprise Sector *Electron. J. Inf. Syst. Eval.* **15** 3–13
[6] Soares S 2012 *The IBM data governance unified process: driving business value with IBM software and best practices* vol 4
[7] Khatri V and Brown C V. 2010 Designing data governance *Commun. ACM* **53** 148
[8] David M 2014 Practical steps for overcoming political challenges in data governance *IDQSummit ECCMA* 1–37
[9] Newman D and Logan D 2006 Governance Is an Essential Building Block for Enterprise Information Management *Gart. Res.* 1–9
[10] Smallwood R F 2014 Defining the Differences Between Information Governance, IT Governance, & Data Governance *Aiim* 1–3
[11] Korhonen, J.J.; Melleri, I.; Hiekkanen, K.; Helenius M 2013 Designing data governance structure: an organisational perspective *GSTF J. Comput.* **2** 7
[12] Jee K and Kim G H 2013 Potentiality of big data in the medical sector: Focus on how to reshape the healthcare system *Healthc. Inform. Res.* **19** 79–85
[13] Mouthaan N 2012 Effects of Big Data Analytics on Organizations ' Value Creation *Nielsmouthaan.Nl*
[14] Morabito V 2015 *Big data and analytics: Strategic and organizational impacts* (Springer International Publishing Switzerland)
[15] Bonenfant, M; Ménard, M; Mondoux, A; Ouellet M 2012 Big data and governance. Research

Group on Information and Surveillance in Daily Life *GRICIS Res. Cent. Univ. Québec, Montréal*

[16] Newman D and Logan D 2008 Gartner Introduces the EIM Maturity Model *Gart. Res. ID Number G00160425* 1–8

[17] Santovena A Z 2013 *Big Data : Evolution , Components , Challenges and Opportunities*

[18] Kord, Davis; Doug P 2012 *Ethics of Big Data* (Sebastopol, CA: O'Reilly)

[19] Tsai J Y, Egelman S, Cranor L and Acquisti A 2011 The effect of online privacy information on purchasing behavior: An experimental study *Inf. Syst. Res.* **22** 254–68

[20] O'Reilly Media 2012 *Big Data Now: 2012 Edition*

[21] Sun H 2011 Enterprise Information Management: Best Practices in Data Governance *Oracle White Pap.* 20

[22] Karel R, Kirby J P, Evelson B, Moore C and Barnett J 2007 Data Governance : What Works And What Doesn ' t

[23] Mario G, Eberthard H, Klaus K, Steve L, Martin O and Michael S 2010 *The art of enterprise information architecture* ed IBM Press (Pearson plc)

[24] Smallwood R F 2014 *Information Governance: Concepts, Strategies, and Best Practices* (Wiley)

[25] International Standard Organization 2005 ISO/IEC 38500:2008 Corporate governance of information technology *Int. Stand.* **2005** 1–36

[26] IBM Institute for Business Value and IBM Strategy and Change 2007 The IBM Data Governance Council Maturity Model : Building a roadmap for effective data governance *Gov. An Int. J. Policy Adm.* 1–16

[27] Isaca 2013 *A Business Framework for the Governance and Management of Enterprise IT*

[28] Cupoli P, Susan E and Henderson D 2014 DAMA-DMBOK2 Framework 1–27

[29] Freitas P A de, Reis E A dos, Michel W S, Gronovicz M E and Rodrigues M A de M 2013 Information Governance, Big Data and Data Quality *2013 IEEE 16th Int. Conf. Comput. Sci. Eng.* 1142–3

[30] Nascio 2009 Data Governance Part II : Maturity Models – A Path to Progress *NASCIO Gov. Ser.* 1–30

[31] Hurwitz J, Nugent A, Halper F and Kaufman M 2013 *Big Data for Dummies* (Hoboken: John Wiley & Sons, Inc.)

[32] Russom P 2008 *The four imperatives of data governance maturity (TDWI monograph series)*

[33] Mohanty S, Jagadeesh M and Srivatsa H 2013 *Big data imperatives: Enterprise 'Big Data'warehouse, 'BI'implementations and analytics* (New York: Apress)

[34] Sathi A 2012 *Big Data Analytics: Disruptive Technologies for Changing the GAme* vol XXXIII (USA: MC Press)

[35] Commonwealth of Australia 2014 *Australian Public Service Better Practice Guide to Big Data* (Australian Government)

[36] Morabito V 2014 *Trends and challenges in digital business innovation* (Springer International Publishing)

[37] Nascio 2008 Data Governance Part I: Managing Information As An Enterprise Asset *NASCIO Gov. Ser.*

[38] Soares S 2013 *A Platform for Big Data Governance and Process Data Governance* (MC Press)

[39] Tallon P P 2013 Corporate Governance of Big Data: Perspectives on Value, Risk, and Cost *IEEE Xplore Digit. Libr.* 32–8

[40] Mcgraw S, Blash D S, Webinar S E I and September S 2008 CMMI on the Web

[41] Nascio 2009 Data Governance Part III : Frameworks – Structure for Organizing Complexity *NASCIO Gov. Ser.* 1–23

[42] Morabito V 2015 Big Data Governance *Big Data Anal.* 83–104

[43] Bertot, John Carlo; Choi H 2013 Big Data and e-Government : Issues , Policies , and Recommendations *Proc. 14th Annu. Int. Conf. Digit. Gov. Res.* 1–10

[44] Michael K and Miller K W 2013 Big Data: New Opportunities and New Challenges [Guest

editors' introduction] *Computer (Long. Beach. Calif).* **46** 22–4
[45] Warden P 2011 *Big Data Glossary: A Guide to the New Generation of Data Tools* (O'Reilly Media)
[46] Cline J S 2014 The promise of Data-Driven Care *N. C. Med. J.* **75** 178–82
[47] Franks B 2012 Taming the Big Data Tidal Wave: finding opportunities in huge data streams with advanced analytics 336
[48] Hurwitz J, Nugent A, Halper F and Kaufman M 2013 Big Data for Dummies *Wiley* 1–5
[49] Raghupathi W and Raghupathi V 2014 Big data analytics in healthcare: promise and potential *Heal. Inf. Sci. Syst.* **2** 3
[50] Webb P, Pollard C and Ridley G 2006 Attempting to define IT governance: Wisdom or folly? *Proc. Annu. Hawaii Int. Conf. Syst. Sci.* **8**

IOP Conf. Series: Materials Science and Engineering **519** (2019) 012024 doi:10.1088/1757-899X/519/1/012024

Information Systems as Support for Competency-Based Assessment Process

M S Contreras

e-Soluciones, Faculty of Engineering, School of System Engineering, Universidad de Cartagena, Colombia

mcontreraso@unicartagena.edu.co

Abstract. Possessing enough and timely information about a process or product and being able to organize and display it efficiently, is a determining factor for decision making in organizations. Educational institutions are not alien to this principle and use different strategies to manage it. Nowadays, the Information and Communication Technologies (ICT) are commonly used to support the evaluation process by competences, to identify strengths and weaknesses and to analyze and reports the results, in order to monitor student performance and apply improvement plans.

This paper shows an analysis of some influential factors in the development of academic reports for basic education institutions, through the implementation of an Information System that provides global and sectional statistics, extracts relevant information on the individual and group performance, simplifies reading and analysis of the data provided, allows the presentation of data in numerical, qualitative and graphical form, highlights the points that move away from the average, among other useful functions.

1. Introduction

An Information System (IS) is a set of interconnected components that support organizations in the decision-making process, through information management. This management includes functions such as collecting, storing, processing, comparing, analyzing, deploying and distributing information [1] [2]. The IS also support the creation of new products and the general control of the organization. Besides, they allow saving time in the development of the tasks and provide added values [2].

One of the most used tools by managers is technology, that has been linked as a dimension of the IS. The Information Technology infrastructure includes the combination of software, hardware, networks, telecommunications and Internet [2]. These elements allow the management and display of information in real time.

Educational institutions, like organizations, have also seen in the information systems an alternative to support their processes in the handling of the large amount of information that they generate daily, making this task easier, faster and more effective. This kind of IS is known as Academic Information System (AIS) [3]. Among the most relevant processes that are developed in the AIS is the assessment of the evaluation and generation of reports that allow a correct monitoring of student performance, in order to implement timely improvement plans and effectively support learning.

One of the ways to measure a student's performance is through the competencies that he/she has developed in the area of study. The term "competence" is related to behaviors, actions or skills that a person can demonstrate in a specific context [4].

Content from this work may be used under the terms of the Creative Commons Attribution 3.0 licence. Any further distribution of this work must maintain attribution to the author(s) and the title of the work, journal citation and DOI.
Published under licence by IOP Publishing Ltd

This paper presents a customized Information System for a non-formal education institution that supports schools in the competency-based assessment process and the challenge of generating quality statistical information.

2. Literature review

2.1. Academic information system
Academic Information System (AIS) is a software tool that allows processing all data of an academic institution quickly [5]. Among the services that are commonly offered in these systems are those related to financial and academic processes. These systems must be built according to the needs of the users [6] and the objective is to achieve the goals of a quality, agile, efficient and accurate service. For this, it is necessary that the information resulting from the AIS be accessible, reliable, consistent, easy to read and relevant [7] [8]. Thanks to Information Technologies (IT), educational institutions have been able to increase their productivity and improve their processes in general [9]. A quality AIS directly provides a competitive advantage [8].

2.2. Software quality attributes
As a part of the development and implementation of any type of software, it is vital to evaluate its quality and performance, for which it is necessary to consider the measurement of 8 attributes (to see their descriptions in Table 1): functional stability, efficiency, compatibility, usability, reliability, security, maintainability, and portability [10]. All these features determine the user experience and the satisfaction level in the use of the system [9].

Table 1. Software quality attributes description.

Attribute	Description
Functional stability	The extent of change in his functions [11].
Efficiency	The relationship between the yield obtained and ideal performance, considering the resources used in a given moment [6].
Compatibility	The ability of a software application to run on different platforms without having to change its structure or format [12].
Usability	Ease of use which a user has can interact with a system, through its interface [13].
Reliability	Be available when needed and behave as expected [11].
Security	Only allow access to information and functions to authorized users [6].
Maintainability	Ability to correct errors or necessary changes to the interface, documentation or code of an application [13].
Portability	Ability to run components or the entire system in an environment for which it was not initially designed [14].

2.3. Competency-based assessment
In Colombia, the specialized entity that offers education evaluation services in all levels (basic, media, technical and professional education), and supports the Ministry of National Education (MEN) in

carrying out State tests and research about the educational quality is the ICFES. Currently, the MEN states that education is aimed at the development of competencies, that is, the know-how in context and, they must be evaluated [15].

Competencies can be classified as generic, specific or transversal. Generic competencies are common to all individuals, regardless of occupation and are related with the necessary abilities to learn, innovate and create [16]. On the other hand, specific competencies are related to private teaching subjects [17]. Finally, transversal competencies are abilities that transcend a specific field, are transdisciplinary competences [18].

3. Methods

This work was developed using the incremental approach or prototyping and was tested in a non-formal education institution that supports the competency-based assessment process in basic and middle education institutions in Colombia, with the purpose of preparing for the State tests at the different levels. Each module was developed according to the general requirements of the evaluated schools, and improvement plans were developed based on the observed results. In addition, the evaluation models were governed by the guidelines given by the ICFES and the MEN. The prototype obtained has been the result of several years of development.

4. Results and Discussion

To support the management of the information resulting from competency-based assessment process in educational institutions was implemented an Academic Information System, called SIAGREC, that contains several modules, described in Table 2.

Table 2. Modules developed in the Academic Information System SIAGREC, to manage reports on competency-based assessment.

Module	Description
Creation of groups and students	Entry of personal information of students, such as full names, ID number, School and grade. This process is done before you start to qualify.
Creation of tests	Entry of information on each test, such as evaluated subjects, number, and order of questions, correct answers. This process is done before you start to qualify.
Reading and processing response sheets	Scanning of the response sheets and processing of the resulting image, to compare the answers given with the correct ones, according to the applied test.
Generation of individual and group reports	In the individual' reports, the score obtained by the student in each evaluated area and its total accumulated is shown. The information is presented in a quantitative and qualitative form. In the group's reports, more detailed and comparative information is presented, by thematic axes and competences, both of the individual scores and of the global averages. In addition, a graphic report is generated that is easier to read and analyze.

Figure 1 shows the flow diagram of the qualification process, once the students have been registered and the information on the applied exam has been entered.

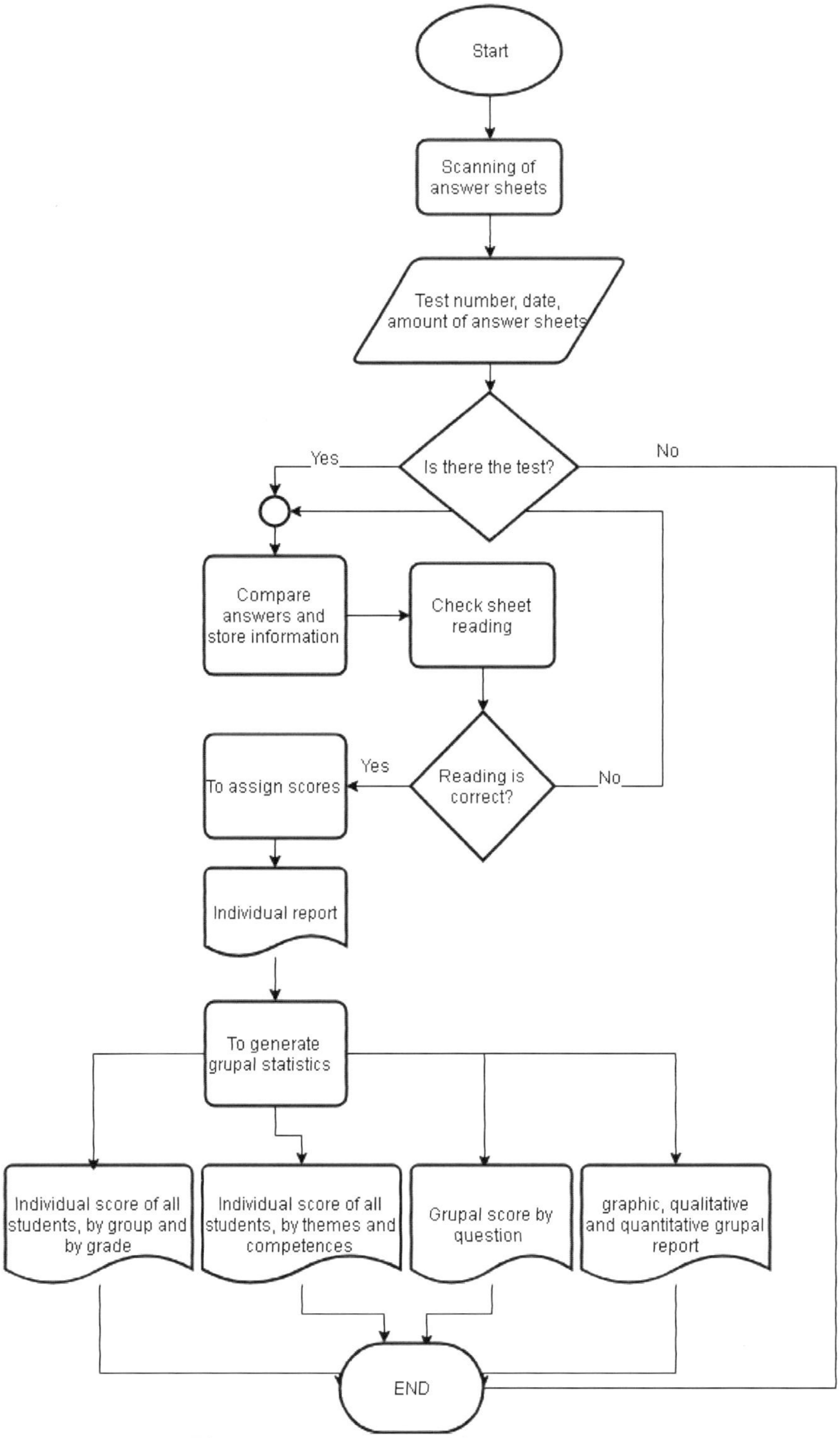

Figure 1. *SIAGREC qualification process.*

Main subjects evaluated are critical reading, mathematics, natural sciences, social sciences, and English. The type of question is a multiple choice with a single answer and the sheet format corresponds to a templated adjusted to each test, where the student fills in an oval manually, according to the chosen option. Figure 2 shows a fragment of the answer sheet.

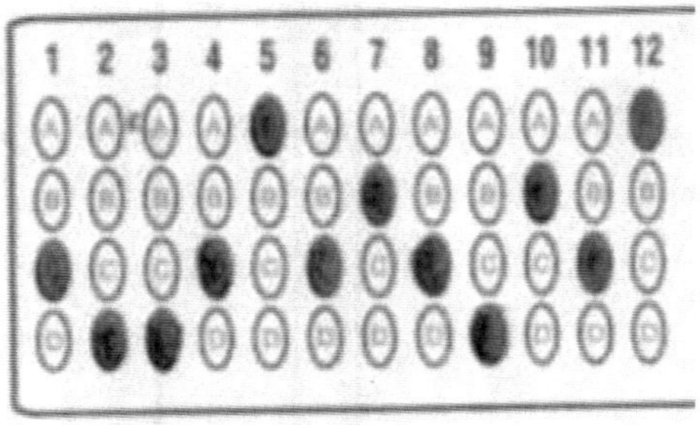

Figure 2. *An answer sheet fragment.*

Reading process is part of the information system but it is done before the qualification process. The most important factor for the correct reading of the leaves is the filling dark' level and background.

4.1. Generated reports

As mentioned, the efficient and timely management of information is a very important factor for decision making and the fulfillment of goals in any organization. In educational institutions, whose purpose is to carry out efficient processes that promote learning and the development of skills, this process takes on more weight.

However, one drawback faced by institutions of this nature is the amount of information that needs to be stored daily, as a result of the application of different tests, especially when the number of students increases. In addition, it is necessary to obtain the results in a short time, in order to achieve improvement plans. In this case, the use of an Academic Information System is an excellent option, and if it is tailored to the institution, much better.

In the present project, a set of reports were developed, some focused-on students and others at the institution.

4.1.1 Individual reports. Figure 3 shows the main data included in this report.

Personal information	General information	Specific information
• Full names • School • ID number • Grade	• Average • Standard deviation • Position in the group	• Average by subject • Performance level achieved in each competence. • Performance level achieved in each theme.

Figure 3. Main data included in the individual report

IOP Conf. Series: Materials Science and Engineering **519** (2019) 012024 doi:10.1088/1757-899X/519/1/012024

In addition to the numerical or qualitative data (performance categories), it was found that a very influential factor to understand the results is to indicate the weaknesses detected in each evaluate the subject.

4.1.2 Group reports. Table 3 shows the different group reports generated in the system and the variables that were considered.

Table 3. Group reports generated in SIAGREC.

Group Report	Variables
List of individual score	Full names, ID number, school, score by subject. Total Average and total standard deviation. Scores are organized from highest to lowest to appreciate the two extremes of the data set.
Performance level list by competency	Full names, ID number, school, performance level by competency.
Performance level list by theme	Full names, ID number, school, performance level by level.
Percentage of correct answers by the question at the group level	The number of each question, subject, the percentage of students that selected each answer option, correct answer, subject evaluated by each question, the percentage of correct answers. This report shows the tendency of success per question and highlights the weaknesses.
Graphic statistical report	Present previous group information but graphically, adding comparisons between subjects, identify the maximum and minimum scores and the performance scenarios relative to the average and standard deviation.

4.1.3 Observed results in the implementation of the SIAGREC system.
The educational institutions that have applied qualified evaluations with the SIAGREC system have obtained outstanding information on the performance of their students, detecting, among other situations, the topics that are most difficult for the students and where it is necessary to reinforce at the group and individual level, the most outstanding students and those who require more support, a continuous record over time that allows them to follow up their processes and develop improvement plans. In addition, they receive specific recommendations on the performance of the competencies by subject. The sum of all information has allowed, over the last 9 years, to feed, improve and adapt the system to align the evaluation process with the standards given by the National Education Ministry.

In terms of the use of the SIAGREC application by the person who performs the qualifying process, an improvement in processing times was achieved. Before implementing the sheet scanning function, an average time of 3 hours and 30 minutes was used to process the tests of 30 students (standard group), measured from the receipt of the sheet pack until they were ready by printing. After the implementation of this function (scanning), the same process was finished in 15 minutes on average. If you consider that, on a date with a large volume of exams, you can get up to 90 groups, it is a considerable saving.

In addition, with the current government policies, the number of grades has been increased in which state assessments are made (previously only eleventh grade and professional students were evaluated, odd grades, technical career students, technology students and professionals are currently being evaluated). This implies that the institutions need to apply pilot evaluations, to prepare themselves, measure their performance and, if necessary, carry out improvement plans.

Finally, we can highlight the improvement in terms of points obtained in the state tests of supported institutions. Although, the results do not only depend on the evaluation process carried out by the application, the opportunity to carry out the improvement plans on time and the given recommendations on specific topics show that they represent a considerable help.

5. Conclusion

Academic Information Systems are a great tool for educational institutions, who can store, process and display relevant information to make decisions about an optimal teaching process, adapted to each group of students, according to their particular needs.

Grouping the information by specific topics and by competence allows identifying with greater ease and dynamism the performance of the students. Besides, with the help of technology it is possible to analyze this information easily and generate different reporting models, in different formats (graphs, qualitative and numerical), through an application that possesses the main attributes of an information system, such as usability, efficiency, reliability, security, maintainability, among others.

References

[1] J V Fonou Dombeu and R Kwuimi 2018 *Afr. J. Inf. Syst.* **10** pp 191-204
[2] K C Laudon and J P Laudon 2014 *Management Information Systems* Edinburgh: Pearson p 645
[3] V R Palilingan and J R Batmetan 2018 *IOP Conf. Ser.: Mater. Sci. Eng.* **306** 012110
[4] E Dwi Hatmanto 2017 *An Evaluation of Competency-based Curricula Implemented in the English Language Education Programs in Five Universities in Indonesia* Melbourne: RMIT University p 411
[5] H P Utomo, A T Bon and M Hendayun 2017 *IOP Conf. Ser.: Mater. Sci. Eng.* **226** 012190
[6] J I Molina and E J Malaikosa 2018 *e-J. Univ.Tribuana Kalabahi* **1** pp 149-62
[7] M Tahvildarzadeh, H Moghaddasi and M Hosseini 2017 *Res.Dev. Med. Educ.* **6** pp 3-11
[8] I A Brahmasari and H Panjaitan 2017 *Arch. Bus. Res.* **5** pp 63-70
[9] F J B Talirongan and A A Hernandez 2017 Issues and Challenges on Academic Information Systems: A Quantitative Study Utilizing Purdue Usability Testing Questionnaire *IEEE 9th Int. Conf.Humanoid Nanotechnol. Inf. Technol.Commun. Control Environ. Manage. HNICEM (2017)*
[10] B Szabó and K Hercegfi 2017 Research questions on integrating user experience approaches into software development processes *IEEE 8th Int. Conf. Cognit. Infocommun. CogInfoCom (2017)* Debrecen
[11] M M Salama 2018 Arquitectural Stability of Self-Adaptative Software Systems. University of Birmingham p299
[12] H J Lee and K L A Yau 2015 *J. Comput. Inf. Syst.* **55** pp 40-47
[13] H Anjum, M I Babar, M Jehanzeb, M Khan, S Chaudhry, S Sultana, Z Shahid, F Zeshan and S N Bhatti 2017 *Int. J. Adv. Comput. Sci. Appl.* **8** pp 249-55
[14] S A Gebregiorgis and J Altmann 2015 *Procedia Comput. Sci.* **68** pp 173-87
[15] ICFES 2013 Alineación del examen Saber 11°. Bogotá p 130
[16] M Sotarauta and T Heinonen 2016 *Triple Helix.* **3** pp 8-27
[17] M Tul, B Leskošek, G Jurak and M Kovac 2015 *HU J. Edu.* **30** pp 268-81
[18] C Langa 2015 *Procedia Social Behav. Sci.* **180** pp 7-12

Design of a web application for the registration and control of the research projects at CORPOSUCRE

C M Pineda-Pertuz[1], C A Mercado-Montes[1] and A D Morales-Acosta[1]

[1]Corporación Universitaria Antonio José de Sucre, Sincelejo, Colombia

carlos_pineda@corposucre.edu.co

Abstract. This project aims to support the University Corporation Antonio José de Sucre-Corposucre, because it will develop an application to keep a record and control of the different documents generated from the projects, emerged of the academic activities carried out in the subjects with an investigative approach, within the Faculty of Engineering of the Institution. This project began with a stage of analysis of the current situation of the faculty of engineering science of Corposucre, in the way of carried out the storage of research projects, which is done in magnetic devices without any backup and control, which can generate deterioration or loss of information. It is important to highlight that with this work the programming methodology Extreme Programming or XP was used with the objective of giving a solution in the less possible time, maintaining a permanent collaboration among all the team members. The result of this work will be a web-based software, which will allow to digitally storing the information of all those research projects that have been developed by the students throughout their professional careers, such as Classroom projects, degree thesis, among others. This in order not only to help its preservation and prevent its deterioration, but also to constitute a source of valuable information for other students and/or teachers who see on this platform a means to draw on, when they need to support their researchs or some work where they are participating. This coupled with this tool will give more visibility to the investigative work of our academic community.

1. Introduction

Information is the basis of all activities carried out in organizations, so it is convenient to design systems to produce and manage it in order to ensure that it is reliable, accurate and available at the right time and the best decisions can be made.

Computer systems have been becoming a fundamental element for all organizations in different sectors of society. These allow to carry out processes more quickly, controlled and organized, improving the levels of productivity of the employees and the competitiveness of the same companies, thanks to the effective treatment of information.

These modern systems involve the intensive use of information and communication technologies (ICT) to capture, process, and distribute data and information throughout the organization. ICT are a set of hardware and software elements and communications applied to the automatic processing of this input, which is as important for organizations as information is.

Content from this work may be used under the terms of the Creative Commons Attribution 3.0 licence. Any further distribution of this work must maintain attribution to the author(s) and the title of the work, journal citation and DOI.
Published under licence by IOP Publishing Ltd

Consequently, an information system will be effective if it provides the necessary information for the organization and do so at the appropriate time, and it will be efficient if it is carried out with the least possible technological, human, temporary and economic resources [1].

All the above is framed within what is known as the new information society, where the automation represents one of its fundamental pillars, radically changing the nature of work, the number of jobs, working conditions and even the organizational model. Embodying more power in the capacity to store, manage, distribute and create information.

The education sector does not escape from this reality and this is why many higher education institutions in Colombia make use of these systems to support their administrative and decision-making processes. According to the Ministry of National education, Strong and reliable information systems contribute to the transformation and modernization of the sector, and are a strategy within the efficiency policy drawn up in the National Government's educational revolution.

Contextualizing a bit in the University Corporation Antonio José de Sucre, although this has many technological tools that serve as support to many internal processes both academic and administrative does not have an application for the registration and control of the research projects developed by the students of the faculty of engineering, that allows to carry out a detailed follow-up of different phases of this process: inscription, reviews, dates of delivery, among others. Currently this work is done in a manual way making it difficult to control and follow up the work done in this regard by students and teachers.

The importance of systematizing the registration and control of research projects lies in the possibility of organizing and consolidating information in order to promote the search for it and thus facilitate the development of successive studies by the community student. In addition, it will allow to conserve digital resources: documents, source files, programs, etc. Thus avoiding its deterioration or loss.

This is how this technological product will be useful because it will significantly reduce the expenses of these projects because it is cheaper to store information in digital format than on paper. It'll improve and expand the information offer to the student population, eliminate temporary barriers between support material and students, favour independent learning and self-learning, and offer new possibilities for the guidance and tutoring of students.

2. Research and development methodology

This research is based on a technological development with a non-experimental qualitative approach, in which it seeks to solve the need of the Faculty of Engineering Sciences of the Antonio José de Sucre University Corporation - CORPOSUCRE in the organization and systematization of its scientific production at Through the proposal of a virtual repository, as the first stage of the research was designed through the XP programming methodology (Xtreme Programming), so when gathering the user stories of the different dependencies that generate scientific content, the platform that could solve the need that the organization has. The user stories act as an open questionnaire that establishes the requirements that the user needs to have the software to meet their needs, so it is used as a qualitative information collection tool. In the future it is expected to carry out a second stage where the repository is implemented to take it to a research with a quantitative approach.

There are numerous methodological proposals that affect different dimensions of the development process. On the one hand, it gives credence to those more traditional proposals that focus especially on the control of the process, rigorously establishing the activities involved, the artefacts to be produced and the tools and notations to be used [2].

For the process of accomplishment of this project we used the Agile development methodology, motivated by the idea of obtaining a quality product in the shortest time possible, giving priority among other things to a good and fluent communication with the customer and a notorious emphasis on delivering visible increments of software rather than handling excessive documentation during the engineering process.

The following illustration shows the most important phases and activities of the XP methodology. This approach is showing its effectiveness in projects with highly changing requirements and when it is demanded to drastically reduce development times, while ensuring high quality in the conception of software architecture [3].

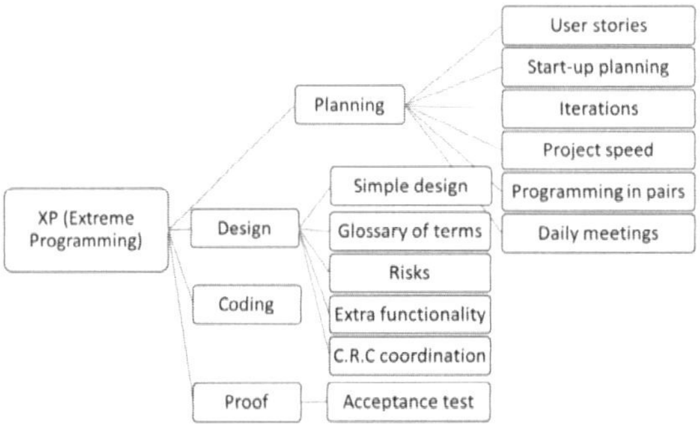

Figure 1. Structure of the XP methodology [4].

The project is divided into several stages, where the first one consisted in the analysis of the requirements of the dependencies of research and social projection of the corporation's engineering program, where the bibliographic research content is produced, so by the. So by the analysis of the requirements using the user stories as instrument of data collection in this phase, a mass of user stories were obtained, where the requirements that would have to comply the software were embodied. Such histories were categorized and assigned to them: a priority based on the importance involved and also an estimated time of duration based on the effort required to be developed. Then a series of iterations were raised, which would include a group of user histories, which should be developed, reviewed and integrated with the rest of the system, proceeding later with the application of the tests of usability necessary for the evaluation of the application, the work team can take note of the possible errors and difficulties encountered by users.

Then a series of iterations were considered, understanding these as the design process of each of the requirements as the platform develops. The iterations would include a group of user stories, which had to be developed, revised and integrated with the rest of the system, proceeding later with the application of the necessary usability tests so that the application was evaluated, being able the work team to take note of the possible errors and difficulties with which the users were found.

As a result of these iterations are obtained the different deliverables that are perceived as increments or advances in the final Product. Computer systems have been becoming a fundamental element for all organizations in different sectors of Society. These allow to carry out processes more quickly, controlled and organized, improving the levels of productivity of the employees and the competitiveness of the same companies.

The education sector does not escape from this reality and this is why many higher education institutions in Colombia make use of these systems to support their administrative and decision-making processes.

3. Proposed solution

The Web application for the registration and control of Corposucre's research projects used free software, highlighting XAMPP as a development package, which incorporates among its tools: PHP, Mysql and Apache.

3.1. PHP

It is a server-side programming language that arises within the current called Open Source. It is characterized by its versatility, robustness and Modularity. As with other similar languages, programs are integrated directly into the HTML Code [5].

3.2. Mysql

It is a relational database management system, being used by many large and popular Websites. Used to create the database, along with tables, where useful information is stored for Users.

3.3. Apache

Open source and Cross-platform HTTP Web server, has wide acceptance on the web and is used by a number of programmers around the world.

These programs are oriented to the development of applications based on the client server Architecture. In which the client is the software that makes the requests to the server, which in this case includes the resources like web pages, files etc. The server is responsible for answering the requests made by the Clients. The latter access the services provided by the server through remote procedure calls using a request-response protocol, such as the HTTP protocol on the WWW [6].

On the other hand, it is worth mentioning that for the development of this project has been used a Web development methodology with access to data base called Model Vista Controller (MVC), whose model represents a software architecture that divides the code in Three components: (i) Model: encapsulates a set of functions that work directly with a database, so that it allows its invocation from other parts of the Application. (ii) view: A view encodes and preserves the final presentation to the user, i.e. it allows to present information and outputs to the user in some defined format, not only in web format; (iii) Controller: A controller is responsible for linking models and views, as well as other modules required for the generation of Results [7].

The following figure illustrates the operation of the MVC Model.

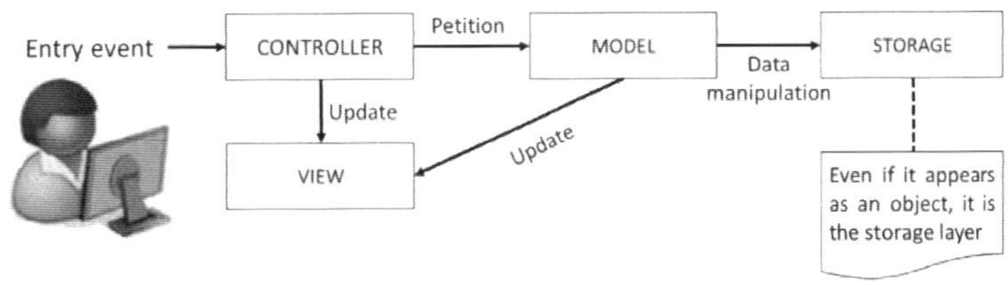

Figure 2. MVC architecture

The flow following the pattern of the MVC architecture is usually as follows:

- The user interacts with the user interface, for example, by pressing a link. This point corresponds, in the pattern, to the VIEW.
- The controller receives (through the INTERFACE) notification of the action requested by the user; That is, the handler manages the event that arrives from the view, produced by a user.
- The controller accesses the model, either in order to query or update data, possibly modifying appropriate to the action requested by the User.
- The handler delegates, to the objects in the view, the task of deploying the user Interface. The view gets its data from the model to generate the appropriate interface for the user, reflecting the changes in this. The model should not have direct knowledge about the VIEW.
- The user interface expects new Interactions.

4. Results

The purpose of this project is to develop a web platform for the registration and control of Corposucre's research projects. By means of this proposal, it is sought that the information is organized to avoid its loss, improve searches and disseminate the information of the projects that are leading in research of the institution.

IOP Conf. Series: Materials Science and Engineering **519** (2019) 012025 doi:10.1088/1757-899X/519/1/012025

Initially, the design of a relationship entity model was carried out, in which the logical architecture of the database is captured, through the identification of a set of objects called entities with their respective attributes or characteristics and the relationships that they exist between them. This modeling process performed with the relationship entity model will allow us to perform the abstraction of the most important data of the system in a simplified and organized way.

The diagram that results from applying the model entity relationship is as follows:

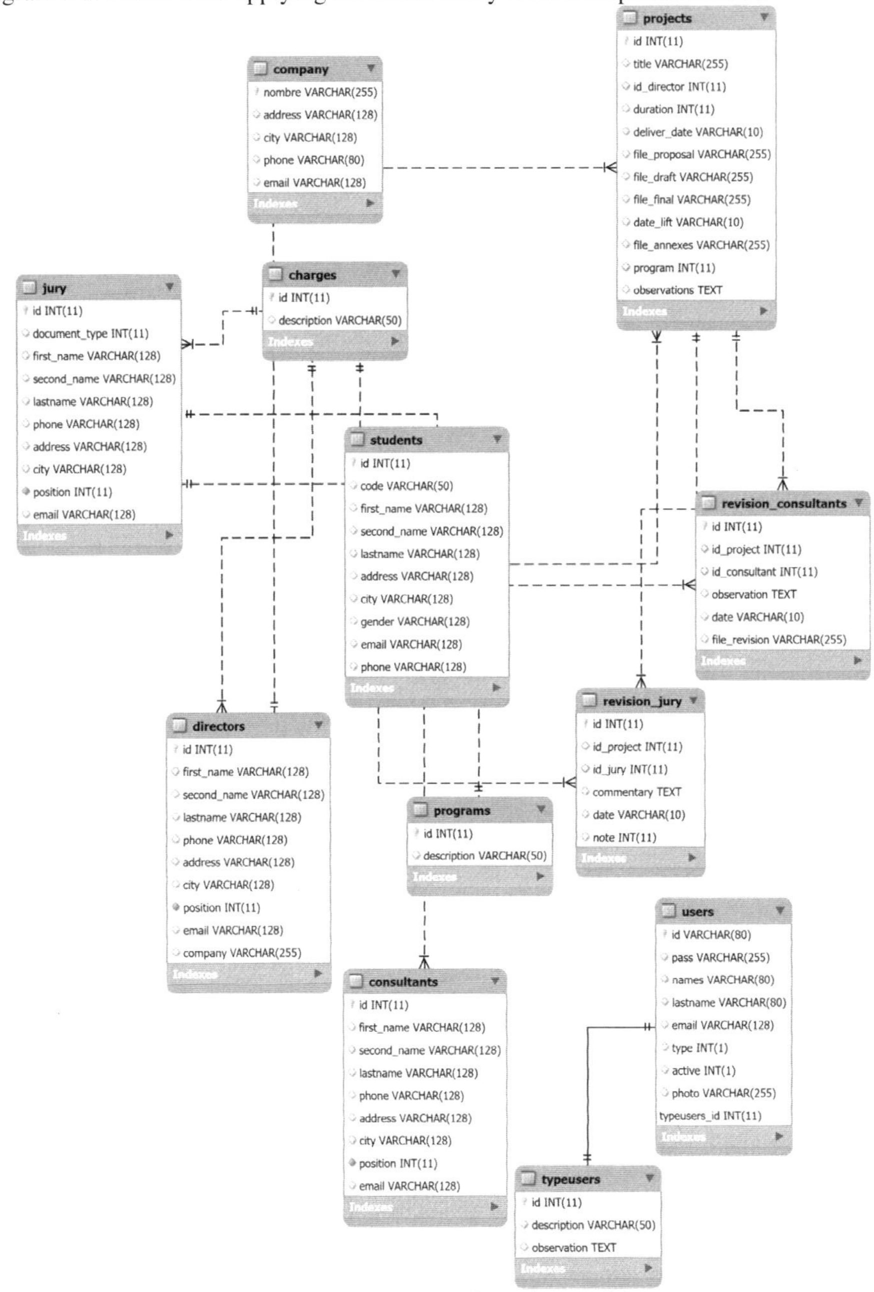

Figure 3. Relationship entity model

In the following images, the application in operation is Evident:

a) Login interface: This website is designed to guarantee secure access to the software, allowing only authorized persons can access to the options offered by the system.

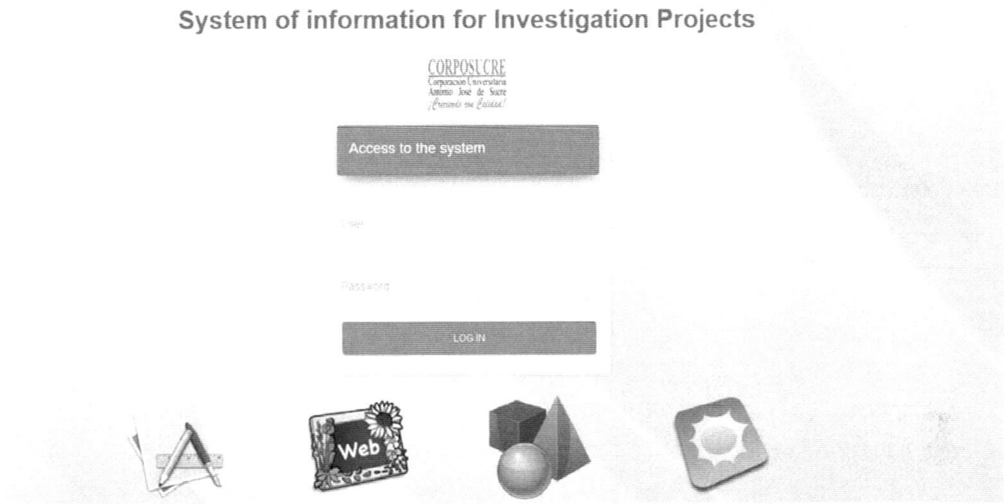

Figure 4. User login

b) Main interface: Provides the user with access to the different modules offered by the program such as: students, agenda, projects, teachers, reviews, among others. Among other things, it also shows a list of the last loaded projects and according to the privileges that the user has, it lets perform different management actions on them.

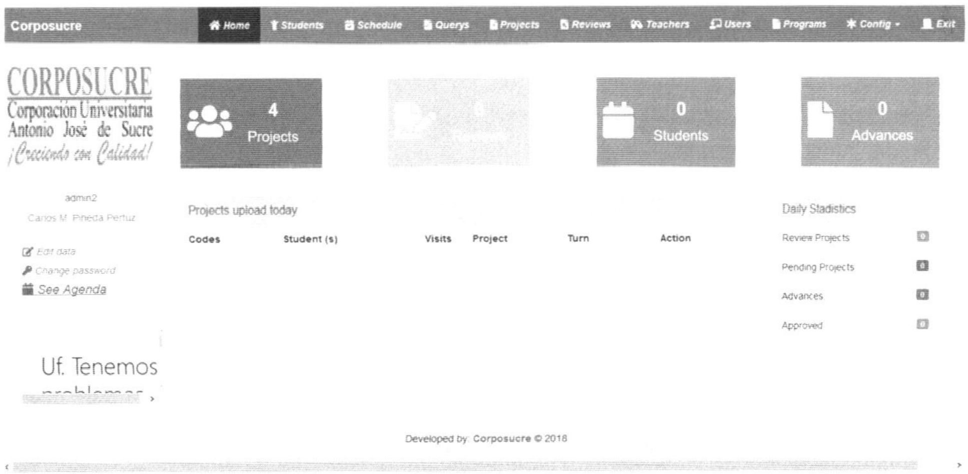

Figure 5. Main interface

c) User management interface: This page is used to register new users to the system, as well as delete, modify and consult existing ones.

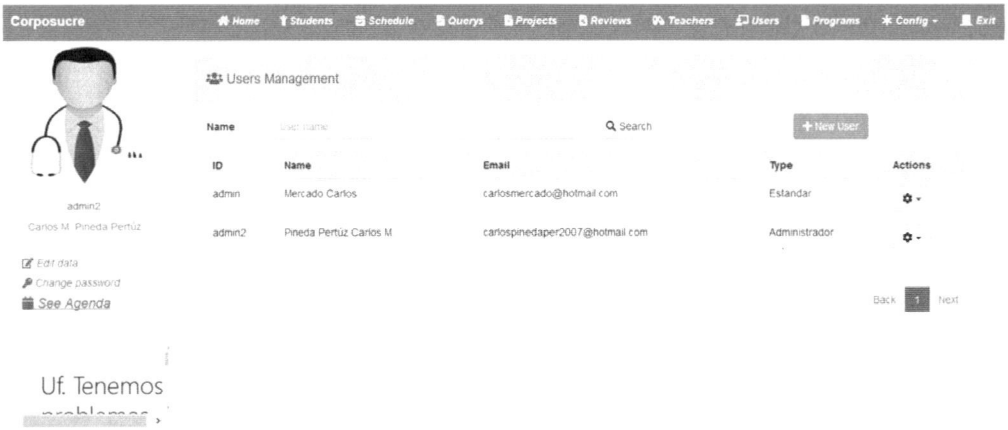

Figure 6. User Signup

d) Project management interface: With this interface the user can perform different actions on the projects registered in the database, such as: make revision, modify, consult, delete among others.

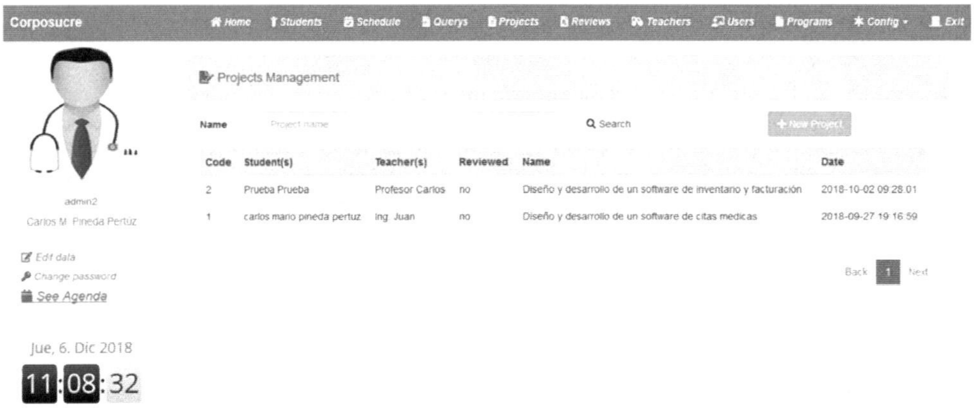

Figure 7. Project Consulting

5. Conclusions

Based on the specific objectives of this research work, the following can be concluded:

Identified the needs presented in the Faculty of Engineering from the collection, analysis and selection of useful information, process that it allowed to understand the current situation of the institution and then propose an optimal solution.

With the implementation of this IT solution will be able to improve the processes of registration and control of the research projects, by making that these are done in a more efficient and efficient way, giving place to that the operations are faster, reliable and Easy to Execute.

This Project has enabled the Agile XP development methodology, something that has turned out to be a novelty for the majority of the members of the work team, since until now they had only faced projects using only the methodologies Traditional.

As future work will be sought to promote the use of this application in other institutional programs, taking into account the positive impact within the faculty of Engineering.

References

[1] De Pablos C López H 2011 Organizacion y transformación de los sistemas de información en las Empresas Madrid: ESIC.

[2] D Wells 2013 Extreme Programming [En línea]. Available: http://www.extremeprogramming.org/.

[3] J Joskowicz 2008 Reglas y Prácticas de eXtreme programming». [En línea]. Available: https://iie.fing.edu.uy/~josej/docs/XP%20-%20Jose%20Joskowicz.pdf.

[4] A jaramillo 2013 Construcción de un aplicativo para el control de activos tenológicos al interior de una organización: beneficios y experiencias, *Lampsakos*, nº 10.

[5] C A 2005 PHP y Mysql: Tecnologías para el desarrollo de aplicaciones web.

[6] I Somerville 2011 Ingeniería del software, Person Education.

[7] Glasserman Morales 2017 Repositorios como soportes para diseminar experiencias de innovación educativa. Síntesis.

[8] Villagrá-Arnedo, Carlos-José 2017 Uso de la herramienta GitHub en la gestión y monitorización de proyectos ABP en cuarto curso del Grado en Ingeniería Multimedia.

[9] Ochoa Agüero, Alexander, Dalier Pérez Águila, and Yali Bles Portú 2015 Repositorio Institucional de contenido educativo de la Escuela Latinoamericana de Medicina Panorama Cuba y Salud 10.2: 2-9.

[10] Villarreal Gonzalo Luján 2017 Interoperabilidad con repositorios digitales: uso de OpenSearch en sitios web institucionales *Conferencia Internacional sobre Bibliotecas y Repositorios Digitales de América Latina (BIREDIAL-ISTEC'17) y Simposio Internacional de Biblioteca Digitales* (SIBD'17) Vol. 7.

[11] López Fernando Ariel 2017 Plan de gestión de datos de investigación: una propuesta argentina del CAICYT-CONICET *VII Conferencia Internacional sobre Bibliotecas y Repositorios Digitales de América Latina (BIREDIAL-ISTEC'17) y XII Simposio Internacional de Biblioteca Digitales (SIBD'17).*

[12] Lahera Yohannis Martí, Grisel del Toro Iglesias, and Yaneivys Gutiérrez Medina *2017 Visibilidad y accesibilidad al patrimonio documental de la Universidad de La Habana: repositorio institucional SCRIPTORIUM* Bibliotecas. Anales de Investigación 12.1: 26-36.

[13] Eléspuru Briceño, Liliana, and Libio Huaroto 2016 Los repositorios institucionales como herramientas para medir los indicadores Altmetrics: experiencia de la Universidad Peruana de Ciencias Aplicadas *(UPC)."* .

[14] Campos, Saúl González, and Luis Felipe Fernández Martínez 2015 Programación Extrema: Prácticas, Aceptación y Controversia CULCyT 15.

[15] Figueroa Roberth G, Camilo J Solís, and Armando A Cabrera 2008 Metodologías tradicionales vs. metodologías ágiles *Universidad Técnica Particular de Loja, Escuela de Ciencias de la Computación.*

Evaluation of the learning strategies of the 10th-grade students in a private school of Bucaramanga

E P Carvajal Valero[1], D M Pabon Gelves[1], N T Castillo[1] and A Tello Rios[1]

[1]Unidades Tecnológicas de Santander - UTS, Colombia

ecarvajal@correo.uts.edu.co

Abstract. This investigation presents the results of the evaluation of the learning strategies of 10th-grade students, to focalize the factors involved in their school learning process. The employed design in this investigation is Quantitative, cross-sectional descriptive. The sample consisted of 30 10th-grade students, to whom the test, Diagnostic Questionnaire for Learning Strategies-CEDEA, a psychoeducative questionnaire (pilotage and validation by experts), and the informed consent to the students and parents. For the questionnaire interpretation, it was taken into account that the scores higher than 60, are considered like strengths and 20 like weak strategies, it was made a relation of the psychoeducative characteristics with the CEDEA variables. The results reveal that old and new students, no matter if they are man or woman, the scores are weak (under 20). In metacognitive-evaluative (Evaluation, planning and self- resources valuation) strategies, information processing strategies (Transference, Memorization, comprehensive, information organization, information acquirement and elaboration), handle of the resources strategies (teamwork and interaction, context control and resources) and dispositional strategies (Perceived self-efficiency, Attention-Active disposition, fitness and mood and anxiety control). In consequence, the data point that the students have a lack of strength in their studying habits, auto regulation in their learning and in the significant learning.

1. Introduction

In the test "Saber 11" (Colombian high school test), the Colombian Institute for higher education promotion ("Instituto Colombiano para el Fomento de la Educación Superior") ICFES. Performs a schools classification in the six categories: A+, A, B, C and D. Which in the schools are classified, and the range of values of the general results index that each of them comprises. The Educative Institution objective of the study, compared with the other educational establishments of Colombia, is on the 2.413 positions, with B qualification; regarding the private education in national level in the place 1718 - qualification B; and in relation with the Santander the place 168- qualification B; in the Bucaramanga´s school the place 52- qualification B, and in the Bucaramanga private schools the place 32, qualification B. Taking into account the situation, and the new educative proposal of the institution, that seeks permanent management, effective and efficient in order to strengthen teachers as leaders of change, and the innovation in their pedagogic procedures, where they develop a series of activities with the aim of enhancing the abilities of students in the direction of meaningful and autoregulated learning. It was desired to evaluate the learning strategies used by tenth-grade students, whose result could help to improve the results on the test "Saber 11" and representative categories of the institution on the school qualification ICFES [1] . With the application of diagnostic questionnaires of learning strategies-CEDEA- in the 10th-grade students, it looks to enhance the role of the teacher from the promotion of

Content from this work may be used under the terms of the Creative Commons Attribution 3.0 licence. Any further distribution of this work must maintain attribution to the author(s) and the title of the work, journal citation and DOI.
Published under licence by IOP Publishing Ltd

meaningful learning and mediation to guide constructivist activities of students. In this way, is provided to the educative system by preparing the students in strategic individuals, autonomous and independent, in the development of the of optimal learning strategies, since the students will have the conscience of what they know and what they don´t know, evaluate what they remember and what they don´t, planning on a systematic way for a determined situation, select appropriate strategies for each situation, identifying if they´re beneficial, modify or replace the strategies for others and lead the research of relevant information in the memory.

The Self-regulation is "Process formed by self-generated thoughts, emotions, and actions that are planned and adapted cyclically to achieve the personal goals" by Zimmerman, [2] . In this way, the self-regulation refers to learning strategies that the students activate when they´re working to achieve the goals that they have set for themselves.

In accordance with Gonzales Fernandez [3] for Zimmerman (1989, 1995) [4] , the students can be considered self-regulated, in the way they're -from a metacognitive, motivational and behavioral point of view- an active participant in their own learning process. It is assumed that the self- regulation in the scholar learning, is an activity that can be trained on the students, achieving they're capable to learn with a superior grade of self-regulation.

Initially the "Cognition control", is the cognitive compound of the self-regulation, also know like metacognition, which deals with the degree of awareness or knowledge of students about their ways of thinking, it covers knowledge, control, and regulation of cognitive processes, content and skills to control those processes in order to organize, review and modify them according to progress and learning outcomes Crus Gutierrez [5] , and that stops Schraw & Moshman , metacognitive activities have three dimensions: Plan, Monitor, and Evaluate.

It implies being aware of the strengths and weaknesses of our own intellectual functioning, and the types of reasoning errors that we usually commit, such awareness would help us to explore our strengths, compensate for our weakness and avoid making common mistakes. The authors Panadero & Tapia, [6] , in their article, explain the phases, process, and acquisition of the self- regulation based on the socio-cognitive model of It implies being aware of the strengths and weaknesses of our own intellectual functioning, and the types of reasoning errors that we usually commit, such awareness would help us to explore our strengths, compensate for our weakness and avoid making common mistakes. The authors Panadero & Tapia [6] , in their article, explain the phases, process, and acquisition of the self-regulation based on the socio-cognitive model of Zimmerman & Moylan [7] , and describe the 3 presented phases: Planning, execution, and self- reflexing.

In the planning phase, is about the initial phase where the student face, for the first time to the task, the student analyze the task, value their capacity to complete it, establish goals and plan. The interest for the task and the motivational orientation, take a crucial role to achieve a good planning and realize the activity adequately. According to Zimmerman and Maoylan [7] aforementioned for Panadero and Tapia [8] , the self-regulation process begins with the analysis of the task where the task to be performed is broken into smaller elements and based on the previous knowledge, a personal strategy is established for its execution By Winne, Self-regulation of learning seen from the models of information processing, [9] . The student establishes his objectives based on the evaluation criteria, which correspond to the standards under which the activity will be evaluated and the level of perfection, is what the student wants to accomplish in the task.

The self-motivating beliefs are also discussed in the planning phase, dealing with values, interest and goals, personal variables that generate and sustain the motivation to carry out an activity. Interact with each other to result in the level and type of motivation that the student will have during the task By Panadero y Tapia [8] .

In the execution phase, in accordance with Panadero and Tapias [8] , it is important that the student maintain concentration and use appropriate learning strategies for two reasons: first, so that they do not diminish their interest and motivation and, second, to reach the learning objectives.

Now in the last phase, self-reflection, according to Panadero & Alonso-Tapia, Theories of educational self-regulation: a comparison and theoretical reflection [10]. The student values his work and tries to explain the reasons for the results obtained that in doing so justifies the causes of his success or failure and, depending on his attributional style, he experiences positive or negative emotions that can influence his motivation and ability to self-regulation in the future. In the research carried out by Elvira-Valdes & Pujol [11], with a sample of 172 volunteer students of both genders, students of the first trimester of the University Initiation Course of the Simon Bolivar University (Venezuela), it was determined that the sampled students present a moderate use of strategies of Academic Self-Regulation, Planning and Control Strategies and present greater use of Reflection strategies, indicating that there is a better execution of the tasks and prompt changes in achieving learning objectives.

Likewise, slight differences were obtained in the use of self-regulatory control strategies for students from Public Institutions with respect to Private Institutions, indicating that these students exercise a greater selection regarding the study environment, study time and effort in the tasks academic. There is also no significant correlation between college entrance scores and high school grades, demonstrating that previous academic performance could be associated with college success, as indicated by several studies where it is shown that if there are deficiencies in the strategies of academic self-regulation, the student may have an inadequate development during the first years of his university education.

This study carried out with high school students in Spain, showed that the subjects make an important use of reference strategies and there is a notable difference between the use they declare to be doing of each other strategies.However, the use of evaluation, writing, planning, regulation and repetition strategies can be appreciated By Del Caño Sanchez, Roman Sanchez, & Foces Gil [12] .

At a national level, the research of Acosta [13] realized out the study called "The learning processes and their incidence in the student desertion in the Pharmaceutical Chemistry Program of the University of Cartagena", within whose objectives was to evaluate the learning strategies employed by the students.

The results indicate that deficits in the learning strategies used are found in all the subjects evaluated and the author concludes that the problem of dropping out has a causal origin in the learning strategies used by the students by Acosta [13] .

2. Methods

The design of the research was cross-sectional descriptive, which responds to the objective of identifying the state of the learning strategies, in the 10th-grade students, showing how they were when the study collected the data. The study sample was formed with a group of 30 10-grade students, all the active students participated in the process, for which the following inclusion criteria were taken into account: Students enrolled and active in the first semester of 2017, old and new students were included. Students who decided to participate voluntarily, with signed informed consent from them and from parents who voluntarily accepted to participate in the research.

Two instruments were used to the data collection: Diagnostic Questionnaire for Learning Strategies-CEDEA, that measured the learning strategies of the students and a questionnaire, designed to identify the psychoeducational characteristics of the students. The Diagnostic Questionnaire for Learning Strategies-CEDEA is based on the fact that all human beings have a great capacity to learn, along with the intellectual potential that each individual develops, essential elements to solve any type of problem they face, so that the proper use of different learning strategies can achieve learning successful. Several studies in the field of cognitive psychology agree that learning strategies interventions, helps the cognitive learning and motivation to learn, likewise, motivational beliefs not only influence the inspiration to learn, but also in the form and quality in which students process the information, selecting and using certain learning strategies by Torrano Montalvo & Gonzalez Torres [14].

However, as indicated by Solis Uribe [15] , all people have the ability to learn, the difference lies in the way and effectiveness in which learning strategies are used, as they correspond to tools for thinking , are resources that help improve reasoning and especially enable the student to develop the ability to "learn to learn." With adequate training and selection of appropriate learning strategies, all students can improve their level of control over learning and performance, so those students with low academic performance can overcome the learning difficulties they present.

According to Gutierrez [5] cited by Solis Uribe [15] , the possibility of learning through learning strategies facilitates meaningful learning, as well as allowing students to establish relationships between what they already know (their own knowledge) and the new information (the objectives and characteristics of the task to be performed). In this way the student learns to use the necessary procedures at the time of studying, he will also know when and why he should use them, and how they will benefit if they use them.

With the instrument of the Diagnostic Questionnaire for Learning Strategies-CEDEA, it was sought to assess the existence and mastery of procedures, techniques, dispositions and other aspects that allow the student to be regularly successful in carrying out the task or in his study according to Solis Uribe [15].

According to what is explained in the technical file by Solis Uribe [15] , within Group A are the metacognitive - evaluative strategies whose function is to consciously control and evaluate the functions, processes and characteristics that influence learning together to the following strategies: Evaluation, Planning, Valuation of own resources.

Within Group B are the information processing strategies and are defined as all those that work with the information that is the subject of learning and are also composed of the following strategies: Transfer, Comprehensive Memorization, Organization of information, Acquisition of Information and Elaboration.

Within Group C are resource management strategies that include control over those external aspects located in the environment that influence learning and presents the following strategies: Group work and interaction, Context control and Resources.

Finally within Group D are the dispositional strategies that are all those strategies that do not manipulate directly with the learning material but that try to favor positive personal conditions of the student that favor learning, is composed of the following strategies : Perceived self-efficacy, Attention - active disposition, Physical and emotional state and Control of anxiety. In this questionnaire it was taken into account that the high scores higher than the 60th percentile considered as strengths that the subject possesses are desirable. When the score obtained is lower than the 20th percentile, it indicates a weakness. The average scores to give statistical significance are not considered.

Respect to the psychoeducational Questionnaire, this instrument was adapted for the adolescent sample that attends 10 baccalaureate degree. The questionnaire has as a background of validity and reliability, its previous use in university students. The questions of the questionnaire come from the instrument built by the psychologists PhD. Leonardo Ortega Murillo y Mg. Patricia D'ıazGordon about the motivation of students in the course of basic psychological processes of the psychology program of the UNAB, presented in a presentation of the National Node of psychological processes, sponsored by the Colombian Association of Faculties of Psychology in 2014. To this same instrument, some questions were modified for its application in an inter-university project with 700 students of psychology, belonging to 17 universities of the country with the purpose of determining psycho-educational characteristics of young people. The questionnaire was also used by Gómez and Arenas (2015), who investigated the learning strategies in a group of nursing students of the UNAB and used the questionnaire to identify the psychoeducational characteristics of these.

The adequacy of the questionnaire, with this sample of 10th-grade students, was made through the process of validity and reliability with pilot test to three 10th-grade students of a private school in Piedecuesta - Santander that was not part of the sample. Likewise, validation was also carried out by expert judges, the rector of the Educational Institution where the sample was applied, the academic coordinator and the psychologist of the same Institution. The application allowed to obtain reliable data, correct ambiguous data in order that the resulting questionnaire was clear and understandable for its later application in the students of the sample. To correct the validation by judges of the psychoeducational questionnaire, the Expert Judgment Template provided by Escobar Pérez & Cuervo

Martínez [16] was used.

The questionnaire is applied with the digital format in Google Forms, where information about the psychoeducational characteristics of 10th-grade students, data on institutional background, questions about academic support, study motivation and study habits was collected. The statistical analysis was descriptive, for the psychoeducational questionnaire through the data obtained through the Google Forms tool. While in the case of analysis on agreement of judges, the SPSS program was used for the Kendall index to establish correlations between them, "Offers the value that makes it possible to decide the level of agreement between the experts.

The value of W oscillates between 0 and 1. The value 1, means an agreement of total agreements and the value of 0, total disagreement, being 1 the desired score " By Cuesta Santos.

3. Results

(i) Sociodemographic characteristics in 10th-grade students.

In the Table 1 we can evidence regarding the distribution of ages there is no difference between the sexes, because 50% of the students (men and women) are in the age range of 14 to 15 years old, the other 50% make up (men and women) ages 16 and 17 years old. Of the 30 students who participated in the research, there was a high tendency of men 56.6% (23.3% between the range of 14 to 15 years old and 33.3% between the range of 16 to 17 years old) and women they make up 43.3% (26.7% between the range of 14 to 15 years old and 16.7% between the range of 16 to 17 years old).

Table 1. Age, Seniority and social stratum distrubution.

Sex	Age %		Seniority %		Stratum %			
	14 - 15	16 - 17	2013 - 2015	2016 - 2017	1	2	3	4
Male	23,3	33,3	23,3	33,3	3,3	3,3	16,7	33,3
Female	26,7	16,7	23,3	20,0	3,3	6,7	16,7	16,7
Total	50	50	46,6	53,3	6,6	10	33,4	50

Source: Recolected data

Regarding the years of seniority that the students have in the school object of study, showed that 53.3% of the students (men and women) are of recent entry (2016 - 2017), while the 46.6% are older. Regarding the socioeconomic strata, 6.6% of the students correspond to stratum 1 (3.3% men and 3.3% women), 10% to stratum 2 (3.3% men and 6.7% women)), 33.4% are from stratum 3 (16.7% men and 16.7% women), with 50% of students corresponding to stratum 4 (33.3% men and 16.7% women). In general, men predominate, recent students (2016 - 2017) and socioeconomic stratum 4.

In the sociodemographic data collected from the families of the students we verify that 40% of the students (men and women) do not have siblings, 36.7% have at least one sibling and 23.3% have 2 siblings. In relation to the people with whom the sample under study coexists, it can be seen that 63.3% of the students cohabit as a two-parent or nuclear family, being understood as those formed by parents and children, 3.3% of students live together only with the father, while 13.3% of the sample resides only with the mother. 6.6% of students live with relatives (uncles and / or grandparents), 10% of the sample live as extended family (father, mother and grandparents, mother and grandparents or mother, grandparents and aunt) and finally the 3 , 3% of the students live as a compound family (mother and stepfather).

Table 2. Number of siblings and people who lives with the student distribution.

Sex	Number of siblings %			Who lives with %					
	Only child	1	2	Nuclear Family	Single parent, Father family	Single parent, Mother family	Relatives	Extense family	Compund family
Male	23,3	20,0	13,3	40,0	3,3	3,3	3,3	3,3	3,3
Female	16,7	16,7	10,0	23,3	0,0	10,0	3,3	6,7	0,0
Total	40,0	36,7	23,3	63,3	3,3	13,3	6,6	10,0	3,3

Source: Recolected data

Table 3. Age and sex distribution for the parents of the students.

Sex	Age %					Householder %	
	30 – 35	36 – 41	42 – 47	48 – 52	> 53	Mother householder	Father householder
Male	1,7	11,7	16,7	13,3	6,7	0	1,7
Female	6,7	15,0	15,0	11,7	1,7	10	0
Total	8,3	26,7	31,7	25,0	8,3	10	1,7

Source: Own construction based on the recollected data.

We reviewed that there is a high population with 31.7% of the parents that are in the age range between 42 to 47 years old (16.7% men and 15% women). Corresponding to the adult phase late, within this same phase is the 25% who are in the range of 48 to 52 years old (13.3% men and 11.7% women) and finally 8.3% belonging to the parents of the students over 53 years old (6.7% men and 1.7% women). Within the intermediate adult phase correspond 26.7% that are within the range between 36 to 41 years old (11.7% men and 15% women) and 8.3% that are in the range of Age between 30 and 35 years old (1.7% men and 6.7% women). It can also be seen that there is a greater female population that is the head of the household (10%) compared to men, which corresponds to 1.7%.

Regarding the academic background of the parents, it was evident that 30% (16.7% men and 13.3% women) are professionals, 11.7% are technologists (3.3% men and 8.3% women), 13.3% are technicians (6.7% men and 6.7% women). The 1.7% that corresponds to the men are with incomplete university studies, the 1.7% corresponding to the women are studying university studies; 3.3% of parents are professionals with some type of specialization (1.7% men and 1.7% women) and 1.7% corresponding to women are professionals with a master's degree. On the other hand, 26.7% of parents studied until high school (15% men and 11.7% women), 1.7% corresponding to men did not finish high school; 6.7% studied up to fifth grade of primary school (1.7% men and 5% women) and 1.7% of men did not finish primary school.

Table 4. Parents scholarship distribution

Sex	Scolarship Incomplete Primary	Primary	Middle school	High school	Tecnic	Technologic	Professional	Incomplete university	Current university	Specialized professional	Master professional
Male	1,7	1,7	15,0	1,7	6,7	3,3	16,7	1,7	0,0	1,7	0,0
Female	0,0	5,0	11,7	0,0	6,7	8,3	13,3	0,0	1,7	1,7	1,7
Total	1,7	6,7	26,7	1,7	13,3	11,7	30,0	1,7	1,7	3,3	1,7

Source: Recolected data

Regarding the academic background of the parents, it was evident that 30% (16.7% men and 13.3% women) are professionals, 11.7% are technologists (3.3% men and 8.3% women), 13.3% are technicians (6.7% men and 6.7% women). The 1.7% that corresponds to the men are with incomplete university studies, the 1.7% corresponding to the women are studying university studies; 3.3% of parents are professionals with some type of specialization (1.7% men and 1.7% women) and 1.7% corresponding to women are professionals with a master's degree. On the other hand, 26.7% of parents studied until high school (15% men and 11.7% women), 1.7% corresponding to men did not finish high school; 6.7% studied up to fifth grade of primary school (1.7% men and 5% women) and 1.7% of men did not finish primary school.

It was also observed that 38.3% of parents are former employees (23.3% men and 15% women). 15% are freelancers (8.3% men and 6.7% women). 13.3% are merchants (10% men and 3.3% women). 10% are administrators (1.7% men and 8.3% women). The other 10% that correspond to women are housewives; 6.7% are drivers (5% men and 1.7% women); the 3.3% corresponding to women perform various jobs, within the population 1.7% is a pensioner own of women and only 1.7% is unemployed that corresponds to men.

Table 5. Parent's civil status.

Civil status	%
Single	3,3
Free union	11,7
Separates	8,3
Divorced	1,7
Married	23,3
New union	1,7

Source: Recolected data

There was a high tendency for parents of family in marital status corresponding to 23.3% of the population, 11.7% are living together in a free union, 8.3% are separated, 3.3% are consider singles, 1.7% are divorced and 1.7% are in a new union.

(ii) Psychoeducational characteristics in 10th-grade students.

The psychoeducational characteristics are composed of several categories such as: institutional background, academic support, study motivation and study habits.

IOP Conf. Series: Materials Science and Engineering **519** (2019) 012026 doi:10.1088/1757-899X/519/1/012026

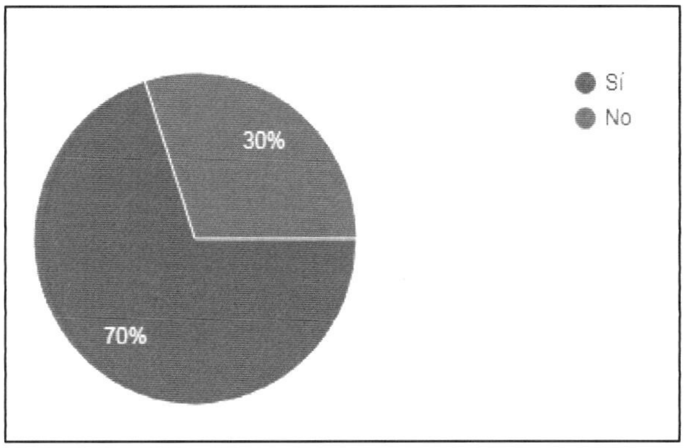

Figure 1. Institutional background

Institutional Background: It is noted that 70% of students come from another school within the baccalaureate, only 30% of the sample, comes from the same school object of study since he started his baccalaureate, that is, sixth grade.

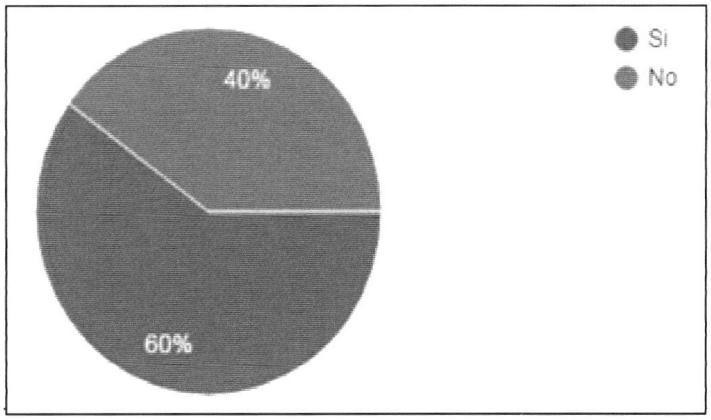

Figure 2. Academic support

Academic support: 60% of students receive some type of academic support to better understand the subjects that are difficult and have good academic performance, while 40% of the sample does not receive any academic support outside the school.

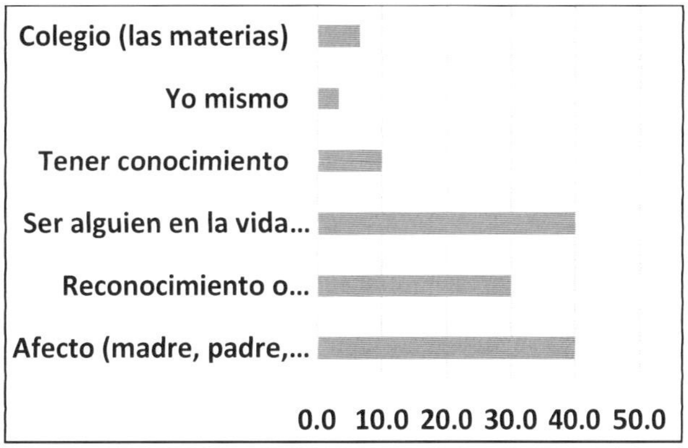

Figure 3. Motivation to study

IOP Conf. Series: Materials Science and Engineering **519** (2019) 012026 doi:10.1088/1757-899X/519/1/012026

Motivation to the study: what motivates to study at present in 40% of the students is the affective from their relatives (mother, father, brothers, girlfriend), also emphasizes that 40% motivates them to be someone in life (get ahead, become professional). 30% of students do it for recognition or rewards (of parents, prizes, money, satisfy tastes, etc.). 10% is motivated by the distinction of own knowledge. 6, 7% of the sample are stimulated by the materials of the Educational Institution and 3.3% are self-motivated.

Table 6. Pshycoeducative characteristics. Questions regarding studying habits.

Questions	Fully agree %	Agree %	Not agree, not disagree %	Disagree %	Fully disagree %
I usually prepare for the evaluations by studying with time	3,3	23,3	46,7	16,7	10
Normally when I do not understand what has been discussed in the course, I look for help	13,3	63,3	23,3	0	0
I tend to study regularly, not only when I have to prepare for an evaluation	3,3	13,3	43,3	26,7	13,3
When I get bad grades, I feel sad, but I continue studying	20	30	26,7	16,7	6,7
The readings that teachers put in their classes have a difficult vocabulary to understand	6,7	16,7	50	16,7	10
I find several interpretations to the same text within the readings proposed in the course	10	36,7	46,7	3,3	3,3
When doing the readings established in the course, I find that everything is important	10	50	23,3	16,7	0
In the development of a reading, I often fail to understand what certain parts of the text refer to	10	36,7	33,3	13,3	6,7
To prepare the subject assessments, it is preferable to rely on the material assigned by the teacher, which in the notes	23,3	30	36,7	6,7	3,3

Source: Results questionnaire psychoeducational characteristics

Study habits: It is evident that there is not much preparation for the evaluations since there are few young people who study regularly, not only for the evaluations as it happens with 3.3% completely agree and 13.3% agree . Otherwise is observed on the majority of the students, that they do not usually study prior to the evaluations, as it happens with the 43.3% that do not assume position in front of it, (neither agreement, nor in disagreement); 26.7% disagreed and 13.3% completely disagree.

Table 7. Most used studying form

Read (High Voice)	10,0
Read and repeat	3,3
Read and analyze exercices	3,3
Read and research	6,7
Read and write	10,0
Summarys	3,3
Draws	6,7
Concept map	3,3
Underlined	3,3
Video	3,3
Internet	26,7
Seek in Relatives, teachers among others	13,3
Take notes	3,3
Transcribe	3,3
Memorize	3,3
Focus and practice	3,3
Silent study	3,3
Manual and theorical	3,3

Source: Results questionnaire psychoeducational characteristics

It indicates the form of study that students use most to learn the contents of the subjects. it is reflected that for the sample there is little search, investigation and deepening of the topics since 26.7% use the internet which is not a reliable source to expand or understand the topics, 13.3% support or resolve doubts with family, friends, teachers or colleagues, 10% read and write as a form of study, 10% are facilitated to read aloud to study and learn, 6.7% read and investigate, 6.7% make drawings as a form of study and for the rest of the sample statistically distributed equally (3.3%) use the following forms of study: reading and Repeat, read and perform exercises, perform summaries, conceptual maps, underlining, use of videos, take notes, transcribe, memorize, concentrate and practice, study in silence without any noise finally also study in a theoretical and manual.

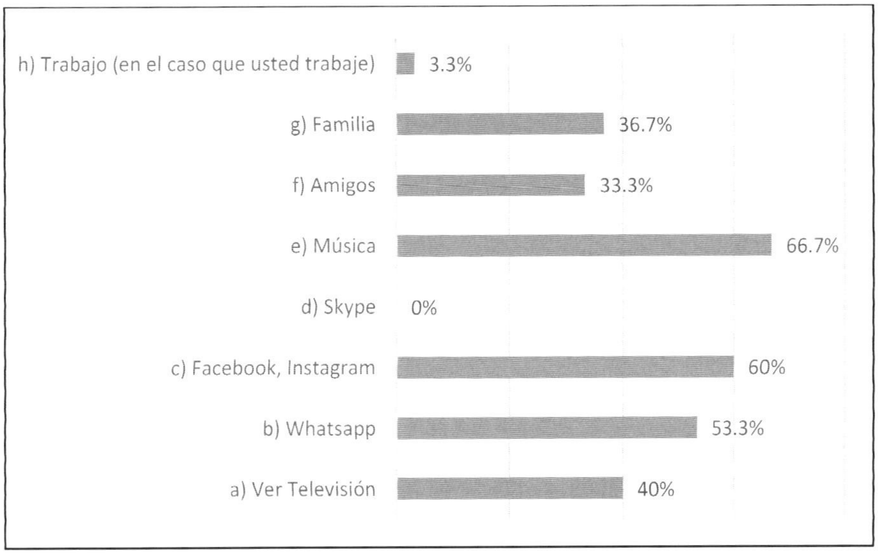

Figure 4. Most frequent activity while they are studying.

The study evidenced that 66.7% of students frequently listen to music when they are studying. 60% of the sample reviews social networks such as Facebook and Instagram, 53.3% reviews WhatsApp, 40% watches television while who studies, 33.3% speak or relate in some way with friends, 36.7% interact with the family while they are studying and 3.3% of the students work at the same time they study.

4. Diagnostic Questionnaire for Learning Strategies-CEDEA

With the test, we evaluated four strategy factors: Metacognitive - evaluative strategies, information processing strategies, resource management strategies and dispositional strategies. The metacognitive - evaluative strategies contain three subfactors such as: evaluation, planning and valuation of own resources. The Information Processing Strategies includes five Subfactors: transfer, comprehensive memorization, organization of information, acquisition of information and elaboration. Resource Management Strategies dominates three Subfactors: group work and interaction, control of context and resources; finally the dispositional Strategies possesses four Subfactors: Perceived self-efficiency, attention - active disposition, physical and mental state and finally control of the anxiety.

For proper questionnaire interpretation, it was taken into account that the high scores higher than the 60 percentile considered as strengths that the subject possesses are desirable. When the score obtained is lower than the 20th percentile, it indicates a weakness. The average scores to give statistical significance were not considered.

In the metacognitive - evaluative strategies, the strengths and weaknesses at the group level were analyzed, as well as between the sexes of the course and between the date of entry of the old students vs those of recent entry, with little strength and greater weakness in the factor being appreciated. Metacognitive Strategies-Evaluative of the sample, indicating that the strategies to control and consciously evaluate the functions, processes and characteristics that influence learning are not the domain in each of the students that show weakness, being few cases in the shows that they are aware to control their learning.

Table 8. Percentage distribution of the strengths and weaknesses that students present in the Metacognitive-Evaluative Strategies factor.

Subfactors	Percentage of students with scores under 20	Percentage of students with scores between 21-59	Percentage of students with scores over 60	Total
Evaluation	50%	23,3%	26,6%	100
Planification	66,6%	20,0%	13,3%	100
Valuation of resources	53,3%	20,0%	23,3%	100

Source: Results questionnaire learning strategies.

In relation to each subfactor, we reviewed the contribution of each one in the use of strategies for learning. The evaluation subfactor (EV) occurs as strength in 26.6% of the opposite sample with the 50% having it weakly. The Planning Subfactor (PL) is a strength for 13.3% and weak for 66.6%. The subfactor Valuation of resources (VR), becomes a strength only for 23.3%, and 53.3% has it weakly.

In the Information Processing Strategies it is evident that the students present greater weakness and insufficient strengths in the Information Processing strategies. Scores lower than 20
are higher than scores greater than 60 for the Subfactors. It is demonstrated that the strategies that work with the information that is object of learning, that is, those that are oriented to make analogies of knowledge, organize and structure the information for its comprehension, highlight the relevant information and integrate the contents with the techniques of information for learning are weak in

students who scored less than 20.

Table 9. Percentage distribution of the strengths and weaknesses that students present in the Information Processing factor

Subfactors	Percentage of students with scores under 20	Percentage of students with scores between 21-59	Percentage of students with scores over 60	Total
Transfer	63,3%	0%	36,6%	100
Comprehensive Memorization	40,0%	23,3%	36,6%	100
Organization of Information	56,7%	30,0%	13,3%	100
Acquisition of Information	50,0%	20,0%	30,0%	100
Elaboration	40,0%	26,6%	33,3%	100

Source: Results questionnaire learning strategies.

In the Table 9, Of the Subfactors characteristic of the Information Processing Strategies. The Subfactor Transfer (T), 63.3% of the students show weakness and 36.6% of the students present strength. That is to say that students with difficulty in the subfactor Transfer do not relate previous knowledge with new knowledge. The Comprehensive Memorization Subfactor (MC) is a weakness at 40% and strength at 36.6%. The Information Organization Subfactor (OI) occurs as a strength in 13.3% of the students, contrary to the 56.7% who present it as a weakness. In the Subfactor Acquisition of Information (AI), it is a strength for 30% of the sample and a weakness for 50%. The Elaboration Subfactor (EL), in the sample 40% presents it as weakness and 33.3% as strength.

It was shown that male students and female students have higher weak scores in information processing strategies, there being no significant difference between men and women.

Table 10. Percentage distribution of the strengths and weaknesses that students present in the Resource Management Strategies factor

Subfactors	Percentage of students with scores under 20	Percentage of students with scores between 21-59	Percentage of students with scores over 60	Total
Teamwork and interaction	33,3%	23,3%	43,3%	100
Context control	46,7%	26,6%	26,6%	100
Resources	50%	33,3%	16,6%	100

Source: Results questionnaire learning strategies.

In the table 10, the Assessment of Resource Management Strategies shows greater weakness and little strength in students. Scores below 20 which shows that those students who obtained low percentile fail to control external aspects located in the environment that influence learning, that is, they cannot avoid distractions that may affect their learning process, such as noise, space, light, heat, etc. For the subfactor group work and interaction (TG), it is a bit higher as strength (43.3%) and lower as weakness (33.3%). The Context Control Subfactor (CC) is a bit higher as a strength (43.3%) and

IOP Conf. Series: Materials Science and Engineering **519** (2019) 012026 doi:10.1088/1757-899X/519/1/012026

lower as a weakness (33.3%). The Context Control Subfactor (CC), is found as weakness in the sample with a score of 46.7% and few students present it as strength obtaining a score of 26.6%. Indicating that those who present it as weak (<20) have no control over the place of study, the environmental conditions of it (light, temperature ...), the layout of study materials, etc. The Subfactor Resources (R) becomes a strength for 16.6% of the students and as a weakness for 50%.

Based on the study, male and female students have insufficient strengths (23.3% - 20%, 6.6% - 20% and 10% - 6.6%) and greater weakness (13.3% - 20%; 23, 3%; 26.7% - 23.3%) in the Resource Management Strategy factor, there being no difference between men and women.

Table 11. Percentage distribution of the strengths and weaknesses that students present in the Dispositional Strategies factor

Subfactors	Percentage of students with scores under 20	Percentage of students with scores between 21-59	Percentage of students with scores over 60	Total
Perceived Self-Efficiency	43,3%	30,0%	26,6%	100
Attention - Active Disposition	30,0%	46,7%	23,3%	100
Physical and psychic state	20,0%	43,3%	36,6%	100
Anxiety Control	23,3%	43,3%	33,3%	100

Source: Results questionnaire learning strategies.

Dispositional Strategies show little strength and greater weakness on the part of the students. It is evidenced, higher percentage of students (men and women), for the Subfactors: Perceived Self-Efficiency (AP), Attention - active disposition (AD), Fitness and Mood(FA), Anxiety Control (CA), with scores below (20%), which shows that students do not master self-motivation, persistence, perception of achievement, voluntary attention, control of nervousness and anguish.

5. Conclusion

From the results of the investigation, we concluded that no differences were found in the results of the tests of learning strategies and questionnaire of psychoeducational characteristics applied to the students. The data obtained were low in the learning strategies without significant differences between men and women, as well as on the dates of entry of students to the educational institution.

Within the sociodemographic characteristics of the students, it is generally found that male students prevail, 56.6% of the sample, 53.3% of the students (men and women) are recent additions (2016 - 2017) and 50% belong to the socioeconomic stratum 4 (33.3% men and 16.7% women). The sociodemographic characteristics of the parents, such as the primary and incomplete baccalaureate levels and the socioeconomic level between 1 to 4, can influence the academic performance of the students. Authors such (Martinez González & Alvarez Blanco, 2005) [17] indicate that the social class and educational levels of the parents can also be factors that condition school failure. These authors affirm that the belonging to one or another social class is determined, above all, by the level of studies and the profession performed by the parents, as well as by the level of income received by the family unit. Pérez-Díaz, Rodríguez and Sánchez (2001) [12] have shown that there is a direct relationship between social class and the resources acquisition such as reference books, computer, etc. This relationship of stratum and education of the parents raised by the authors, is confirmed by the academic performance of the sample of students of the present project.

Regarding to learning strategies, students with weakness in the Metacognitive - Evaluative Strategies are identified at a general level. It means that students, both men and women, whether they are from old (2013-2015) or recently admitted (2016 - 2017) in the institution, have learned little to control and consciously evaluate the functions, processes and characteristics that influence learning. From the theoretical point of view, this aspect of the strategies is related to self-regulation in the student, Pintrich & De Groot establishes its objectives based on the evaluation criteria, which correspond to the standards under which it will be evaluated. the activity that the student wants to achieve in the task. The students in the sample are not aware of their learning process. Have little time organization and do not identify the forms of study that are best adapted to facilitate their learning. This also implies gaps in the routine and studying habits.

The weakness of the Information Processing scores in students is notorious, so they do not manage to work with the information that is the object of learning. Likewise, there is evidence of weakness in the students, to orient themselves to make knowledge analogies, organize and structure the information for their comprehension, highlight the relevant information and integrate the contents with information techniques for learning. According to Panadero and Tapia (2014) [8] the strategic planning, is to develop an action plan and choose the right strategies to succeed in the task. Zimmerman (2008) [7] , indicates that planning is a self- regulatory process par excellence, and is also a predictor of the success that will be achieved in the task: the longer planning time, the better results. By having weak information processing strategies can affect their expectations of self-efficacy producing demotivation and negatively affect their results. There is no difference between men and women independent of seniority in the institution. Little identify the form of study that best adapts to facilitate their learning for example the use of conceptual maps, underlining, identification of main and secondary ideas, summaries, drawings, perform exercises, take notes, etc.

In Resource Management Strategies, they are weak in students. That is, they can not control external aspects in the environment that influence learning because they can not control distracting factors such as noise, lighting, networks, music, space, temperature, among others, from the environment at the time of study, so that not being controlled by the student, affect their learning in some way. The resource enviroment control, is achieve creating an enviroment with the least number of distractions, and that facilittes the development of the task, increasing its effectiveness. Corno, likewise according to Panadero and Tapias (2014) [8] , it is important that the student keep focused and use proper learning strategies for two reasons: first, so that they do not diminish their interest and motivation and, second, to achieve the learning objectives.

The position of young people with respect to dispositional strategies, is of little control over the study for self-motivation, persistence, perception of achievements (goals), voluntary attention, control of nervousness and distress. Attitudes towards learning and motivation are taken by parents and teachers, indicating that students do not learn because they are not motivated or are not motivated because they do not learn. Motivation directly influences the type of cognitive processes and learning strategies that the student sets in motion when faced with a learning task by Gonzalez Torres [14] . The author Ausubel (1983) also explains that motivation is necessary for learning, for the author the motivation for achievement is made up of various dimensions such as the cognitive impulse, the affiliative drive and the motivation for improving the self.

The research showed that the students in the sample have few conditions to achieve meaningful learning, they do not organize the information, they present insufficient relationships between previous knowledge and the new ones, since the important thing is not how the information is presented, but how the information is presented. New information is integrated into the existing knowledge structure. The students show little disposition towards learning, presenting little interest and belief in the study.

References

[1] ICFES I C 2015 *Guía de Interpretación y Uso de Resultados de las* (Bogotá)

[2] Zimmerman B 2011 *Routledge* **49** 64

[3] Fernández A G 2001 *de Asociaciones de Psicología* **30**

[4] Zimmerman B 2001 *Aprendizaje Autorregulado y Académico. Nueva* (York: Springer-Verlag)

[5] Gutiérrez A C La metacognicion como un proceso de autorregulación del aprendizaje significativo, en matemáticas de sexto grado de educación primaria (Monterrey, nuevo león, méxico)

[6] Panadero E and Tapia J 2014 *Canales de Psicología* **450** 462

[7] Zimmerman B and Moylan A 2009 *Routledge* **299** 315

[8] Panadero E and Alonso-Tapia J 2014 *Elsevier* **11** 22

[9] Winne P 2001 *Lawrence Erlbaum Associates* **153**

[10] Tapia J A and Panadero E 2010 *Infancia y Aprendizaje* **385** 397

[11] Elvira Valdés M P L 2012 *Latinoamericana de Ciencias Sociales, Niñez y Juventud* **367** 378

[12] Sánchez M D C, Sánchez J M R and Gil J F 2000 *Revista de Psicodidáctica* **0**

[13] Acosta M S 2009 *Los procesos de aprendizaje y su incidencia en la deserción estudiantil en el Programa de Química Farmacéutica* (de la Universidad de Cartagena. Psicología desde)

[14] Montalvo F T and Torres M C G 2004 *Revista electrónica de Investigación Psicoeducativa* **1** 34

[15] Uribe G S 2011 *Ficha técnica Cuestionario Diagnóstico Estrategias de Aprendizaje* (Bucaramanga: Impact PSY)

[16] Pérez J E and Martínez A C 2008 *Avances en medición* **27** 36

[17] González R A M and Álvarez Blanco L 2005 *Aula Abierta* **127**

IOP Conf. Series: Materials Science and Engineering **519** (2019) 012027 doi:10.1088/1757-899X/519/1/012027

Development of prototype educational software for children with hearing disabilities for kids

L J Polo-Amador, A Tello-Rios, N Tavera-Castillo, J D Castrillon-Gomez and W Gomez-Sanchez

Unidades Tecnológicas de Santander (UTS).

lpolo@correo.uts.edu.co

Abstract. At present, there are different factors that lead to creating this auditory difficulty as a medical diagnosis. Becoming an impediment to the integral development of the people since a greater effort must be made for their formation and communication in the daily life. In this way, that in the process of academic training for these people, who have this type of difficulty should be based on the use of new methodological strategies, generating new learning scenarios for the best class development. It will contribute in that space given that its condition of absolute deficiency of listening makes it difficult to understand a specific context, in this case in Spanish, this fundamental matter being a communication mechanism in their daily lives and in the framework of interpretive and argumentative competences. For this subject and given the condition of the student, is considered important to accompany the learning with ICT tools that allow the process to be more dynamic, from the virtual context where the student will directly interact with a learning content in a didactic way understanding in a better context this information that is taught in a classroom with a traditional methodology. Among the main advantages of using these computing resources in class is: having a homogeneous group with respect to the knowledge that corresponds to the degree they are studying and improving their lexicon ability for communication in their daily lives as they go to enrich their vocabulary and understand some words they did not understand. One of the many reasons of the distractions today, are the new technologies, the mobile devices like cellphones, videogames, audio and video players, among other things which don't allow keep the needed focus in class.

Through the use of the developed web application it will dynamically offer a virtual space to interact with the contents of the Spanish subject to the student improving their communication and understanding the context where they live in their daily lives. Applying technological tools searching to solve this problem. The development of this research starts by generating a tool in this case a web application that, in a dynamic way, provides spaces and training activities in the basic contents of primary in Spanish, offering the student a variety of levels in those who can improve knowledge and apply the acquired notions in their training process. Additionally, you can keep track of your progress and score in the levels you are surpassing. By De la Torre S [1].

1. Introduction

During the development of this research we are looking the improvement of the learning in the Spanish area. In some countries, having a disability can more than double a child's schooling options, compared with those that do not. In Burkina Faso, having a disability increases the risk of children not receiving

Content from this work may be used under the terms of the Creative Commons Attribution 3.0 licence. Any further distribution of this work must maintain attribution to the author(s) and the title of the work, journal citation and DOI.
Published under licence by IOP Publishing Ltd

school education by two and a half times. It is therefore not surprising that in many countries children with disabilities make up the vast majority of those who do not attend school. For example, in Nepal, 85% of all out-of-school children are disabled. By Discapacidad Colombia, s.f [2] .

During the development of this research we are looking the improvement of the learning in the Spanish area. In some countries, having a disability can more than double a child's schooling options, compared with those that do not. In Burkina Faso, having a disability increases the risk of children not receiving school education by two and a half times. It is therefore not surprising that in many countries children with disabilities make up the vast majority of those who do not attend school. For example, in Nepal, 85% of all out-of-school children are disabled. By Discapacidad Colombia, s.f [2] . For this purpose, the construction of a web application is proposed, which in an interactive and dynamic way is a support tool in the training process in the main notions of the subject of Spanish. The website allows, among other things, the student's access, since it can be installed in an easy and intuitive way locally, so that its use can be from the computers of an institution. Or, it can be installed in a private hosting that would allow its use from anywhere.

This ease of use is also generated by the simple programming language that will be used, being this PHP, which comes by default in paid hosting plans or locally through XAMPP.

The design will be very visual with basic interactions to make easier the process of any user, independent of disability or academic level, and to strengthen inclusive education, visual assistance with sign language and on-screen text will be presented.

This is how it seeks to provide inclusive education with technological tools that are pleasant to the student and easily accessible. Obtaining as a benefit the strengthening in this area of knowledge.

2. Methods

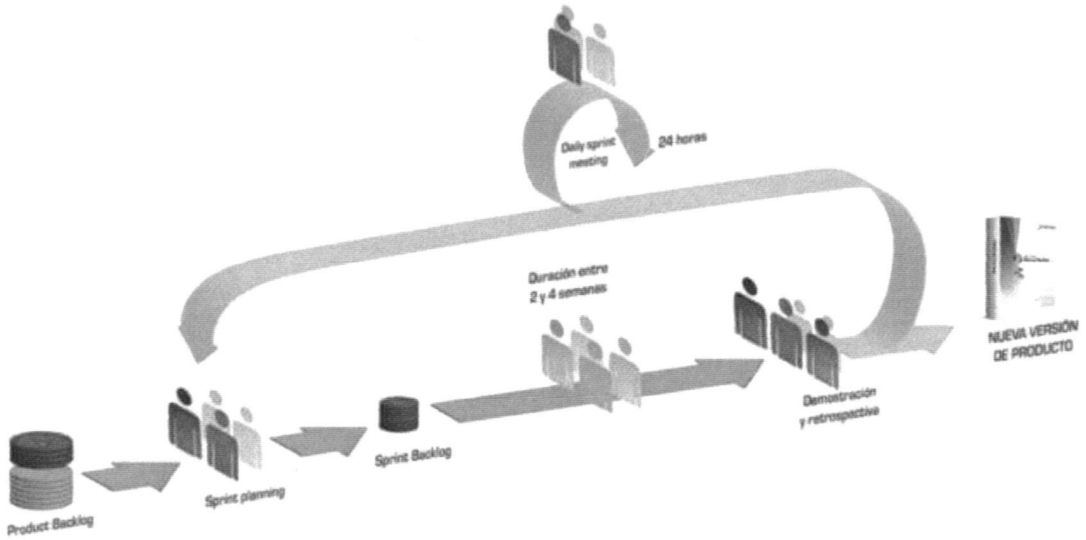

Figure 1. Methodology of development

The development is done iteratively and incrementally. Each iteration, called Sprint, has a pre-established duration that on this occasion will be one (1) week, resulting in a software update with new features ready to be used. In each new Sprint, the functionality already built is adjusted and new features are added, always prioritizing those that provide the most business value. By Dyba, T. Dingsoyr . [3]

Product Backlog: Set of requirements called stories described in a non-technical language and prioritized by business value, or what is the same, by return of investment considering its benefit and cost. The requirements and priorities are reviewed and adjusted during the course of the project at regular intervals. By Fowlery [4]

On this point of the process, requirements indicated by the user are raised. It is in this initial phase, when the information of agendas and questions proposed for the levels that will be scheduled on the website and will be presented to students is collected.

Sprint Planning: Meeting during which the Product Owner presents the backlog stories in order of priority. The team determines the number of stories that can be committed to complete in that sprint, for a second part of the meeting, decide and organize how you will get. By Schwaber. [5]

On this point of the process, requirements indicated by the user are raised. It is in this initial phase, when the information of agendas and questions proposed for the levels that will be scheduled on the website and will be presented to students is collected.

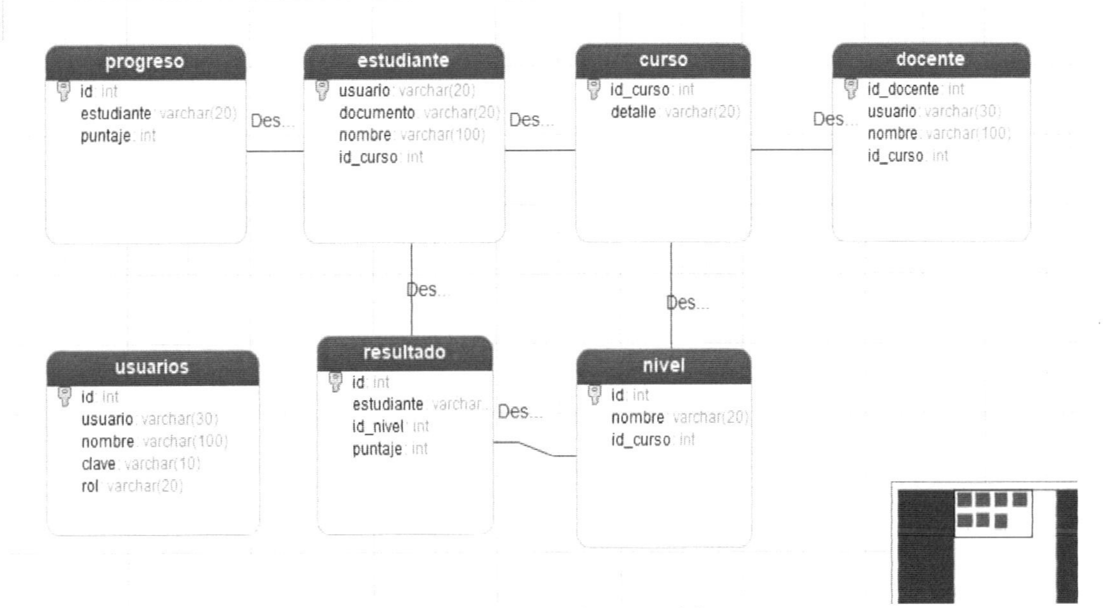

Figure 2. Data base model

In this way it is considered by the development team that the most optimal programming language is PHP with JavaScript. And the design is developed in HTML and CSS with the Bootstrap framework. The text editor that allows us to link all these tools with ease and productivity that has been chosen is Sublime Text. Then, you can see an image that demonstrates this process.

Sprint: Iteration of fixed duration during which the team works to convert the stories of the Product Backlog to which it has committed, in a new version of the fully operational software. Poppendieck M [6].

In each Sprint of seven (7) days duration, twenty levels will be delivered with questions related to the one-year syllabus. Sign language videos, multiple choice questions and text interpretation are presented.

IOP Conf. Series: Materials Science and Engineering **519** (2019) 012027 doi:10.1088/1757-899X/519/1/012027

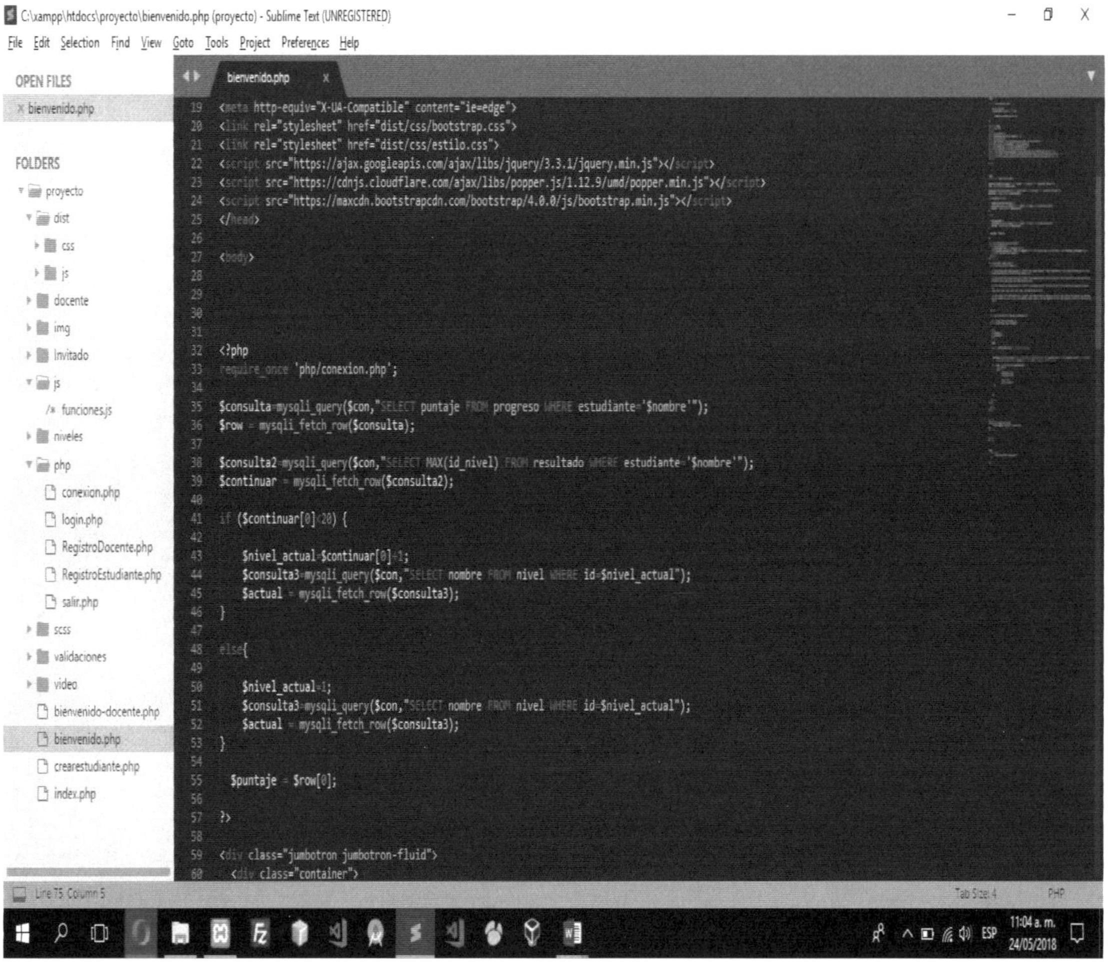

Figure 3. Source code

Table 1. Sprint Levels

Week number	Level	Grade
Week one	1 to 20	1st
Week two	1 to 20	2nd
Week three	1 to 20	3dr
Week four	1 to 20	4th
Week five	1 to 20	5th

Sprint Backlog: List of tasks necessary to carry out the stories of the sprint. (Softeng, s.f.) The list that will take as a base each Sprint, is the one corresponding to the table 1, where the thematic one corresponding to each course is indicated, and work will be done in the delivery of the levels taking into account these questions.

Daily sprint meeting: Daily meeting with 15 minutes max of duration, where the team get synchronized and work on the best way. Each one comment what they do the last day, and what is going to do today and if have any issues.

Every meeting is focused on the main thematic of the level planted, the presentation of the questions

IOP Conf. Series: Materials Science and Engineering **519** (2019) 012027 doi:10.1088/1757-899X/519/1/012027

and the feedback which allows accelerate and improve the development and designs times.

Demo and restropective: Meeting held at the end of the sprint where the team presents the stories obtained through a demonstration of the product. Later, in the retrospective, the team analyzes what was done well, what processes could be improved and discusses how to perfect them.

The weekly delivery of each course with twenty (20) levels, is accompanied by software tests, in which it is expected to participate in the solution of the levels and evaluate the presentation performance.

In Scrum, the team focuses on building quality software. The management of a Scrum project focuses on defining what are the characteristics that the product should have to build (what to build, what not and in what order) and to overcome any obstacle that could hinder the task of the development team.By Cockbun. [7]

The Scrum team consists of the following roles:

Scrum master: The development team leader, in this case corresponds to Juan David Castrillon Gomez. He works directly with the Product Owner.

Product Owner: Representative of shareholders or clients, for the times of this project we take as reference the director: engineer Leidy Polo.

Team: Professional team, represented by the authors of the project.

3. Results

Web application: The purpose of this project since its conception was to develop an interactive web application for teaching Spanish to children with hearing disabilities. This web, connects to a database worked in MySQL, it presents a graphical interface that allows the management and interaction of each user and the separation of them with roles and the result obtained in this project can be seen below.

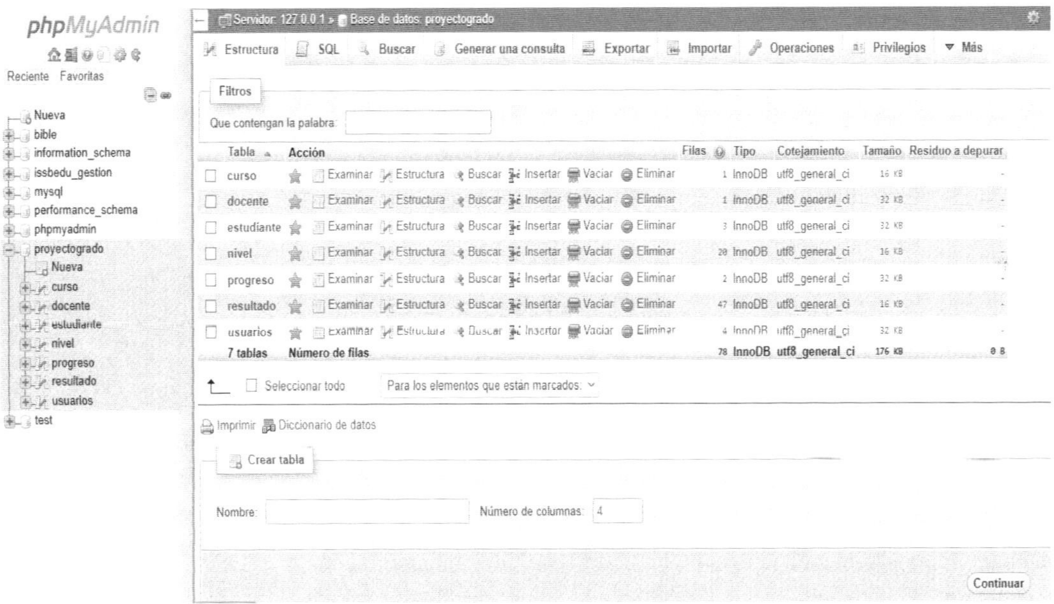

Figure 4. Data base model.

Through phpmyadmin, you can access the graphical interface where you can perform all the own management of the database. Since the creation of the same and its tables, deletion and data insertion processes. Through this tool you can also export and import the database together with its content.

Additionally, phpmyadmin has a console for managing the database through SQL code. In the following Figure 4 we can see the console.

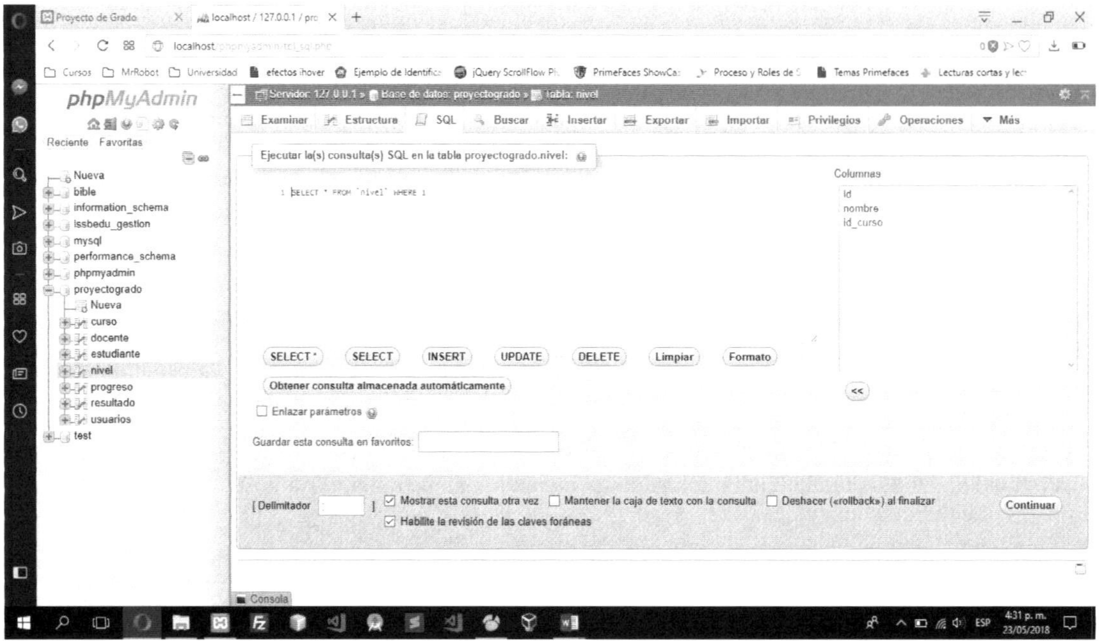

Figure 5. Data base administrator

The next result generated by the project refers to the web page itself. Which initially presents an interface to log in, register in the system as a teacher or access a nice game as a guest.

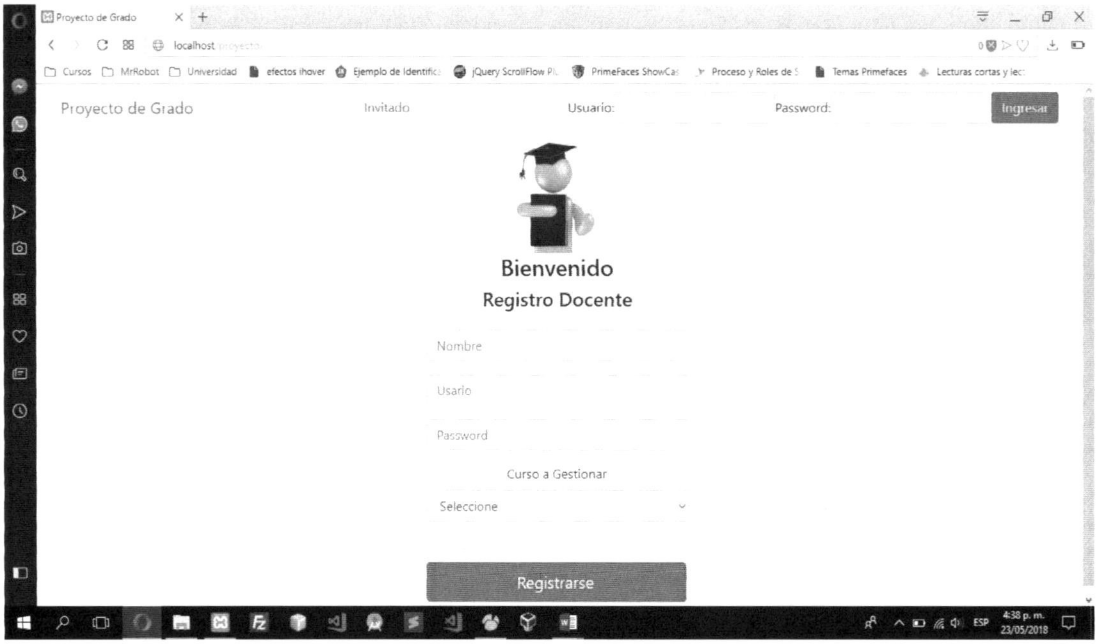

Figure 6. Log in screen.

From this home screen, the user can log in. In this process, the system validates what type of role the person trying to access has.

For this purpose, two roles have been registered, one as a teacher and the other as a student. Next, we can see the start screen that receives a user that has been validated with the role of teacher. Clearly exposing three basic aspects of management for this case.

(i) Student registration, where you can create new students who will later use the application.
(ii) Find news from students to know what levels they have passed and know their score. In this search part, you can perform this process both by student name and by document.

Review the highest scores, in the descriptive table, you can quickly see which student leads the most progress.

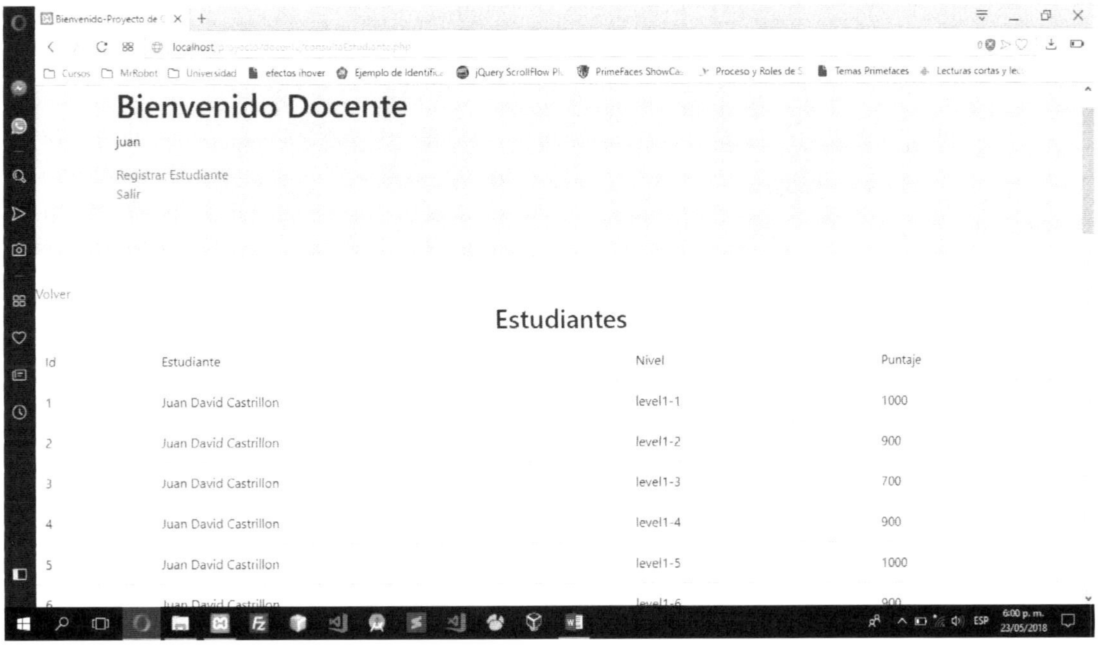

Figure 7. List of students.

Then, the student's access to continue their activities is evident. In this screen you will always have the top banner visible where you can return to the initial screen or close session.

The questions have a multiple choice and a score that will reduce its value if the answer you choose is incorrect. Sometimes, these questions will be accompanied by stories or video instructions, with the use of sign language.

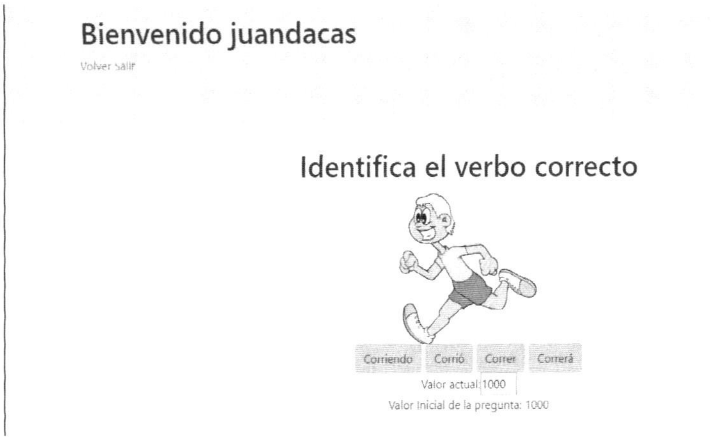

Figure 8. Interactive action.

Figure 9. Reading comprehension, interactive activity.

The no usage of the application to a 13 students same, where we can find the following results.

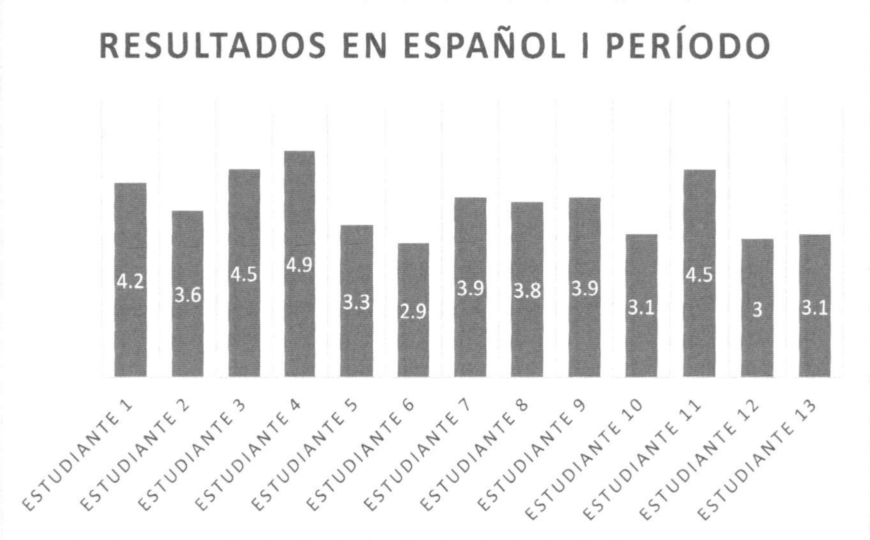

Figure 10. First period results.

IOP Conf. Series: Materials Science and Engineering **519** (2019) 012027 doi:10.1088/1757-899X/519/1/012027

After the implementation and the application usage during the second period in Spanish. We could see these results.

Figure 11. Second period results.

4. Discussion

Through the development of this research project, it was possible to generate measurable results of great importance in the field of inclusive education, directly in Spanish taught in primary basic education. It is evidenced by way of conclusion, how important it is to offer learning alternatives for students with hearing disabilities should not limit their educational process.

This is how you get to a web product that, in a simple, dynamic and visually pleasing way, presents hearing disable student in basic training in the subject of Spanish. Given that, through multiple choice questions, with videos in language of signs and activities of reading comprehension levels and forms in an inclusive manner and based on principles of equal formation. This is because the curricular content of the subject is the same for all students and is also implemented in the web application, but with visual aids thinking about this population.

In the levels faced by the student as user of the application, there are questions that have a purely cognitive focus and academic training purposes. To arrive at this type of deliverable, there was the problem of choosing and writing the questions and solutions for each level, but this was faced and solved with teacher information gathering. Where the main shortcomings of this group of students were evident. Through an academic research, the best alternative was presented in each presentation of the website.

Through simple computer tools and application of programming languages, it is concluded that inclusive education is a broad field to discuss, but with the use of ICT tools can be offered tools that support and enrich the training process.

A social approach is created that is not very casual from software engineering, system or its derivatives and that in turn creates awareness and improvements to the development of the country, thinking of education as an alternative of growth in all possible areas of a nation.

5. Conclusion

The project "Development of prototype software for children with hearing impairment", developed here, leaves great possibilities open mainly in three fields. The first one refers to the creation of spaces for inclusive education, adaptation of classrooms and academic training sites and special training for teachers who will directly face the challenge of educating a young person with a hearing disability.

The second field is the one about extending the project, an alternative is creating a web service, allowing access through mobile applications, or the creation of an app for devices. It is also recommended continuing the creation of more levels and thus extend the durability of this tool, given that its development was in php language without using any framework and can be taken as a basis for learning this language.

And finally, the third field, corresponds to generate software, web tools or mobile applications that contribute to inclusive education in new academic areas, such as mathematics, science, languages, history and other subjects that provide training to people with hearing disabilities in the different educational levels.

References References

[1] V D T S V 2010 *Estrategìas creativas en la enseñanza universitaria. Obtenido de Universidad de* Barcelona: España

[2] (sf D C La educación inclusiva para niños con discapacidad Obtenido de http://discapacidadcolombia.com/index.php/articulos-discapacidad/educacion-inclusiva

[3] Dyba˚ T and yr T D *Softw. Technol* **2008**

[4] Fowler M, Beck K and Brant J 1999 *Refactoring: Improving the Design of Existing Code* (Addison-Wesley)

[5] K S, M B and RC M 2001 *Agile Software Development with SCRUM* Prentice Hall

[6] M P and T P 2003 *Lean Software Development: An Agile Toolkit for Software Development Managers* Addison Wesley

[7] Cockbun A and Williams L 2000 "The Costs and Benefits of Pair Programming". Humans and Technology Technical Report

IOP Conf. Series: Materials Science and Engineering **519** (2019) 012028 doi:10.1088/1757-899X/519/1/012028

Prototype for identification of vehicle plates and character recognition implemented in Raspberry pi

E Arrieta-Rodríguez[1], L F Murillo[1], M Arnedo[1], A Caicedo[1], M A Fuentes[1]

[1] SINÚ University sectional Cartagena de indias-Colombia

investigacionsistemas@unisinucartagena.edu.co

Abstract. This paper presents the development of a prototype for the identification of license plates and character recognition implemented in a RASPBERRY PI 3 device. The device is used as a program controller, allowing to make a real time video from an integrated camera. From the video an image of the license plate is extracted, which allows to recognize the characters from it. Subsequently, the information of the license plates, current date, time, and camera code is integrated into a server which can be accessed from a web application. With this it can be achieved a systematic control of vehicles income and output.

1. Introduction
Artificial vision (AV) is a field of artificial intelligence (AI) that tries to create technological systems of optical recognition through methods that allow acquiring, processing and analyzing real-world images in order to produce information that can be treated by a machine [1].

An application of AV are the automatic recognition systems of vehicle plates (ALPR), these are composed by hardware and software allowing the reading of the vehicle plate and the recognition of the alphanumeric characters through optical character recognition techniques (OCR) [2].

This work presents the development of a prototype for the identification of vehicle plates and recognition of the characters it has, through the RASPBERRY Pi 3 electronic device. Its main functionalities are; automatic plate identification, character recognition using OCR, generation of the plate text and automatic sending of information to the server. The user interacts with the system through a web application that displays the information registered in the database, such as the date, time, photo and plate of the vehicle [3].

2. Related work
There are different investigations and projects related to the processing and analysis of images applied to the recognition of vehicular plates, now we presented the most relevant investigations.The work entitled "sistema de visión artificial para el reconocimiento de placas en vehículos particulares" developed by students of the University of San Buenaventura from Bogotá in 2008, this system allowed to identify the six characters of the license plate of the vehicle through the processing of the digital image, and in turn records the recognition data in an editable text file that contains additional information such as time and date of identification. The acquisition of the images was made through a webcam. Various image processing techniques were used such as color filters, morphological operations, growth of regions among others in order to achieve the location of the plate and the segmentation of the characters of the same. This work recommends the use of neural networks for future work [2]. Another important work is the "Sistema de visión artificial para la detección y lectura de matrículas" develop by the department of systems engineering and automation of the School of Industrial

Content from this work may be used under the terms of the Creative Commons Attribution 3.0 licence. Any further distribution of this work must maintain attribution to the author(s) and the title of the work, journal citation and DOI.
Published under licence by IOP Publishing Ltd

IOP Conf. Series: Materials Science and Engineering **519** (2019) 012028 doi:10.1088/1757-899X/519/1/012028

Engineering of the University of Valladolid. It arises from the idea of developing a system for the detection and recognition of license plates in automobiles. The main objective is to be able to use it in the control of access to parking, the camera when detecting the vehicle takes the image. It will consist of three blocks: location of the plate of the vehicle registration, treatment of the same and recognition of the characters. This algorithm was created to obtain satisfactory results regardless of the size and rotation that the plate may have. For the development of this work, the OpenCV open source library based on the C ++ programming language was used [4].

Finally, we can mention "3LPR" which is an automatic license plate recognition system (LPR system) used to identify and register the vehicles that access or leave a parking lot, achieving a great control of vehicle license plates, and therefore greater security. The system of automatic license plate recognition is not only focused on parking, but can be used in all those facilities that need to control, monitor and have a record of all vehicles that pass a certain access. Examples are the private garages of companies, shopping centers, tolls, hospitals, etc. [5].

There are several models of video cameras that incorporate recognition of vehicle plates, such as the camera HIKVision IP model DS2CD4A26FWDIZS / P that with a resolution of 2 megapixels and the use of a specialized algorithm ALPR, is used as a security element in accesses to parking lots and vehicle circulation areas. Its cost is usually relatively high; some can cost more than 4 times the price of a normal camera with similar characteristics.

3. Methods and Materials

For the realization of this work a series of tools and materials were used.

3.1. Rasp Raspberry PI 3

It is a complete computing system with good characteristics, which is commonly used for the development of embedded systems, due to its relatively low cost, its low power consumption, and its processing power, allows to build dedicated systems to solve problems that previously they required the use of computers.

Figure 1 shows a Raspberry PI 3 that has pins that are unbuffered, that is, they do not have protection buffers, taking into account the above, you must be careful with the magnitudes (voltages, intensity, etc.) when you connect some components to them to avoid damaging the board, the different pins it has: Power pins, DNC (Do Not Connect), normal GPIO and special GPIO [4].

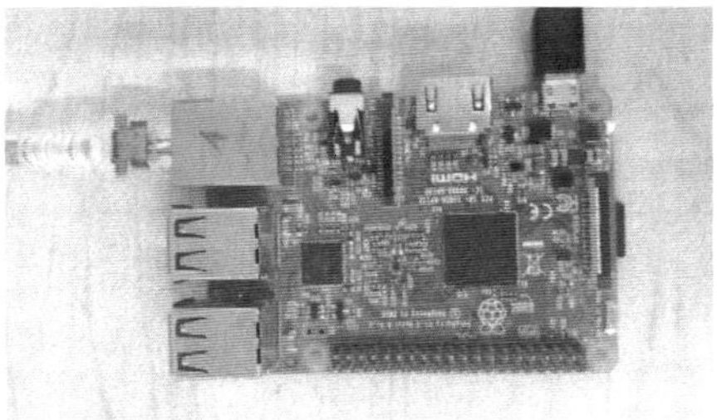

Figure 1. *Raspberry-pi-3-model-b*

This device in its web platform has several Operating Systems, which can be installed and configured in the optimal way for a good performance in the field that is required to use, some of the OS are: Raspbian (the most used), Ubuntu Mate, Snappy Ubuntu Core, Windows 10 IOT Core, OSMC, LIBREELEC, Pinet, RISC OS, Weather Station and Ichigojam RPI

3.2. Open ALPR

OpenALPR is a framework for image processing; modification and application of algorithms in images. Tesseract OCR is an open source software for character recognition that uses machine learning algorithms [7]. This library of recognition of open source automatic vehicle plates written in C ++ with links in C #, Java, Node.js and Python. The library analyzes images and video sequences to identify license plates. The result is the text representation of the characters of the license plate [5].

This library already has the basic parameters configured for the detection of vehicle plates, which allows training for the recognition of plates with the formation of any country. This library, being open source, allows you to make modifications directly in your coding to make improvements or customize it for our needs.

OpenALPR relies on other libraries for its operation, these being:

- OpenCV: this a framework for image processing; modification and application of algorithms in images.
- OCR Tesseract: is an open source software for character recognition that uses machine learning algorithms.

Automatic License Plate Recognition (ALPR) is a mass surveillance method that uses optical character recognition in images to read vehicle license plates as a special set of hardware and software components.

The automatic license plate recognition process is divided into two main steps: plate detection and plate recognition. The first step is to detect the location of the plate in the entire frame of the camera. When a plate is detected in an image, the plate segment is sent to the second step, which is the recognition, which uses an OCR algorithm to determine the alphanumeric characters on the plate [2].

Three additional steps that are commonly used in pattern recognition algorithms are defined:

- Segmentation: detects and removes each region of interest in the image.
- Feature extraction: extracts a set of characteristics or variables from a region.
- Classification: classifies each image region into "plate" or "no plate" in the plate detection step [8-9].

3.3. Open ALPR

For the recognition of vehicle license plates, OpenALPR was implemented, which analyzes images and video sequences to identify vehicle plates, this introduces in its encoding libraries that is based on computer vision techniques, powerful, such as OpenCV and Tesseract OCR, which are open source and handle automatic learning algorithms. A web application was developed to capture data and manage information. Figure 2 shows the detection of the region that contains the license plate by applying OpenCV functions.

Figure 2. *Plate capture*

Then apply a segmentation to obtain characters in the registration and recognition or identification of the car registration number by tesseract-ocr. In Figure 3 the interface where the vehicles that have entered during the day is listed, in the column "Photography" the image captured in the Raspberry and other data is loaded.

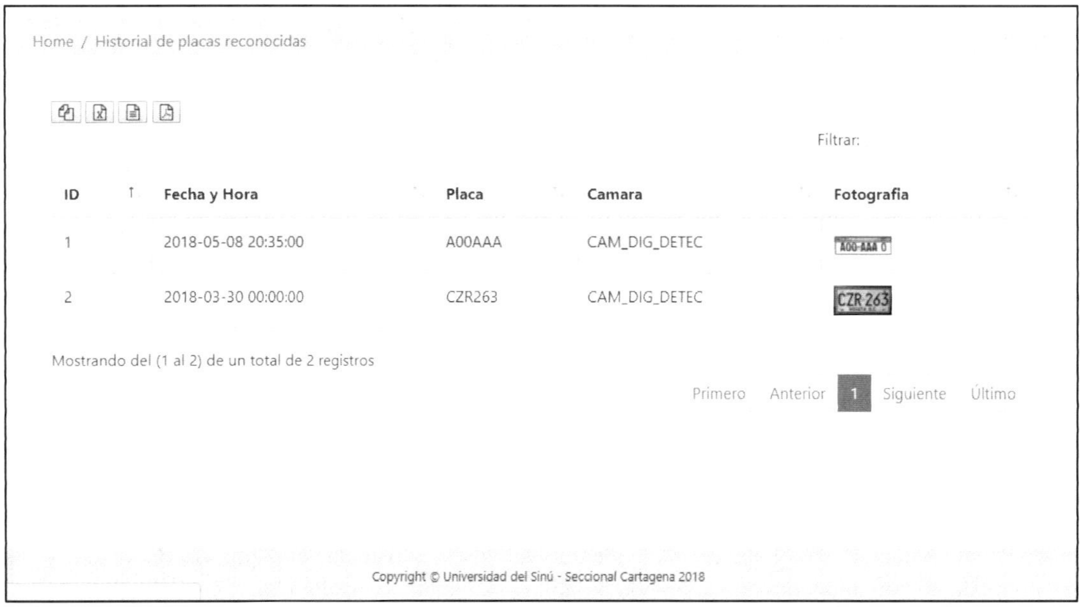

Figure 3. *List of plates*

4. Results

For the elaboration of the algorithm, several libraries that specialize in artificial vision were tested. The library selected was OpenALPR that allows to implement artificial neural networks for image processing, the process of retraining of the neural network was done with a new set of images of vehicular plates.

The tests were carried out with batches of 16 plates, of images taken at different angles and at different times of the day. Each plate is formed by six characters of which 3 are letters and 3 are numbers. Three conditions were presented in which the algorithm was tested, these are: Diagonal image, Front (normal) image and deteriorated plate image.

The result of the diagonal image tests indicates that these plates could not be recognized by the algorithm because the plate could not be extracted from the complete image, normally this occurs because the images have an angle of inclination greater than 30°.

In the case, in which the image is taken from the front, a percentage of success was obtained 81% of the time, the remaining 19% that presented an error, there was an average of two erroneous characters er plate and with a maximum of three error characters on a single plate. The deteriorated plates were recognized by the ANPR algorithm and the image plate was obtained. In the recognition of the OCR some characters are recognized, this is due to the fact that no uniform tone is found on the plate.

Additionally, a tendency to produce errors was identified because the algorithm was not trained with the characters of the plate divided between letters and numbers, so the error increases with the letters that have similarity with numeric characters, for example, A and 4, O and 0, I and 1 among others.

5. Conclusions and recommendations

A prototype of vehicle plate identification and recognition was built, automating the process of information registration through the OpenALPR library that is installed from the Raspberry PI device. Additionally, a user interface was developed to facilitate the queries of historical data of the vehicles that have entered.

For future work it is recommended to improve the training process of the character recognition algorithm, testing with a larger set of images containing vehicles with Colombian license plates. To discard the recognitions that do not correspond to vehicle license plates, it is recommended to indicate to the algorithm how the vehicle license plates are made in Colombia (3 letters - 3 numbers), and if it is a motorcycle it is formed by three letters, two numbers and a letter at the end.

References

[1] Cindy J and Diego G 2008 *Desarrollo de un sistema de visión artificial para el reconocimiento de placas en vehículos particulares Ingeniería Mecatrónica* (Universidad de San Buenaventura, Bogotá)
[2] Dhiraj Y Gaikwad 2014 A Review Paper on Automatic Number Plate Recognition (ANPR) System (*International Journal of Innovative Research in Advanced Engineering*)
[3] Daniel L, R S Naureen M, Shervin E, and Jason S *A Review Dominar OpenCV con proyectos prácticos de visión artificial.*
[4] Roberto M 2014 Sistema de visión artificial para la detección y lectura de matrículas *(Departamento de Ingeniería de Sistemas y Automática,* UNIVERSIDAD DE VALLADOLID)
[5] i+D3 Equipamientos Tecnológicos SL 2018 *3LPR Sistema LPR reconocimiento de matrículas automático*," i+D3,. [Online]. Available: https://imasdetres.com/sistema-anpr-reconocimiento-automatico-matriculas/.
[6] RASPBERRY PI FOUNDATION "*Raspberry Pi 3 Model B*," raspberry pi. [Online]. Available: https://www.raspberrypi.org/products/raspberry-pi-3-model-b/.
[7] Miguel A 2003 *Qué es Python Desarrollo Web*, [Online]. Available: https://desarrolloweb.com/articulos/1325.php.
[8] Guido van R, 2009 *El tutorial de Python* (Python Software Foundation)
[9] Campos A 2014 *Desarrollo De Software De Reconocimiento De Matrículas De Coche* (Universidad Politécnica de Valencia, Gandia)

Architecture of a Technology Platform for sustainable tourism management under the NTS-TS 002 standard

F García Ramírez[1], J Puello Beltrán[1]

[1]Tecnologico Comfenalco-Cartagena. Faculty of Engineering. Systems Engineering Program.

fgarcia@tecnologicocomfenalco.edu.co

Abstract. The importance of the tourism sector in Colombia has focused on aspects related to competitiveness and sustainability given the impact of tourism activity in the different destinations where it is developed. In order to maintain this advantage, it is essential to strengthen the management of sustainable tourism, optimizing the benefits for users, the destination and the local economy, while minimizing impacts on the environment. In this sense, the governmental demands oriented towards the implementation of these policies in the sector, as indicated by the Sectorial Technical Norms (NTS), specifically, the NTS-TS 002, requires the support of a technological management tool that allows the controlled development of the activities and indicators established by the regulations.

Based on this need in the tourism sector, a research project was developed based on the proposal of a management model of the NTS-TS 002 standard, was articulated with a software architecture that later will be specified in a web application to support the diagnostic process, monitoring of the indicators included in the standard and which allow accommodation and lodging establishments to be subsequently certified.

1. Introduction

The concept of sustainable development has been consolidated in the Brundtland report, being that which "allows attention to the needs of present generations without compromising the possibility of future generations to satisfy their own needs" [1].

Sustainable development worldwide has become a relevant topic, oriented to activities such as the use and renewal of natural resources, analyzing the effects of the generation of waste and pollutants on the environment and the economic distribution of regions, as well as natural preservation. This concept for some authors is defined as a challenge mainly for hotel establishments, considering the multiple initiatives that must implement the process gets slow [2].

Sustainable development consists in obtaining that the human being reaches to cover his basic necessities, without the attempt deteriorates the environment in which it is developed; however, achieving both the preservation of the environment and the satisfaction of the necessities is a disjunctive for some authors since in occasions the search of an economic balance brings as consequence significant impacts in cultural, environmental and social [3].

Hotel establishments are a key link in the value chain of the tourism sector, given their impact on the strategies implemented in the framework of corporate social responsibility, seen from the environmental, socio-cultural and economic dimensions, which is why they should include between

Content from this work may be used under the terms of the Creative Commons Attribution 3.0 licence. Any further distribution of this work must maintain attribution to the author(s) and the title of the work, journal citation and DOI.
Published under licence by IOP Publishing Ltd

their tourism activities strategies aimed at generating effective policies that result in the profitability of the organization in the long term and strengthen their relations with the environment with the aim of growing their business in the long term, understanding the impacts of their activity on local society or on the staff of the company. Among the potential benefits that may derive from these policies, the following should be highlighted: positive discrimination by the client, who increasingly incorporates social responsibility criteria in their purchasing decisions; greater reputation and brand image, greater competitiveness, cost savings in some cases, customer loyalty and even greater economic performance, this last point being a particularly relevant debate in the hotel sector, together with the uncertainty of whether there is a sustainable tourism sub-sector with a sufficiently attractive market to justify the creation of a specific offer[4].

The responsibility to maintain the balance between the dimensions of sustainability commits the government entities, who develop strategies aimed at motivating tourism service providers to generate strategies to implement sustainable practices in their businesses. This is how the recognition to the management granted through a certification is used as a tool to support the sustainable practices, as well as to satisfy the necessities of clients and the generation of the continuous improvement. Sustainability certification is a mechanism with which it is possible to measure qualitatively and quantitatively the performance of tourism through its operating practices[5].

In Colombia, different studies have been carried out that demonstrate the need to adopt a commitment to sustainable development: in the city of Bucaramanga, management variables were identified that influence service quality in the development of sustainable tourist destinations, through the adaptation of a measurement instrument for hotel service managers based on evaluation models identified in the literature and contrasted with the Focus Group of hotel experts in the region [6]. In the city of Bogotá, a study was conducted on the results of the hotels that implemented the NTS-TS 002 standard, and it was found that the greatest value in the sustainability requirements of the standard has been the decrease in the consumption of water and energy services and, therefore, their associated costs [7]. In Santa Marta, a study showed that the accommodation and lodging company subject to the investigation is closer to proactivity than to legal reactivity, since it begins to carry out practices of social responsibility that it is not obliged to carry out by law [8].

With respect to legal regulations, the Government of Colombia began a legislative process with Law 300 of 1996 - General Law of Tourism [9][1], which established some guidelines for the establishment of norms that regulate tourism sustainability; later work continued with the issuance of Law 1558 of 2012, which made it mandatory to implement sectorial technical norms [10] ; and finally Resolution 0148 of 2015, which establishes mandatory compliance with the Sectoral Technical Quality Standards related to sustainable tourism through a self assessment process documented and supervised by the Ministry of Industry and Commerce [11].

Taking into account the opportunity of applied research that the scenario described above entailed, support was given to the development of a project oriented towards the design of a software architecture for a technological tool that supports the process of management of indicators and activities established by the sectoral technical standard NTS-TS 002 [12], oriented to accommodation and hosting establishments at the national level. This project was articulated with the research group in Software Engineering and New Technologies - GISNET assigned to the programs of Systems Engineering and Electronic Engineering, with the institutional support of the Technological Comfenalco-Cartagena.

2. Methodology

For the development of the project objective, the following stages were planned:

Stage 1:

A. Design of the Indicator Management Model for Standard NTS-TS 002.
B. Software Architecture Design

Stage 2:

C. Design of the Indicator Management Tool for the NTS-TS 002 Standard.

D. Implementation in a real controlled environment
E. Evolution and Improvements

The results evidenced in the article correspond to Stage No. 1.

3. Stage Results 1.

3.1 Indicator Management Model for Standard NTS-TS 002

Taking as references Standard NTS-TS 002 [12] and the Implementation Guide issued by the Ministry of Industry, Trade and Tourism [13], a proposal was designed for an indicator management model for standard NTS-TS 002, as show in Figure. 1:

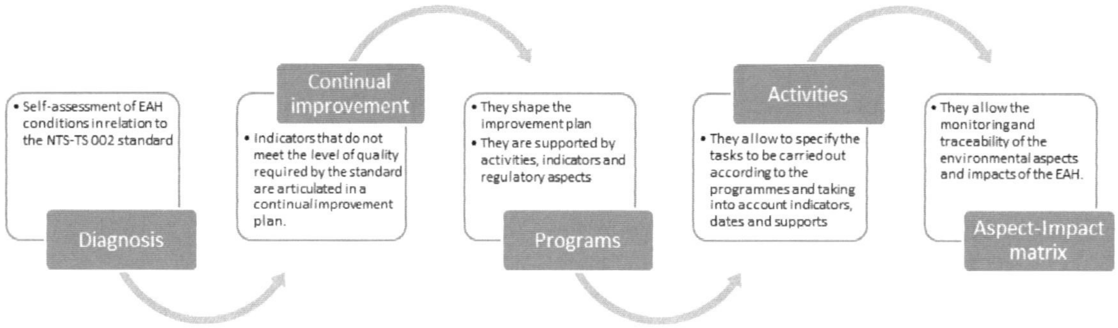

Figure 1. Indicator Management Model of Standard NTS-TS 002

According to the figure, the proposed model includes the processes of:

A. Diagnosis → oriented towards the self-evaluation of the conditions of the establishment of accommodation and lodging (EAH) in relation to the fulfillment indicators of the norm NTS-TS 002.
B. Continuous improvement → the process of managing the results of tourism sustainability indicators that are below the minimum quality conditions, coming from the environmental, socio-cultural and economic aspects contemplated in the standard and which in turn provide input for improvement programmes and plans.
C. Programs → process in charge of programme management within environmental, socio-cultural and economic aspects, as shown in Figure 2 [14].

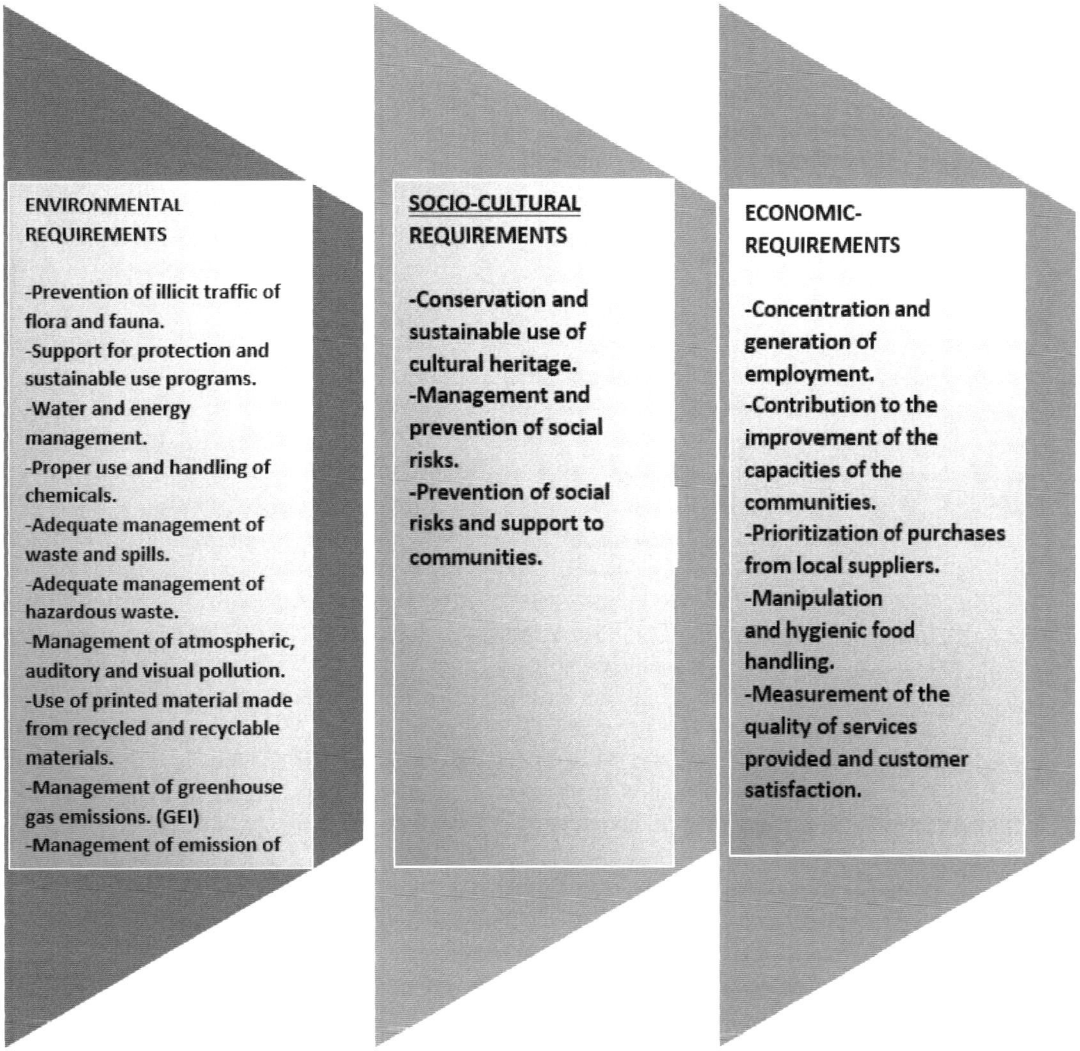

Figure 2. Aspects and programs covered by the NTS-TS 002 standard

D. Activities → process in charge of the management of the activities, including the tasks of assignment and description of the same, according to the indicators, aspects and programs to develop in the establishment of accommodation and lodging.
E. Aspect-Impact Matrix → process oriented to the management of the traceability of the environmental aspects and impacts of the accommodation and lodging establishment. The relationship between these indicators is best shown in Figure. 3 [15]:

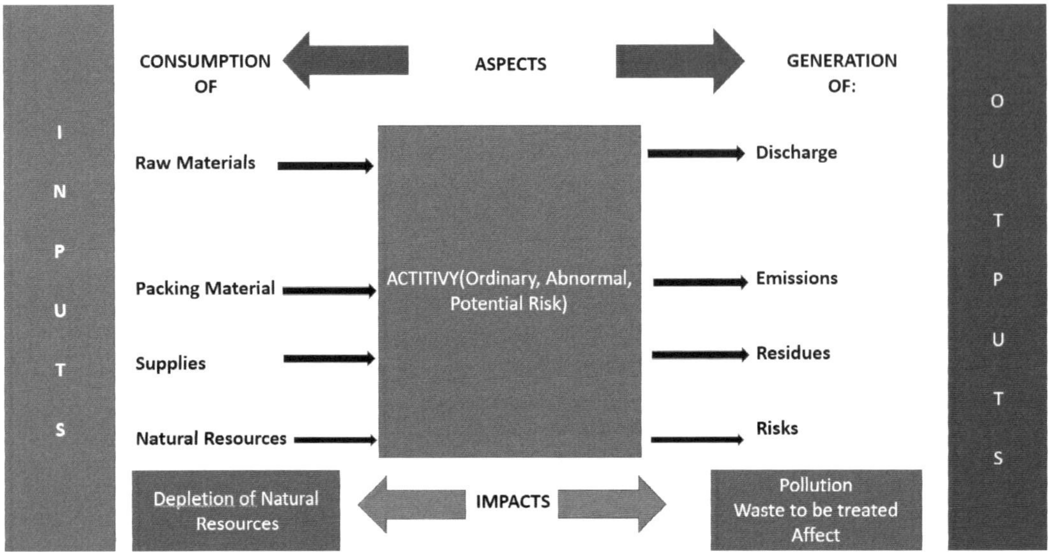

Figure 3. Relationship between environmental aspects and impacts according to NTS-TS 002.

3.2 *Software Architecture for the NTS-TS 002 standard management tool*

From the model proposed for the management of the indicators of the NTS-TS 002 standard, a software architecture was designed for the subsequent development of a web tool.

Figure 4 shows the proposed software architecture, which is shown as PHP programming language, which has many advantages for the development of web solutions, such as a large community of developers, several frameworks at both front-end and back-end level; as far as data management is concerned, MySQL has been selected, which is also an open and free product that has a trajectory and recognition in the software development industry.

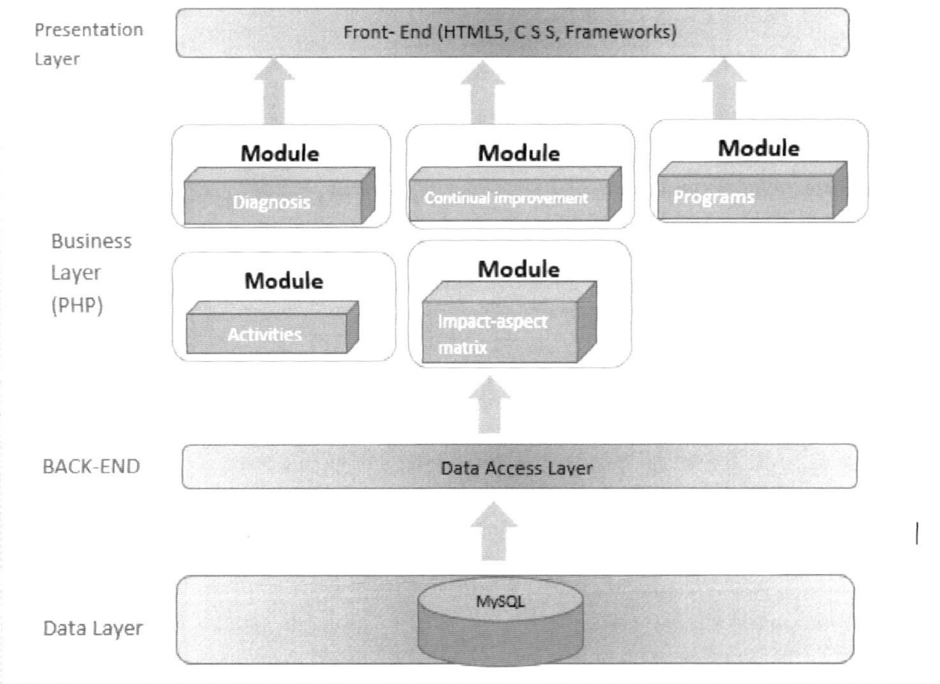

Figure 4. Proposed software architecture for the NTS-TS 002 indicator management tool.

As part of the design exercise, an estimation was made of the expected functionality flowchart for the web application to be developed, based on the previously proposed software architecture, as shown in Figure. 5:

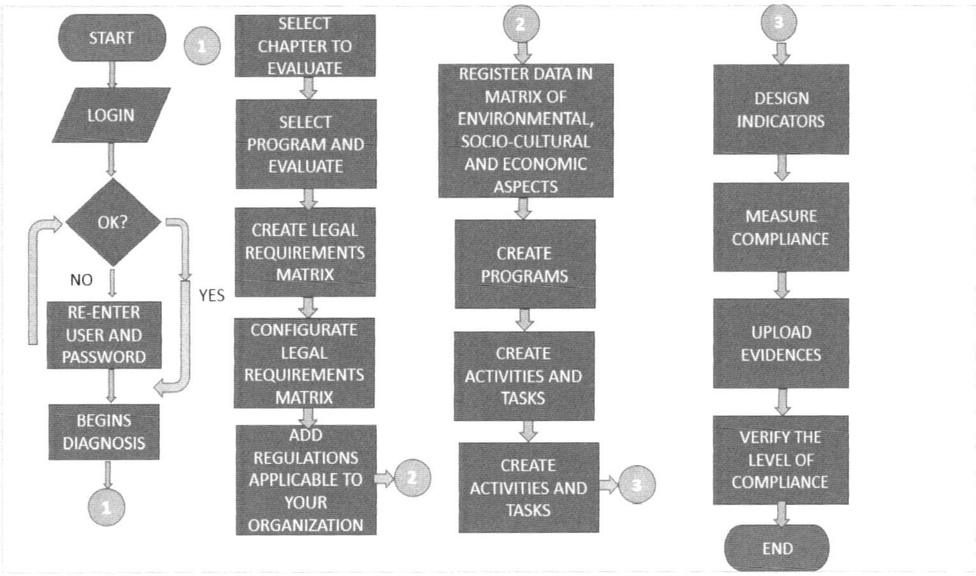

Figure 5. Proposed flow chart for the NTS-TS 002 indicator management tool.

4. Conclusions

The obligatory adoption of the sectoral technical standard NTS-TS 002 for accommodation and lodging establishments (EAH), promulgated by national government regulations, is a challenge for companies in this sector, especially the smaller ones, which require technological support, proposed by a tool that allows them to manage in a suitable and productive way, the different indicators that must be adopted to achieve tourism quality certification, allowing them to compete in the same conditions as major tourism service companies. As future challenges of this research work, is to implement the proposed architecture in a web application that can be installed in a controlled manner in accommodation establishments and lodging, allowing it to submit to validation and verification in relation to the requirements of the standard NTS TS 002.

References

[1] Brundtland G 1987 *Our Common Future: Brundtland Report.*

[2] Ginneken R 2016 *Sustainability challenges and opportunities arising from the owner-operator split in hotels* International Journal of Hospitality Management Vol **54**: p. 35-42.

[3] Corredor E 2008 *El turismo sostenible y endógeno: una nueva actividad económica en Venezuela* Universidad, Ciencia y Tecnología Vol **12**.

[4] Fernández M & Cuadrado R *2011 La responsabilidad social empresarial en el sector hotelero: revisión de la literatura científica* Cuadernos de Turismo Vol **28**.

[5] Diaz I & De La Fuente H 2013 *Análisis de los factores determinantes de la calidad percibida del servicio prestado por una cooperativa de ahorro y crédito: una aplicación basada en modelos de ecuaciones estructurales* Revista Ingeniare.

[6] Monsalve C & Hernández S 2015 *Gestion de la calidad del servicio en la hoteleria como elemento clave en el desarrollo de destinos turisticos sostenibles: caso Bucaramanga* Revista EAN Vol **78**.

[7] Aldana M 2014 *Análisis de los resultados obtenidos en hoteles de Bogotá, a partir de la Implementación de la norma NTS-TS 002* (Teoria y Praxis Investigativa) Vol **9**-2.

[8] Peña D & Serra A 2012 *Responsabilidad Social Empresarial en el Sector Turistico. Estudio de Caso en empresa de alojamiento de la ciudad de Santa Marta, Colombia* (Estudios y Perspectivas en Turismo) **21**-6.
[9] Ministerio de Industria, Comercio y Turismo 1996 *Ley 300 de 1996.*
[10] Ministerio de Industria, Comercio y Turismo 2012 *Ley 1558 de 2012.*
[11] Ministerio de Industria, Comercio y Turismo 2015 *Resolucion 0148 de 19 Enero de 2015.*
[12] ICONTEC 2006 *Norma Tecnica Sectorial Colombiana NTS-TS 002.*
[13] COTELCO, MINCOMERCIO, FONTUR 2014 *Guía para la implementación de la NTS TS 002 en establecimientos de alojamiento y hospedaje.*
[14] Camara de Comercio de Ibagué. Camara de Comercio de Ibagué. [Online].; 2016. Available from: https://www.ccibague.org/archivos/rnt/instructivo-NTS-TS.pdf
[15] Ortiz Urquijo VaOSH. Repositorio Universidad Javeriana. [Online].; 2015 [cited 2018 10 15. Available from: https://repository.javeriana.edu.co/bitstream/handle/10554/17961/OrtizUrquijoVictorHugo2015.pdf?sequence=1.

IOP Conf. Series: Materials Science and Engineering **519** (2019) 012030 doi:10.1088/1757-899X/519/1/012030

Integration of learning objects for adaptative learning

M Rojas-Contreras and O Portilla-Jaimes

Universidad de Pamplona, Pamplona, Norte de Santander, Colombia

mrojas@unipamplona.edu.co

Abstract. The integration of learning management systems with learning object management systems allows to increase the availability, reuse, integration and adaptability of educational contents according to the characteristics and preferences of the clients. However, given the new trends in the integration of learning objects, software architectures and recommendation systems, it is of high added value to optimize the services offered to customers through an integrated learning management system aimed at the automatic construction of Adaptive training processes in higher education. The main result is the description of an integral system of virtual education based on the competences required by the students, which integrates learning objects (OA) through a semantic OA recommender system and then the system sequencing the OAs to respond to the required competences. The software architecture of the system is structured in two layers: The layer of OA discovery and adaptability and the layer of generation of adaptive training processes, which have as a design principle the web services in order to guarantee the interoperability of the layers.

1. Introduction

Advances in information and communication technologies have allowed innovating and adapting the business models of organizations in different areas of society, in addition, the implementation of cybernetic systems to optimize processes and generate value to through services and new products to these organizations. Education does not escape such disruption, which requires the generation of new conceptions and new paradigms such as university 4.0, which consists of the application of industry 4.0 in the field of higher education. According to [1], industry 4.0 is characterized by the transformation of the productive apparatus worldwide through cyber-systems. In particular, one of the segments in which the impact of this paradigm is evident is that of virtual education, specifically with the integration of learning object management systems into learning management systems, this integration strategy has optimized the way in which the elaboration, distribution and access processes of educational contents are executed through the reuse and integration of learning objects (OA), which are stored and managed in repositories of learning objects (ROA) [2].

On the other hand, according to [3] not all people learn in the same way, which implies a personalization and adaptability of the way to teach each student, due to this, the design of a software architecture is described to take an adaptive teaching process, which is based on web services technologies and recommendation systems. Specifically, the architecture for the adaptive integration of OA is designed in 2 layers, the first layer is semantics, which is composed of the necessary components to deliver personalized recommendations to each student. Each recommendation is made up of a set of ranked OA according to a numerical value calculated by a mating mathematical function, which indicates to what degree each OA is suitable to study a certain topic, considering the particularities defined in the profile of each user (student). Next, the integration layer is designed, which is responsible

Content from this work may be used under the terms of the Creative Commons Attribution 3.0 licence. Any further distribution of this work must maintain attribution to the author(s) and the title of the work, journal citation and DOI.
Published under licence by IOP Publishing Ltd

for carrying out the automatic construction of adaptive training processes in Higher Education. In order to carry out such construction automatically, an intelligent integrating semantic service is required, which receives a training request expressed in competences (by a client application) and from that request performs an automatic composition process that generates the necessary training blocks. The semantic integration service is implemented as a web service, which incorporates planning and learning techniques in artificial intelligence. The use of artificial intelligence allows you to automatically generate, intelligent and personalized blocks of training sequenced and adapted according to the learning style, the level of knowledge, the psychological and emotional characterization of the student.

2. Methodology

The process of building the software architecture is supported by methods taken from software engineering and supported by artificial intelligence techniques. The methodological approach used in the design of the proposed architecture is the deductive, which starts from the approach of a software architectural design solution and continues with the decomposition of the architecture into layers or subsystems, in a specific way, the architecture is broken down into two layers the layer of OA discovery and adaptability and the generation layer of adaptive formation processes, in turn, each layer is structured into components. Figure 1 presents the component diagram associated with the proposed architecture.

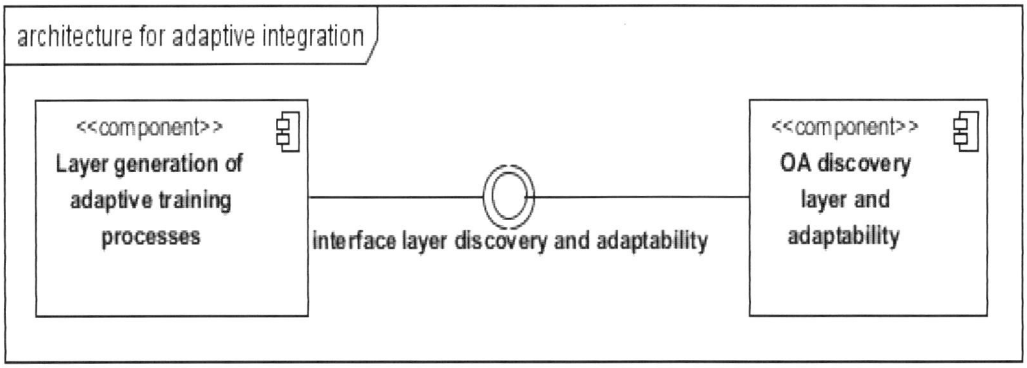

Figure 1. Architecture for adaptive integration

2.1. OA discovery layer and adaptability

According to [4], the discovery layer of OA and semantic adaptability is structured in six components as shown in Figure 2.

2.1.1. OAs recommender system. It is defined as a hybrid semantic recommender, which involves a knowledge-based component, a collaborative component, and a content-based component. The fundamental role of the OA recommender system is to build a set of OA recommendations, which it considers adequate to teach a topic to a student (for which the semantic integration service requested the recommendation). The recommender makes an assessment of the elements of the recommendation set, in such a way that he recommends the objects in an orderly manner according to a numerical value that represents the degree of suitability that each LO has to teach a certain competence or subject.

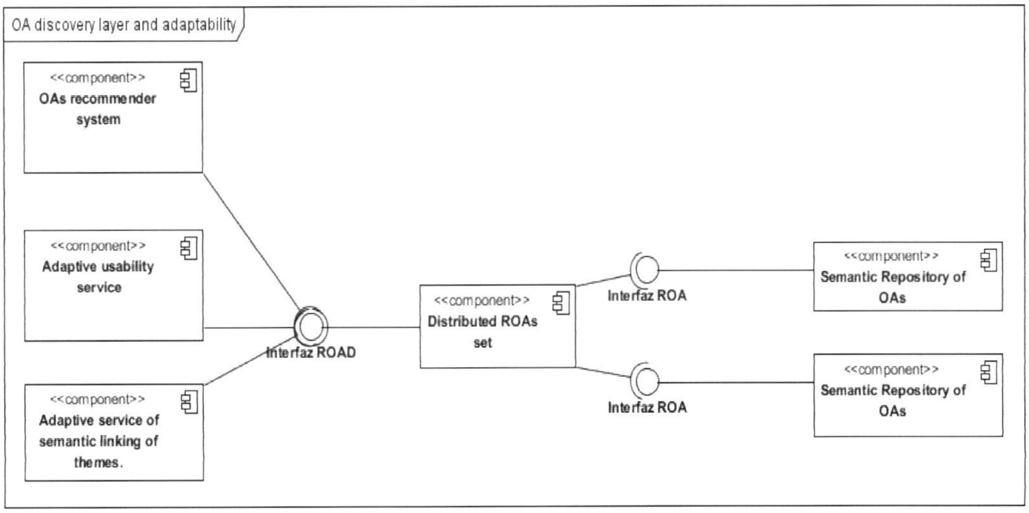

Figure 2. OA discovery layer and adaptability

2.1.2. Semantic Repository of OAs. It is an OA storage medium related to the metadata associated with the educational contents that have been submitted to a certification process. The fundamental element of the repository is an ontological model presented in [4], which was designed based on the LOM standard. This ontology contains the necessary elements, so that the recommender can make the requested recommendations.

2.1.3. Distributed ROAs set. It is made up of the different repositories of OA distributed worldwide that offer OA search services. According to [5] the distributed repositories allow to improve the levels of reuse and integration of the LOs. These sets of ROAs are used by the recommender system, only in the case that they do not have OA certificates that comply with the requested recommendation.

2.1.4. Adaptive usability service. This service is composed of a repository with the information of the user profiles, in which the knowledge regarding the preferences of each user and the use of OAs that each one makes is stored. In addition, this intelligent agent that has a web service that offers information about a user's profile, requested by whoever requires it. For the proposed architecture, the client of said web service is the recommender service, who uses this knowledge to personalize and adapt the recommendations. The information registered in the user profile comes from the dynamics generated in the system.

2.1.5. Adaptive service of semantic linking of themes. This component is designed as an intelligent agent, which is responsible for providing the recommender system with the different topics semantically linked to the topic requested in the recommendation. This agent is based on the use of an ontology of semantic linking of topics, which is useful for any application that requires it.

2.1.6. Adaptive quality service and performance evaluation. Component designed as an intelligent agent that manages all knowledge related to the quality and performance evaluation of each of the OAs that are certified. For architecture, this agent must be implemented with intelligent techniques that allow to permanently re-evaluate the quality and performance of each one of the OAs (learn), according to the dynamics of the system. That is, this agent must incorporate artificial intelligence that allows him to learn to differentiate that OA has higher quality than another for a certain user at the time of making the recommendation.

2.1.7. Adaptive service of pedagogical aspects. This component is responsible for calculating the contribution of pedagogical personalization in the function of utility of each object to be recommended, according to the student's learning style and other characteristics of their profile. The implementation of this intelligent agent requires artificial intelligence in such a way that it can determine, through experience, the didactic and methodological strategies that are most convenient for the student. The foregoing implies that an OA with suitable didactic and methodological strategies for the student requesting the recommendation will show greater contribution of pedagogical personalization for the calculation of the mating mathematical function.

2.2. Layer generation of adaptive training processes

This layer is responsible for performing the automatic construction of adaptive training processes. In order to carry out such construction automatically, an intelligent integrating semantic service is required, which receives a training request expressed in competences (by a client application) and, based on said request, perform an automatic composition process that generates the necessary training blocks. The semantic integration service is implemented as a web service, which incorporates planning and learning techniques in artificial intelligence. The use of artificial intelligence allows it to automatically and intelligently generate sequenced and personalized training blocks according to the learning style, the level of knowledge, the psychological and emotional characterization of the student. The semantic integration service is structured in two processes as shown in figure 3.

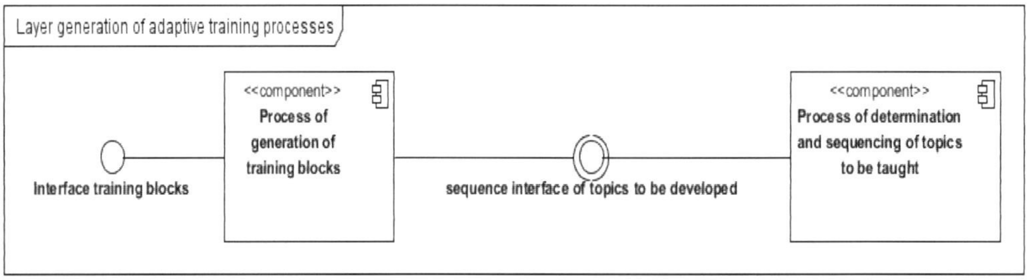

Figure 3. Layer generation of adaptative training processes

2.2.1. Process of determination and sequencing of topics to be taught. In this process, the integrating service uses an ontological domain model based on rules, which allows you to infer the different topics that must be taught to ensure that the student really learns the requested competence. In addition to determining the different subjects that must teach to achieve the competition, it is necessary that the integrating semantic service determines the order in which said topics should be taught.

2.2.2. Process of generation of training blocks. This process is responsible for integrating into each training block several existing OA that are registered in the semantic repository that has the semantic layer. To do this, the integrating semantic service uses the sequence of topics to be taught (obtained in the previous process), requesting a semantic layer recommendation for each topic that belongs to the sequence of topics to be taught. For each of the recommendation requests, the semantic layer will provide a recommendation that will serve as an input to the intelligent algorithms for generating training blocks to include the OA that best suits the student for whom it is being formed.

3. Results

The software architecture described solves a problem of integration of systems that act in isolation, specifically, the OA repository systems store educational contents packaged with metadata that are used according to the needs of the courses, later manually, the LMS add selected OAs to the course that will contain the educational content. The presented architecture integrates and automates the processes of selection and integration of OAs and additionally performs the integration process according to the profiles of the students based on the requirements they need to develop the competencies of a training process. The presented software architecture is structured in two layers, the semantic layer and the integration layer, which use theoretical foundations of artificial intelligence and software engineering for their construction.

To assess the suitability of the proposed architecture, a panel was convened, made up of three experts in the area. Each of them was presented with the evaluation criteria presented in Table 1.

Table 1. Evaluation test for the proposed architecture

	Test
C1	Depth level of the architecture description
C2	Represents a viable solution to the problem of customization and adaptivity of teaching automatically
C3	It represents a viable solution to the problem of integration and reuse of digital teaching contents automatically
C4	The proposed architecture incorporates updated and appropriate technologies
C5	The work presented is in accordance with the current requirements in TEL
C6	The presented work includes the most relevant aspects in customization, adaptation and integration of digital contents
C7	The work is novel regarding the of arte state existing
C8	Architecture is adequate in terms of mechanisms of representation and exchange of knowledge
C9	The proposed architecture is independent of specific platforms and technologies for its implementation

The results obtained in the evaluation of the adaptive architecture are shown in table 2.

Table 2. Results obtained in the evaluation of experts

	C1	C2	C3	C4	C5	C6	C7	C8	C9
Expert 1	4.1	5.0	4.8	5.0	5.0	4.9	4.5	4.0	5.0
Expert 2	4.5	4.6	4.7	4.8	5.0	5.0	4.8	3.8	5.0
Expert 3	3.2	4.2	4.4	5.0	5.0	4.8	4.4	3.7	5.0
Average	**3.93**	**4.6**	**4.63**	**4.93**	**5.0**	**4.9**	**4.56**	**3.83**	**5.0**

The above results make it clear that the presented work is new, is in accordance with the current requirements regarding Technology for Improvement of Learning TEL, covers the relevant aspects in terms of adaptability and personalization of the teaching and also provides a feasible mechanism to perform integration automatic digital content.

4. Conclusions

Artificial intelligence techniques constitute a field of high importance for the generation of value in the design of systems integration solutions, as is the case of the proposed architecture, which contributes to the reuse and integration of OAs automatically, complemented with features that bring and align virtual learning environments to the current needs of higher education institutions, specifically in the construction of training processes based on the individual characteristics or profiles of each of their students or clients through recommendation systems.

The teaching-learning processes are being influenced by the current disruption generated by the fourth industrial revolution, which implies that virtual learning environments should involve architectures such as the one presented here that not only integrate and reuse existing educational content, but also optimize the teaching process through the personalization and adaptation of contents according to the particularities of each student.

References

[1] Joyanes L 2017 Industria 4.0: El camino a la cuarta revolución industrial (Barcelona, Alfaomega)

[2] Rojas M, Montilva J, Hurtado M 2013 Diseño de repositorios de objetos de aprendizaje como estrategia de reutilización e integración de contenidos en modelos de educación virtual. *LACCEI 2013.* Cancún México.

[3] Openheimer A 2011 ¡Basta de historias!: La obsesión latinoamericana con el pasado y las 12 claves del futuro. Ed. Debate.

[4] Portilla O, Aguilar J 2015 Plataforma de gestión de la nube de fuentes de conocimiento aplicando el paradigma ODA. *Technical report, CEMISID*, Universidad de Los Andes, Mérida.

[5] Montilva J, Rojas M, Orjuela A 2011 RDOA-WS: Repositorio distribuido de objetos de aprendizaje soportado con servicios web. *Revista avances en sistemas e informatica.* Volumen 8 Número 2

Use of ICT Tools using the PBL Methodology as a Student Learning Strategy

S A Sanchez [1], C A Mercado[1], J F Serpa[1], J S Julio[1] and A J Almanza[1]

[1]Corporación Universitaria Antonio José de Sucre, Sincelejo, Colombia, Engineering science faculty, Grupo de investigación GINTEING

sergio_sanchez@corposucre.edu.co

Abstract. The community of Altos de la Sabana is a population composed of families of limited economic resources and victims of the internal armed conflict in Sucre, located in low-income housing. In this urbanization is the Institution of Higher Education of the Holy Shroud. Given that at present it does not have technological and innovative tools for academic use, information provided by the rector and the academic coordinator of this institution, as well as being considered as a vulnerable community by the national government, which gives them a priority position when implementing educational and social projects; The main objective of this project is to encourage the students of the secondary school of the institution to use technological tools to solve problems in their environment. This research has a quantitative approach and the design of the research focuses on field research and with a pre-experimental method. For the execution of this project, 3 moments were concentrated; initially, an initial test was carried out to validate knowledge in technological tools and the self-directed profile, then some training was carried out using the problem-based methodology and finally a final evaluation to measure acquired competences. The results of the post-test, in contrast to the initial test, revealed that 65% of students in grade 11 of the institution demonstrate an empowerment in the use of ICT tools and concepts in the field of basic analog-digital electronics. This is how ICTs are based on the education system, as they are facilitating tools in the teaching-learning process.

1. Introduction

The speed of adoption and penetration of digital technologies advances rapidly in the different countries of Latin America, access to digital, genders, age groups and the gap between urban and rural areas, is increasing. By 2015, 81% of households in developed countries had access to the Internet, compared to 34% in developing countries and 7% in least developed countries [1]. Other studies from the Economic Commission for Latin America ECLAC, indicate poverty rates in the region of 28% in the countries that comprise it, while the DANE, shows an indicator of poverty for the department of Sucre. 44.7%, Sucre being the fifth department with the highest percentage of poverty, preceded by Chocó 62.8%, La Guajira 53.3%, Córdoba 46.6% and Magdalena 44.8% [2].

Reflecting these inequalities in education, being the main challenge for Latin America, the universalization of basic education and the media, incorporating social sectors, improving quality and massifying higher education. The new teaching systems configured around information and communication technologies must redefine learning techniques to lead them to a more flexible type of teaching processes [3].

Content from this work may be used under the terms of the Creative Commons Attribution 3.0 licence. Any further distribution of this work must maintain attribution to the author(s) and the title of the work, journal citation and DOI.
Published under licence by IOP Publishing Ltd

With the passing of the years, technology is a fundamental part of the activities of the human being, it is necessary to have knowledge about ICT tools. Since the world has modernized to such a scale that people must adapt to these technological changes, with these tools anyone can develop activities in less time, in addition to providing the society with many forms of communication, education and jobs. According to the Ministry of National Education of Colombia, the use of technological tools is necessary to bring students' knowledge to the world around them, so it is necessary that they have access to this type of tools for the development of competent skills [4].

Master classes taught in basic and secondary institutions increasingly discourage students, teachers need new resources and tools to generate more didactic classes. The Ministry of National Education proposes an educational revolution to transform the Colombian educational system in terms of magnitude and relevance through strategies that expand coverage with equity criteria, improve the quality of the education system and increase the efficiency and productivity of the sector. Two of the 37 projects that support these postulates include the incorporation of new technologies and methodologies in the country's education. In coverage, higher education and quality in basic and secondary education, with elements such as connectivity and information technology, use of television programs and educational radio.

The OECD, in its educational aspect, is concerned with the quality of education, seeking, analyzing and promoting effective educational environments, based on the results of the students [5]. An in-depth analysis is not necessary to verify that the most innovative environments use new technologies to support their methodology, in order to favor the child's integration into the current real world. The effectiveness in the learning process is much higher when the new technologies are integrated.

UNESCO in its global report on education, points out that virtual learning environments constitute an entirely new form of educational technology and offer a complex set of opportunities and tasks to educational institutions around the world, the environment of Virtual learning defines it as an interactive computer program of a pedagogical nature that has an integrated communication capacity. Virtual learning environments are, therefore, a relatively recent innovation and the result of the convergence of information technology and telecommunications [6].

In Colombia ICT has taken a big boom in the educational field since the Ministry of Information Technologies (MinTic) has made a variety of projects that allow the teacher the flexibility to carry out the teaching-learning process, one of the projects with the highest national recognition is the digital live plan (2014-2018), which supports the strengthening of the education sector, all this is supported by different programs that aim at the following objectives; contribute with the TIC to the reduction of the poverty, to generate employment in our country and to form micro, small and medium companies (Mipymes) with technological tools [7].

Master classes taught in basic and secondary institutions increasingly discourage students, teachers need new resources and tools to generate more didactic classes. The Ministry of National Education proposes an Educational Revolution to transform the Colombian educational system in terms of magnitude and relevance through strategies that expand coverage with criteria of equity, improve the quality of the education system and increase the efficiency and productivity of the sector. Two of the 37 projects that support these postulates include the incorporation of new technologies and methodologies in the country's education. In coverage, higher education and quality in basic and secondary education, with elements such as connectivity and information technology, use of television programs and educational radio.

Based on the research conducted by Pérez, C titled; Sustainable Social Entrepreneurship in Vulnerable Communities: Altos de la Sabana case in the city of Sincelejo; it has been possible to determine that the residents of the Urbanization are vulnerable people, because they are a population that was deprived of their homes because of different types of violence. This Urbanization is part of the 1 million free housing program of the national government [8]. Within the free housing program, an educational facility was built with the capacity of 22 classrooms that allowed 800 students to be accommodated. From the research carried out by the authors mentioned above, it can be deduced that the great majority of students of the Altos de la Sabana Educational Institution have little communication with their parents, since they are working in the informal sector, which leads to bad remuneration. , and instability, which by effect forces them to leave a lot of home to survive, evidencing that it is a community with a high degree of social vulnerability that affects in one way or another the educational performance of many children and young people in this area.

On the other hand, an interview was conducted with the rector Lodis Madera of the Altos de la Sabana Educational Institution, where she manifested the lack of technological resources and the way in which this problem affects the students' learning, since they do not have a classroom appropriate in which interact with technological tools and thus awaken skills that allow improving the teaching-learning processes of the student and teacher role. Currently the institution has a population of approximately 2200 students, with a ratio of 15 computers and two videobeam, evidencing a high technological gap by the number of students and teachers that the institution has.

This research is done in order to provide students with a way to mitigate the technological gap presented by this institution, through didactic prototypes, which will help them to improve their skills and be at the forefront of the technologies used today. in day. Within the public policies of the Colombian government, ICT is a way to strengthen education systems.

Interestingly, vocational guidance is a very important issue when training students with a strategic and assertive professional projection to the current era, which usually has a labor market in constant change. But none of this is possible without professional intervention for the discovery of talents or attitudes relating to an area, in the case of this project computer science. In this way increase the rise of computer professionals passionate about the profession and thus avoid academic desertion in the university stage, which is often the fate of many young people who due to ignorance of their talents, lack of advice and training, end up frustrating their dreams.

Society today demands from educational institutions a greater link with the processes that define the future of the country. As a consequence, great tasks are posed in efficiently fulfilling the training of students capable of facing new problems; consolidating a broad provision regarding the use of information and communication technologies, fostering progress in the generation of knowledge.

Many projects that have emphasized the use of technological tools using the ABP methodology and highlight the lack of mechanisms to take students to improve their learning process in a dynamic way that allows to get out of what is commonly given in the classroom. It is in this way that its authors emphasize the use of the aforementioned methodology and its great influence in the apprehension of knowledge, relying on information and communication tools.

A bibliographic review was carried out at national and international level, where the following antecedents with affinity to the research were found, as evidenced in table 1 [9-13].

1.1 Information of Communication Technologies

Information of Communication Technologies makes part of the most significant basis of the new education, and it represents the new scenario to comprehend the educational phenomenon that implies the new educational and pedagogical culture offering to this purpose a valuable support for its use strengthens the information interaction and exchange in appropriate surroundings and in various contexts that help the movement of competences and knowledge among students, breaking the traditional classroom developments and favouring the students' cooperative work [14].

1.2 Learning based on projects

This learning model requires for the professors to be creators, guiders and facilitators who stimulates the students to learn, discover and feel satisfied with all the knowledge they have acquired, thing that can be achieved if the teaching process, based on projects, is applied the best way. In other words, the learning process can be seen as accumulative, self-adjusted, guided, collaborative and individual [15].

1.3 Innovating technological tools

These tools constitute a learning laboratory about robotic and technological sciences from an integrating perspective that does not separate the other knowledge fields such as arts, design and style. These allow students to achieve new knowledge and develop technological abilities in the fields of basic and digital electronic, building with their own hands projects ranging from lamps, alarm bells and toy cars to handcrafted boats controlled by remote controls [16].

Table 1. Bibliographic review of antecedents of investigation.

Project author (s)	**Year of implementation or development.**	**Project's name**	**Methodology of the project**
Javier Calleja Pascal.	University of La Rioja., Academic year.	ABP and the use of ICT: a proposal to improve education.	The methodology that will be implemented is ABP (Project-based learning). Belted by a constructivist model.
Vanesa Ausín, Víctor Abella, Vanesa Delgado y David Hortigüela.	University of Burgos, Faculty of Education, Department of Education Sciences.	Project Based Learning through ICT. An Experience of Teaching Innovation from the University Classrooms.	Project Based Learning, ABP or PBL, which is part of the active methodologies. Field experimental model.
Ana García, Varcálcel Muñoz Repiso, Verónica Basilotta Gómez Pablos	Dept. of Didactics, Organization and Research Methods. University of Salamanca.	Project-Based Learning (PBL): evaluation from the perspective of P-education students.	Based on the ABP, (Project-based learning), supported by the descriptive methodology.
Vicent Estruch, Josep Silva	Dept. of Computer Systems and Computing. Polytechnic university of Valenci	Project-based learning in the Computer Engineering career.	Focused on the ABP methodology, (Project-based learning), supported by innovative pedagogical methods.
Gómez Trigueros, Isabel María.	General Didactic Department and Specific Didactics Social Sciences Area University of Alicante	The didactics of Geography and ICT: new teaching and learning methodologies.	Active methodologies such as Problem Based Learning (PBL), supported by collaborative and cooperative didactic strategies.

1.4 Hackids tools (Arduino set)

These tools constitute a laboratory for learning about robotics and technology sciences from an inclusive perspective that does not separate sciences from other areas of knowledge, such as art, design and fashion. They allow young people to develop knowledge and technological skills in the field of basic and digital electronics, building with their own hands projects that range from flashlights, alarms, toy carts and even craft boats made with recycling objects, managed by remote control [17].

These Are Elements Composed by a great variety of electronic components as sensors, engines, leds, resistors, Arduino panels, etc. whereby research can be generated.

2. Methodological framework of the investigation.

2.1 Design of investigation.
The focus of this research is quantitative, as well as the wait, the information, the analysis, the information collected. According to Hernández, Fernández and Baptista, the quantitative approach uses data collection to test hypotheses, with the numerical base and statistical analysis, to establish behavioral patterns and theories [18]. The research is field and with pre-experimental method. The selected sample is of 40 students with ages between 13 and 17 years, degree of the high educational institution of the savannah of the municipality of Sincelejo.

2.2 Technological resources.
The technological means with which the objectives of the project will be achieved are:

- Computer.
- Internet connection.
- Printer.
- Digital camera.
- Academic database.
- Office automation applications.
- CAD software for electronic design.
- Programming software.
- Procedures of the investigation.

2.3 Procedures of the investigation.
For the development of this research data is collected in three phase.

Initially for the first diagnostic phase of the investigation, we proceed to the application of an instrument that measures the degree of appropriation of knowledge that students have of the average of the Educational Institution Altos de la Sabana, in terms of analogical electronics concepts and digital; frequency of use of social networks in terms of academic part and self-directed profile.

In the second phase begins with the transfer of knowledge about the use of innovative technological tools, following the methodology based on projects, where students learn to build their own knowledge by applying the tools and the needs of their social environment.

In the third phase, an application is developed through the use of innovative technological tools Hackids or other digital electronic components, where young people develop a project together with CORPOSUCRE research seed students, showing their skills acquired in the transfer phase of the project. knowledge. Also elaborating an ex post analysis, through the application of the same diagnostic instrument applied in the initial phase, which allows to reflect the degree of knowledge acquired by the students, in terms of the use of the tools.

3. Results

This research will work in several moments, a pre-evaluation and a post-evaluation in order to encourage the use of ICT in the media of the high school of La Sabana in the municipality of Sincelejo. From the general analysis of the results of the initial test, we can affirm that the vast majority of students do not have knowledge about the technological tools, concepts and basic topics of electronics. They do not use the web frequently for educational purposes, but to check social networks such as (Facebook, Instagram) and YouTube. This could be a reason why students have academic deficiencies because they do not use the teaching tools properly. Another fundamental reason why students show flaws is that the vast

IOP Conf. Series: Materials Science and Engineering **519** (2019) 012031 doi:10.1088/1757-899X/519/1/012031

majority do not have Internet and computers. Important and fundamental tools for independent learning of each one.

For the second time, training sessions were held over a period of 5 months between the month of June and October of 2018, where 6 students and two teachers of the electronic engineering programs and system of the Antonio José de Sucre University Corporation, they worked on concepts related to analogue and digital electronics using the project-based methodology. The training cycle is evident in the following figure 1.

Figure 1. Trainings in technological tools using the PBL methodology.

For the third time the students carried out different projects focused on needs related to their environment, punctually in security, energy, unemployment, water, among others. Below, some elaborate rapid prototypes are evidenced in figure 2;

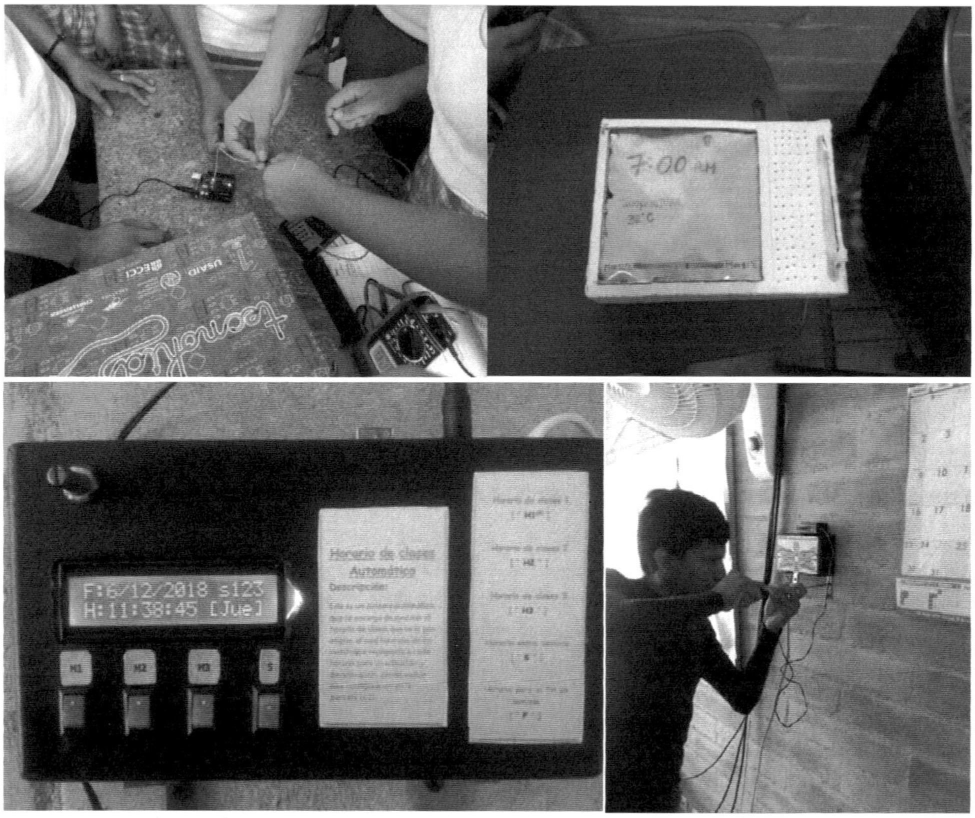

Figure 2. Design of rapid prototypes using the PBL methodology.

The contrast of the initial test and the final test related to the attitudinal part and concepts related to ICT is shown below in figure 3, figure 4, figure 5 and figure 6. for the test related to the ICT concepts, it is valued if the students answered correctly (good), incorrectly (bad) and does not know (does not know). In the table 2, the type of response of the attitudinal part is specified.

Table 2. Likert response of the attitudinal part

Minimization	Answer
(S)	I always feel and think like that
(CS)	I almost always feel and think like that
(O)	Occasionally I feel and think like that
(CN)	I almost never feel and think like that
(N)	I never feel and think like that.

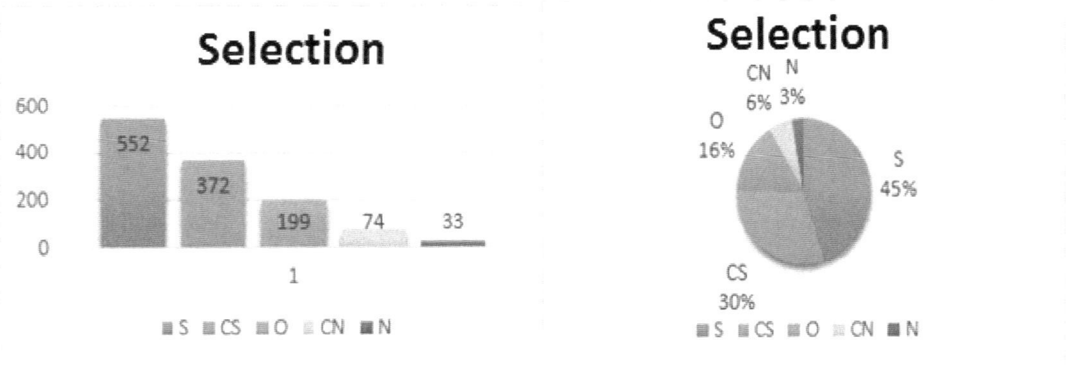

Figure 3. Results of the initial attitudinal test

75% of the students showed a positive result in the questions related to decision making, social value, attitudes, motivation and identify alternatives for solving problems.

Figure 4. Results of the final test of the attitudinal part

85% of the students showed a positive result in the questions related to decision making, social value, attitudes, motivation and identify alternatives for solving problems.

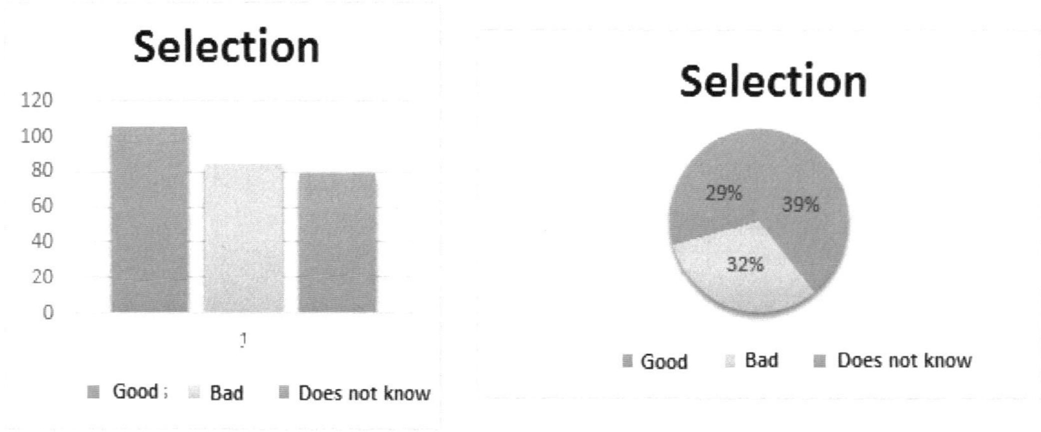

Figure 5. Results of the initial test part of concepts related to ICT

61% of students showed significant shortcomings in basic electronics concepts and also in terms of recognition of electronic components and units of measurement.

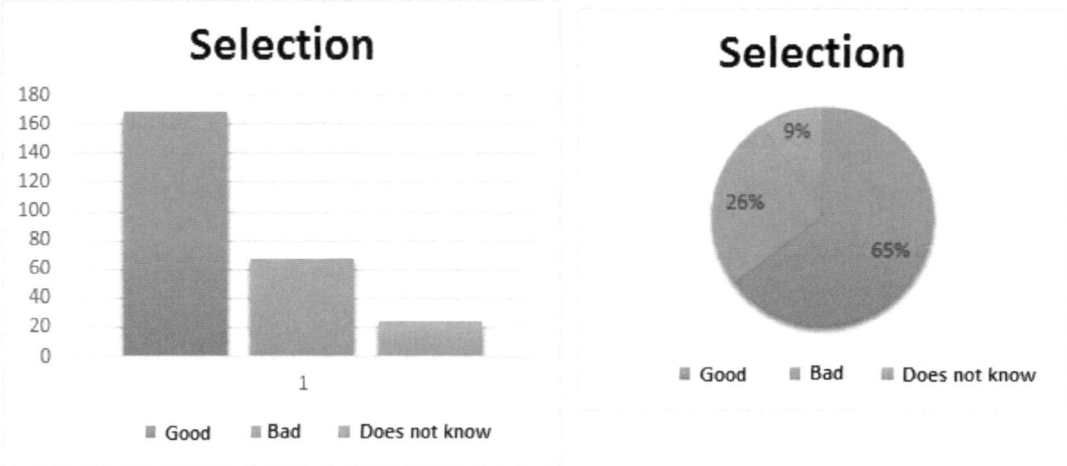

Figure 6. Results of the final test part of concepts related to ICT

65% of the students had the knowledge when answering the questions referring to basic electronics concepts and also in terms of recognition of electronic components and units of measurement.

4. Conclusion.

In a conclusive way we can affirm, from the results of this investigation, the benefits that can contribute to the educative institution of the savannah of the municipality of Sincelejo are the following:

- Democratization of culture, through technologies currently live in a globalized world in which we can access large amounts of information from anywhere in the world
- Appropriation of ICT for social inclusion, as a challenge to promote suitable scenarios of orientation of technologies for the approach of vulnerable populations to these tools.
- The ICT allow generating new teaching methodologies for teachers, abolishing the monotony of the master classes that have been working for a long time and generating more didactic spaces for students.

- With this type of tools, students can generate a sense of belonging for their environment, managing to solve problems in their communities.
- This class of modules mitigates technological illiteracy, generating skills and abilities that allow reducing the social, economic and labor gap in young people who are opposed to adapting to the use of the tools that surround them.
- The methodology of problem-based learning used in this research can be very positive for students, with it they can acquire skills related to creativity, problem solving and critical thinking, which will allow a better performance in the classrooms. It will also support the development of skills and life skills.
- Through this research, collaborative work is encouraged in students from classrooms, fostering values and good school life in a society increasingly deteriorated by individual behavior and social competitiveness.
- It would be pertinent to replicate this type of project in other highly vulnerable educational institutions, which do not have this type of technological tools.

With this research, it was possible to encourage and encourage the use of ICT using Arduino technology as a teaching and learning strategy in the students of the Altos de la Sabana Educational Institution, improving the teaching-student role. Finally, it is intended to replicate this project in another institution.

5. References

[1] PNUD 2015 Panorama General. Informe sobre desarrollo humano Consultado en agosto 2018.
[2] DANE 2015 Cuentas Nacionales. Consultado en Septiembre 2018.
[3] Salinas J 2006 Cambios metodológicos con las TIC. Estrategias didácticas y entornos virtuales de enseñanza-aprendizaje. Consultado en septiembre 2016.
[4] Ministry of National Education 2009, *año de la pertinencia educativa. Altablero.* December 2008 - January 2009; 48.
[5] OECD 2005 *Teachers matter: Attracting, developing and retaining effective teachers:* OECD Publications 13 p. Report No 1. Paris, France.
[6] UNESCO 2018. *World Education Report: UNESCO's reports*; Report No 1. Paris, France.
[7] MinTIC [Internet] 2018. Bogotá D.C. *El plan Vive Digital 2014-2018.* Available from: https://www.mintic.gov.co/portal/vivedigital/612/w3-article-19654.html
[8] Departamento Nacional de Planeación (DNP) 2014. *Informe al Congreso de la República,* Colombia. Presidente: Juan Manuel Santos.
[9] Pascual, J. C. 2018 *ABP y el uso de las TIC: una propuesta para la mejora de la educacion.* Universidad de la Rioja, https://biblioteca.unirioja.es/tfe_e/TFE002987.pdf.
[10] Vanesa Austin, V. A. 2016 *Aprendizaje Basado en Proyectos a través de las TIC.* Una Experiencia de Innovación Docente desde las Aulas Universitarias. scielo, https://scielo.conicyt.cl/pdf/formuniv/v9n3/art05.pdf.
[11] Ana García Valcárcel Muñoz, V. B.-P. 2017 *Aprendizaje Basado en Proyectos (ABP): evaluación desde la perspectiva de alumnos de Educación Primaria.* Revista de Investigación Educativa, http://revistas.um.es/rie/article/view/246811.
[12] Vicent Estruch, J. S. 2015 *Aprendizaje basado en proyectos en la carrera de la ingenieria informatica.*bioinfo, http://bioinfo.uib.es/~joemiro/aenui/procJenui/Jen2006/prDef0089_70efdf2ec9.pdf.
[13] Gómez Trigueros, I. M. 2016. La didáctica de la Geografía y las TIC: nuevas metodologías de enseñanza y aprendizaje. Rua, http://rua.ua.es/dspace/handle/10045/56570.
[14] Aguilar M. 2012 *Aprendizaje y tecnologías de información y comunicación: Hacia nuevos escenarios educativos.* Revista latinoamericana de ciencias sociales, niñez y juventud ; 10 (2):

801 - 811.

[15] EPALE 2018. *Plataforma electrónica dedicada a la enseñanza para adultos en Europa [Internet]. European commission; Aprendizaje basado en proyectos (ABP)*; February 22 [cited 2018 August 15]. [En línea]. Available from: https://ec.europa.eu/epale/es/resource-centre/content/aprendizaje-basado-en-proyectos-abp.

[16] Enda Santander SM and Solano Naula AE 2016. *Innovación pedagógica en relación con la tecnología educativa del subnivel medio de educación general básica [Degree work]. Machala*, Ecuador: UTMACH. Unidad Académica de Ciencias Sociales; [cited 2018 August 18]. Available from: http://repositorio.utmachala.edu.ec/handle/48000/8182.

[17] Candelas-Herías FA, Garcia GJ, Pomares J, Jara CA, Delgado Rodríguez A, Mateo Agulló C, Mira Martínez D and Pérez Alepuz J. 2015 *Experiencias sobre el uso de la plataforma Arduino en prácticas de Automatización y Robótica [Electronic resources]*. Alicane, España: Universidad de Alicante. Departamento de Física, Ingeniería de Sistemas y Teoría de la Señal; 2015 [cited 2018 August 18]. Available from: http://rua.ua.es/dspace/handle/10045/48815

[18] Sampieri RH, Collado CF and Lucio PB 2014 *Metodología de la investigación*. Vol 3. México: McGraw-Hill; 2006. p 850

Development of a web application for consultation and online payments of technological services in higher education institutions

C J Gómez-Sierra and B J López-Bustamante

Fundación tecnológica Antonio de arevalo, Tecnar, Cartagena, Pedro de Heredia Avenue, 49A Street # 31-45, Sector Tesca

cesar.gomez@tecnar.edu.co

Abstract. Currently services for online payments using gateways, have been displacing traditional payments. In this way, in addition to providing ease to the user, it also increases revenues from the use of these technologies. All electronic payment methods make life easier and make processing much cheaper, more efficient and more friendly for the end user, who with just a couple of clicks can process any request from your Smartphone, Tablet or Laptop. At present, the marketing processes of Tecnar's Technological Office services are carried out manually, as well as the offer to the external community in general. The process of paying for a service, vital to guarantee the income and sustainability of the business unit, does not have an online payment platform. For this reason, it is necessary to develop an information system that allows executing consultations and online payments.
A dynamic website will be delivered in a first phase using the PayU platform, to pay for computer maintenance services. Prior to this, there will be an investigation of other sites that use electronic commerce to promote their products. Finally, the development of the site will be carried out, implementing a database in MySQL, developing the site in HTML5 language through the BoilerPlate framework and the CSS3, JavaScript and PHP languages. In addition, implementing the Scrum methodology, which allows us to incorporate a set of good practices to work as a team and obtain better results. Preliminary effects include the creation of the database, the administration module and the PayU gateway registration.

1. Introduction

E-commerce can be defined, in a broad sense, as any form of transaction or exchange of commercial information based on the transmission of data over communication networks such as the Internet. In this sense, the concept of electronic commerce not only includes the electronic purchase and sale of goods, information or services, but also the use of the network for activities before or after the sale, such as: advertising, information search (about products, suppliers, etc.), negotiation between buyer and seller, customer service before and after the sale, etc. [1].

Undoubtedly, the rise of information and communication technologies and all their dissemination and development at an academic, social, cultural and business level means that every day people prepare and adopt the changes that are generated from the implementation of services and electronic commerce on the web (online payments). In this context, it is important to know how and when to develop strategies and digital marketing campaigns to publicize the services offered, such as WhatsApp marketing, Sendinblue and Acumbamail, accompanied by a business model on the Internet for its propagation as the B2B (Business to Business), B2C (Business to Consumer) or C2C (Consumer to Consumer) [2].

Content from this work may be used under the terms of the Creative Commons Attribution 3.0 licence. Any further distribution of this work must maintain attribution to the author(s) and the title of the work, journal citation and DOI.
Published under licence by IOP Publishing Ltd

Today, companies are forced to innovate and implement changes in their organization to take full advantage of the opportunities offered by electronic commerce; otherwise, they will be destined for bankruptcy. Therefore, the effective application of the strategy of transforming traditional businesses to electronic businesses, turns out to be a very important factor to obtain a competitive advantage over others.

Based on the above, the development of a Web Application for consultation and on-line payments of Technological Services in Higher Education Institutions is projected, with a user-friendly interface that can be used to pay for computer maintenance services in a first stage through of the PayU payment gateway implementation [3].

The website that will be delivered will have a Responsive & Clean appearance and a high degree of usability. Features very typical of simple programming languages such as: Php, Html5, MySQL, Css3 and JavaScript

2. Methodology implemented

The development of this project is done with the Scrum collaborative framework, which is used for complex or simple implementations and belongs to the outstanding agile methodologies that handle iterations and incremental lines. This framework is usually used for software projects, although today, thanks to its adaptability, it can be used in different contexts. The Scrum life cycle is made up of three (3) very important roles, such as the Product Owner (owner of the product), the Scrum Master and the Development Team [4].

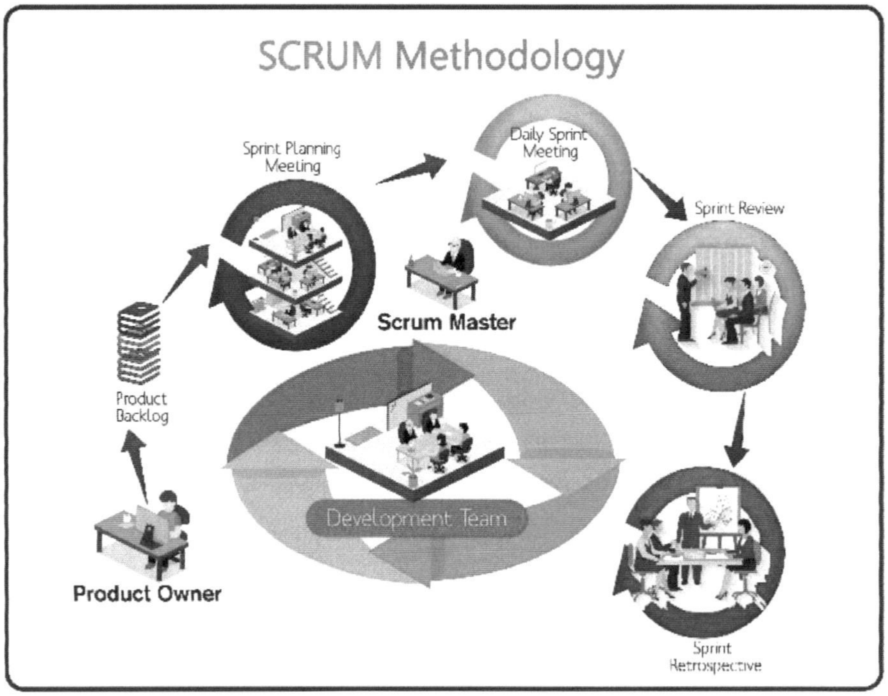

Figure 1. Scrum methodology

2.1. Product Owner

It is the person who represents the clients, something similar to the requirements engineer. This leader is in charge of raising all the information and knowing first hand all the needs of the client and thus be able to transmit them to the Development Team and the Scrum Master, so that they can build on that need.

2.2. Scrum Master

He is the leader of the work team, but he is not the person who gives orders or who says how things should be done, basically he is a moderator that helps the Development Team to understand the need of the client and the need that the Product Owner has raised.

2.3. Development Team

They are all people trained to solve the client's needs, building the project. These developers can be testers or analysts and must comply with the iterations (each sprint). This work team is represented by the author or owner of the project.

2.4. Product Backlog

Document defined by the Product Owner and contains all the needs, requirements and ideas of the client (these are called stories) that must be met. At this point, all the maintenance plans that will be offered on the website will be collected and created to be presented to the end user.

2.5. Sprint

Iteration whose duration is fixed beforehand and the equipment must comply with the provisions of the Product Backlog. Sprint is carried out every 20 days for this project.

2.6. Sprint Planning Meeting

Meeting to present by the Product Owner the stories captured in the Product Backlog, in order of priorities. In this initial stage, we present the design of the database to be implemented and the Sprint periodicity that will be 20 days from the first meeting. In addition, this meeting also reaches the conclusion that the programming language to be used is Php and JavaScript. The design and the interface will be made with Html5 and Css3 through the BoilerPlate Framework and the Sublime Text 3 editor, followed by an image that demonstrates this development.

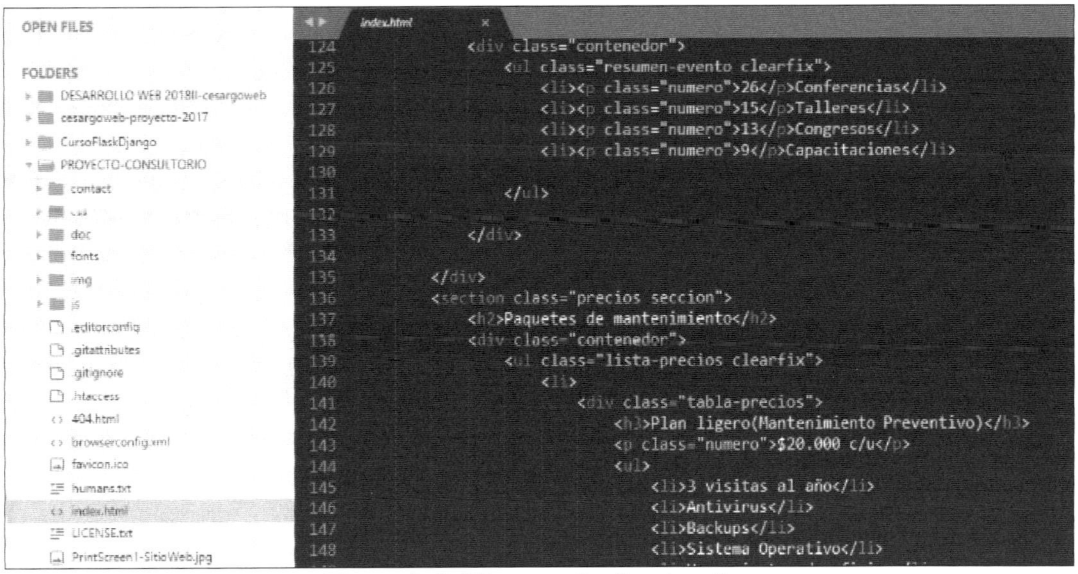

Figure 2. Sublime text 3 editor interface

2.7. Sprint Backlog

List of features that are obtained from the Sprint Planning Meeting. These are the requirements that must be built in each Sprint, which for this project is 20 days.

2.8. Daily Sprint Meeting
Daily meeting between the Scrum Master and the Development Team, maximum 15 minutes to track what exists in the Sprint and ask some specific questions: what was done yesterday, what is being done today, what will be done tomorrow and what problems will be they found.

2.9. Sprint Review
Meeting that takes place at the end of the Sprint to verify the fulfilment of the goals and objectives of the Sprint that is being worked on to guarantee the delivery of the final product.

2.10. Sprint Retrospective
Meeting that takes place after delivering the product in which the results of the previous Sprint are analysed and to see if there is something to improve or some shortcomings.

3. Dynamic website

Before starting the development and implementation of this project, it was necessary to do a study and survey of information in the city of Cartagena, to measure the percentage of use of online payment gateways and tools that are being used by Higher Education Institutions.

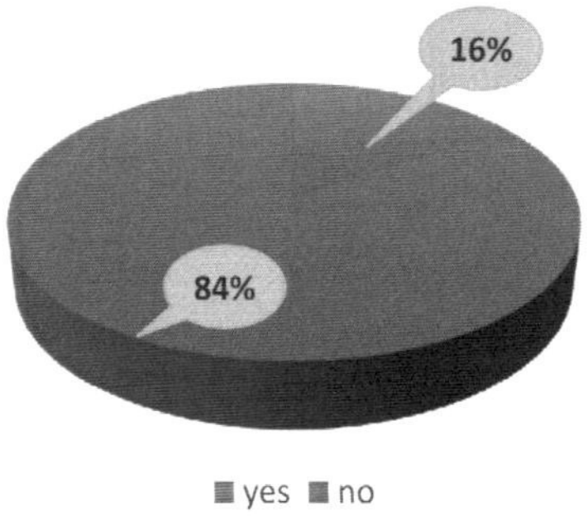

Figure 3. Online Payment System in Universities of Cartagena

Taking as a sample a number of nineteen (19) universities, it was evident that the data is not very encouraging in terms of the use of virtual payment systems. In figure 3 we can see that in the city of Cartagena, only 16% of the universities studied, have online payment and billing system, which shows that traditional collection systems such as payment in bank branches are still being implemented or windows of educational institutions and this registers 84% of the total intervened.

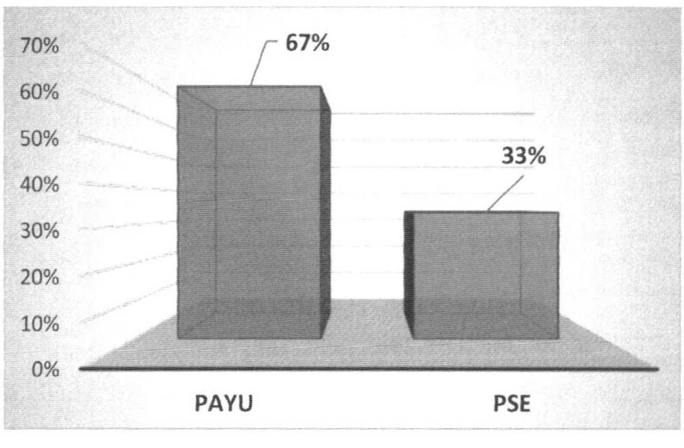

Figure 4. Payment Platforms in Universities of Cartagena

IOP Conf. Series: Materials Science and Engineering **519** (2019) 012032 doi:10.1088/1757-899X/519/1/012032

It can be observed in Figure 4, that the payment platform preferred by the Institutions that implement online payments is PayU, with a 67% share. This is explained by its technological versatility and by the easy way to be integrated into the information system of each entity.

In addition, this research shows that 100% of the institutions involved use their online payment method for tuition and continuing education such as specializations, extension courses, diploma courses and seminars, among others. Today in Cartagena there is no Higher Education Institution that promotes and sells technology services through its website. For this reason, the implementation of this online payment system through a dynamic website using the PayU payment gateway is relevant for the region.

This is the reason why we see the need to develop a site in which people, entrepreneurs, private and public educational institutions and strategic allies, can pay online for the technological services that are being promoted from the Tecnar technological office by middle of a PayU payment gateway. That is why we say that this project is relevant in our region, adaptable and scalable to other Institutions of Higher Education.

The website (consultoriotecnologico.tecnar.edu.co), which is anchored to the domain and server of the Institution, consists of a totally friendly interface, with a high degree of usability, with the application of standards and good programming practices. That W3C offers us and also with a Responsive & Clean design that provides the ease of navigation on any Android and iOS technology mobile device [5].

Figure 5. Homepage

The technological services that in this first stage of the project will be offered to be able to request and pay online are the different preventive maintenance plans to computers, at a regional level.

Figure 6. Option to pay

The design and structure of the site is supported in the Html5 language by means of the BoilerPlate Framework, which provides all the labels for the development of the interface, where any person could navigate without any inconvenience, a main banner with the name of the center is presented technological and navigation to social networks to achieve interaction with the user, a navigation bar with 4 (four) home links, who we are, services and contact us. The module to choose and pay for services online is in the central part of the main page or index. After creating the entire structure, the styles and the final appearance are defined with the CSS3 language, which gives us a unique aspect for visualization [6].

The dynamic part is worked with the Php programming language that connects with the MySQL database where all the tables that will contain the information of the visitors and people who make transactions are housed. This database is fully administrable with the PhpMyAdmin software, with which all the tables and fields that are worked on are created and modified [7].

For this project a previous investigation was carried out on which platform would be the most appropriate to connect to the website for the ease of our clients. It was found that several payment gateways operate in Colombia such as: Enlineapagos, Pagoagil, PayU, Pagosinteligentes, Ev1enlinea, Paypal, 2CheckOut, Interpagos, among others. According to the analysis carried out, it was decided to implement the PayU platform due to its important scope and characteristics. This gateway receives national and international payments, customers receive notifications to their registered email or cell phone immediately after their transaction, registration to the PayU account is free, the client can integrate their banking agreements, very low processing fees, anti-fraud control, monitoring of ongoing transactions, technical and operational support 7 * 24, debit card payments, credit card payments, cash payments, bank payments and PayU's option (in which the customer can pay the purchase up to 14 days without interest), among others [8].

Table 1. Comparison between payment systems

Payment systems	Advantage	Limitations
PayU	Security Anti fraud module Ease of integration Variety of methods World presence	Network dependency Returns
PayPal	Email association Security Speed by mail Immediateness of money Free to the buyer	Money limit High Commissions
Pse	Real time confirmation Real time identification Cost savings Decongestion	Registration in banks Second banking code

In Table 1, we can observe some advantages and limitations on which we base ourselves to choose the PayU catwalk as the basis of this project, because its structure is quite robust and has a high degree of security. Its antifraud module does not allow thefts or false transactions that harm the end user and is implemented almost everywhere as a payment system that provides a variety of methods for the ease of people [9].

For the action of making a consultation or online payment, the user must go to the central part of the site, where you will find 3 (three) maintenance packages and by choosing the desired plan by means of an input, you can use the pay button connect to PayU's database and online payment gateway to complete your transaction. Below is a detail of the website interface implemented [10].

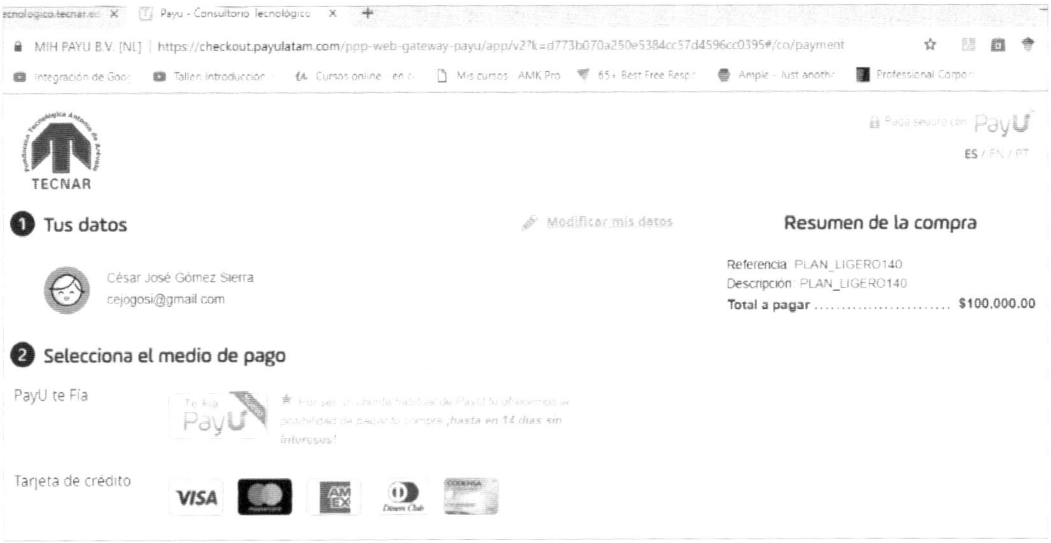

Figure 7. Payment finalization

4. Conclusions

With the development of this project, the implementation of an online payment system is achieved, which facilitates the achievement of the line of business of the Technological Office of the Antonio de Arevalo Technological Foundation and allows the possibility that many Higher Education Institutions, they also take ownership and can implement it as a model of additional generation of economic resources. This web page, whose interface is clean and totally friendly, allows through a system of entry of the number of computers that the user needs to be intervened, to have the total value calculated and make the payment with just a couple of clicks.

The project Development of a Web Application for consultation and online payments of Technological Services in Higher Education Institutions, is implemented in a scalable way, because it can be easily updated and also integrate new products to promote with their respective calculations and payments. The scalability of this software project also opens up the possibility of developing its mobile application for computers with Android and IoS operating systems.

IOP Conf. Series: Materials Science and Engineering **519** (2019) 012032 doi:10.1088/1757-899X/519/1/012032

5. References

[1] http://rii.cujae.edu.cu/index.php/revistaind/article/view/156/140
[2] Carter, L., & Belanger, F. (2004). The influence of perceived characteristics of innovating on e-government adoption. Electronic Journal of E-government, 2(1), 11-20.
[3] Możdżyński, D. (2017). The conceptions of new payment methods based on revised payment services directive (PSD2). Information Systems in Management, 6.
[4] Hossain, E., Babar, M. A., & Paik, H. Y. (2009, July). Using scrum in global software development: a systematic literature review. In Global Software Engineering, 2009. ICGSE 2009. Fourth IEEE International Conference on (pp. 175-184). Ieee.
[5] Natda, K. V. (2013). Responsive web design. Eduvantage, 1(1).
[6] Gauchat, J. D. (2012). El gran libro de HTML5, CSS3 y Javascript. Marcombo.
[7] Romano, J. M. G. (2015). Desarrollo de sitios web con PHP y MySQL. Obtenido de http://www. lsi. us. es/cursos/cursophp/apuntes/tema1. pdf.
[8] Chattopadhyay, A., Chatterjee, R., & Saha, A. A Study to Gauge Consumer Orientation towards E-Wallet Service Providers In Urban India. THE JOURNAL OF ARMY INSTITUTE OF MANAGEMENT KOLKATA, 17.
[9] Carvajal Escobar, K. H., & Ricaurte Puentes, L. F. (2018). Formulación de un plan de negocios para el desarrollo y comercialización de un aplicativo móvil encaminado a la gestión de reservas de canchas deportivas en el sector de Chapinero y Teusaquillo.
[10] Robles Salazar, J. M., López, M., & Lizek, G. (2018). IMPLEMENTACIÓN DE E-COMMERCE Y PUNTO DE VENTA ODOO PARA ABASTOTAL. COM. Instituto Tecnológico de Colima.

IOP Conf. Series: Materials Science and Engineering **519** (2019) 012033 doi:10.1088/1757-899X/519/1/012033

System of information for the management in the optics of the city of Riohacha la Guajira

J Quintero[1] **and S Romero**[1]

[1] Engineering Faculty, University of La Guajira, Riohacha, La Guajira- Colombia.

jquintero@uniguajira.edu.co

Abstract. The objective of the research was to develop an information system for the management in the optics of the city of Riohacha, which allows the information to be obtained in a reliable, timely manner, facilitating the whole process carried out in these companies, for the management and provision of services to their patients. The system includes an administrative management module and a sales and inventory module. This research is of applied projective type, with non-experimental, cross-field design, on the information management needs that are presented in the opticians of the city of Riohacha. The result of this investigation was the development and implementation of a system to strengthen the process of administrative management of the optical of the city of Riohacha, this software is the result of an analysis of the prerequisites of the processes of the optics of the city, the design and development of the system, tests and finally the commissioning of the System in one of the analyzed optics: the Iris Optics of the Guajira.

1. Introduction

The theme developed in the present investigation has the title: Information system for the management in the optics of the city of Riohacha la Guajira, where they will study the requirements with which a system must comply to give an adequate management to the information that corresponds to the area of visual health in the city of Riohacha, to develop a software tailored to meet the standards and needs that are presented in the opticians, since it is very important for this type of companies to store and give a good use of the information of the patients they attend, treatments, formulas, data of the EPS and companies with which they have contracts.

During the beginning of the investigation, a background search was carried out related to the research topic, in the environment to which the proposed information system is directed, at the local level these types of projects have not been developed, which is why we take as references the data that we find corresponding to some information systems that exist in Colombia to carry out the tasks within clinics and opticians, such as AgilMED which is used in more than 400 optical and more than 30 clinics nationwide, Oftalvisión professional products Specialized software applied to Visual Health, of which we took into account the information shown in the documentation that describes the functionalities offered.

The criteria that would be taken into account to develop the information system and the main development tools that currently exist in the market were also observed. Likewise, the methodology of the basic life cycle for software development was applied: An analysis was made of the requirements of the process carried out in the optics, the design of the interfaces, queries, reports and databases for the process of management of the optics, the implementation of the information system, the realization of the tests of the designed system and finally the start-up of the management system for optics.

Content from this work may be used under the terms of the Creative Commons Attribution 3.0 licence. Any further distribution of this work must maintain attribution to the author(s) and the title of the work, journal citation and DOI.
Published under licence by IOP Publishing Ltd

From the methodological point of view, the quantitative approach was used; the research is Projective, applied with a non-experimental, cross-field design. As you can see the results of this research, was the implementation of the Management System in the Optical in the city of Riohacha

2. Situation and Context

In this new knowledge and information society, certain traditional models and concepts are no longer valid, since the world is currently governed by a new order, the changes introduced by technology in our way of life make it necessary to rethink traditional models in all the fields being one of them the computer science and the systems, foundations of this process of transition [1].

All this has made the management of information is increasingly dependent on technology, as the increasing volumes of it are handled and its clearly multimedia nature require a treatment with increasingly sophisticated means. Access to networks such as the Internet via personal computers or the complexity of banking systems and fully computerized air reservations are clear proof that without technology, the use of information would be impossible at present. It is for this reason that software and Information Systems are important concepts when talking about Technologies. These are present in all information processes, since these functions are carried out with increasing intensity by computers.

Software is playing an increasingly innovative role in the Knowledge Society, enabling solutions for companies and introducing significant changes in the behavior of end users, both at home, at work and the implementation of information systems in the organizations, provide the possibility of obtaining reliable, timely, effective information and great competitive advantages.

According to the above, the opticians are organizations that are within the health sector, they provide care to people with visual problems, the lenses are diagnosed and prescribed according to the problem. Currently in the District of Riohacha there are around seven (7) Opticians, which provide various services to its users; some of them manage software tools, but they do not have information systems that allow them to have an integral treatment of the information. Because in some cases they have software created to handle the clinical part and that is not considered optimal for the management of the administrative part, sales and inventory; that is why some of these companies are looking for a way to obtain software that is useful for the management of all their areas.

With this research, we seek to create a customized information system that satisfies all the needs and requirements of the optical systems of the city of Riohacha, oriented to the web and adaptive or responsive.

3. Theoretical bases

3.1. The information systems

An information system (IS) is a set of elements oriented to the treatment and administration of data and information, organized and ready for later use, generated to cover a need or objective.

These elements will be part of any of the following categories: people, data, activities or work techniques, material resources in general. Generally, computer and communication resources, although not necessarily.

"Information system is a set of interrelated components that gather, store, and distribute information to support the decision making and control of an organization" [2].

For [3] the information systems of an organization are made up of the human being and material things, in charge of the information treatment of a business through the access of data, its development, accumulation and subsequent exit.

It can be any organized combination of people, hardware, software, communication networks and information resources that store, retrieve, transform and disseminate information in an organization [4].

It is a conjunction of elements such as human talent, hardware, software, networks that work harmoniously, organized, synchronized and integrated to process data, ordering, transforming and analyzing them, through the different organizational subsystems to produce information.

It is important to clarify that in all the above definitions all the elements interact to process the data and give rise to more elaborate information, which is distributed in the most appropriate way possible in a given organization, according to its objectives, and this applies as much the manual processes as well as the automatic processes. It is commonly believed that all information systems do their processing automatically.

Many information systems are initially manual systems that are then converted into computerized systems. In this research, when we talk about Information Systems, we will always refer to computerized information systems, so our concept of this type of system is: "Automated information system, is a set of elements, such as: software, hardware, database, telecommunications, people and procedures specifically configured to collect, store, and process data to be converted into timely and relevant information for decision making ", [5].

3.2.Components or elements of an information system

The information systems are composed of the following elements:

Software.

It is the logical component, it is constituted by the computer programs, the routines and instructions that make up the information system.

Hardware.

Computer equipment used to carry out input, processing and output activities, and may be keyboards, devices among others

Database.

It is an organized set of data and information

Telecommunications.

They are the electronic transmission of communication signals that allow organizations to create networks of computer systems. Since they make it possible for people to communicate between different channels and media, and thus facilitate work and access to information

People.

They are the most important element of most computer-based information systems. Information systems personnel include all individuals who administer, operate, program and maintain the system. They are all those who use the system to obtain results, such as financial executives, manufacturing operators, etc.

Procedures.

They are the strategies, policies, methods and rules for the use of automated information systems.

3.3. Activities of the Information Systems.

[6], considers that: "the purposes of information systems, like that of any system within an organization are: process entries, maintain data files related to the organization and produce information, reports and other Departures".

According to Quintero&Romero, information systems perform the following main activities based on the computer:

• Information entry: process in which the system takes the data it requires to process the information, through work stations, keyboard, diskettes, magnetic tapes, barcode, etc.

• Storage of information: it is one of the most important activities that a computer has, since through this property the system can remember the information stored in the previous session or process.

• Processing of information: this feature of the systems allows the transformation of the source data into information that can be used for decision making, which makes it possible, among other things, for a decision maker to generate a financial projection starting from of the data that contains a statement of income or a balance sheet in a base year.

• Exit of information: it is the capacity of an IS to extract the processed information or data of entry to the outside. Typical output units are printers, plotters, magnetic tapes, diskettes, voice, etc.

• Feedback: it is here, where it serves to make changes in input or processing activities. And it occurs when there is the presence of errors or problems, and you have the need to correct input data or it can also be when you want to modify or improve a process. This activity is of great importance for administrators and decision

3.4. Classification of Information Systems.
Consider that information systems are developed with different purposes, which depend on the needs of the company. Here is a table that summarizes all the categories according to their purposes [7]:

Table 1. Classification of Information Systems.

According to the organizational levels	Departmental Systems Institutional Systems Inter-Organizational Systems
According to Functional Areas (Central Mission Activities)	Accounting Systems Finance Systems System of Manufactures (Operations / Production) Marketing and Sales Systems
According to the type of support they provide	Transaction processing system **(TPS).** Management information systems **(MIS)**. Executive information systems **(EIS).** Decision support systems **(DSS).**
According to Computational Architecture	Centralized based on mainframes Centralized based on server network Distributed (Client-Server, Web-Intranet, Service-Oriented) Independent PC
According to the Activity Level Supported	Strategic Tactical / Management-Medium Operational

3.5. Software life cycle
Software life cycle. It is the process that is followed to build, deliver and evolve the software, from the conception of an idea to the delivery and withdrawal of the system. The different intermediate phases that are required to validate the development of a software are defined, that is, to guarantee that the software meets the requirements for the application and verification

3.5.1. Types of software life cycles
3.5.1.1. Linear life cycle. It is the simplest of all models. It consists of decomposing the overall activity of the project into separate stages that are carried out in a linear manner, that is, each stage is carried out only once, after the previous stage and before the next stage. From the point of view of management, it also requires that it be known from the outset, with excessive rigidity, what will happen in each of the different stages before starting it. The latter also minimizes the possibilities of errors during coding and minimizes the need to require information from the client or the user.

3.5.1.2. Pure cascade life cycle. This model of life cycle was proposed by Winston Royce in the year 1970. It is a cycle of life that admits iterations, contrary to the belief that it is a sequential life cycle like the linear one. After each stage, one or several revisions are made to check if the next one can be passed. It is a rigid model, not very flexible, and with many restrictions. Although it was one of the first, and served as the basis for the rest of the life cycle models.

3.5.1.3. Life cycle in V. This cycle was designed by Alan Davis, and contains the same stages as the life cycle in pure cascade. Unlike the former, two feedback covers were added between the analysis and maintenance stages, and between the design and debugging stages.

3.5.1.4.. Classical systems development lifecycle. According [8], the life cycle method for the development of systems is the set of activities that analysts, designers and users carry out to develop and implement an information system. The life cycle method for system development consists of the following Stages and each of them of different phases: Planning, Analysis, Design, Implementation, Testing, Implementation or Installation and Maintenance.

3.6. Optometry.
Optometry refers to the set of processes that allow measuring the limits of vision, visual acuity and chromatic sense. It also studies the retraction of the eye, that is, the deviation of the light rays. The objective of optometry is to allow a correction of vision through adapted glasses or contact lenses. All of these measurements are made through an optometer. In France, the practice of optometry is not defined by a legal framework defined in the Public Health Code. Opticians and professionals who make glasses use it illegally to adapt contact lenses.

3.7. Optics.
It refers to everything that is linked to the vision. An optician can be a store dedicated to selling products such as eyeglasses, whether for augmentation or sun, or a person who has a qualification that enables him to perform in this sector. Optics is, on the other hand, the technique for making lenses and other devices that allow people with conditions such as astigmatism or nearsightedness to see better.

3.8. Management Processes in Optics
In general, optics are set aside and appointments are canceled, patients are taken care of, their personal data and those of their EPS are entered, the patient's medical history is assessed by optometry, lenses and frames are prescribed and sold.

4. Methodology
In accordance with the objectives set and the intentionality of the researchers, this research supports their epistemological actions in the qualitative approach and the systemic approach; this agrees with what [9] says: In summary, the universe is the system of all systems. For this reason, it can only be understood and effectively controlled if a systemic approach combined with the scientific method is adopted. Consequently, the systemic approach provides an adequate analytical and conceptual framework for the study of systems, where interactions are the fundamental aspect for their understanding.

The type of research is Proyective, since its main objective is to propose an information system for the management of information in optics [10], which tells us that projective research proposes to provide solutions to a situation determined from a process of inquiry, which involves exploring, describing, explaining and proposing alternatives for change.

Likewise, the design of the research, refers to "where and when the information is collected, if they are living sources and is collected in its natural context the design is field", such criteria are Applicable to the present investigation, in which the information and data necessary for the realization of it, is collected directly from the opticians located in the city of Riohacha. Consequently, the research is of the field, that is, the data of interest are collected directly from reality through the specific work of the researcher. On the other hand, the design of the research is non-experimental, that is, it is an investigation where the independent variables are not intentionally varied, which for this research is the Information System for the management of opticians. At the same time, the study is of transectional or transversal design, since the observation of the phenomenon is carried out in a single moment, without evaluating its evolution over a period of time [11].

4.1. Población

Similarly, the population of a research represents the unit of observation, by which statistical information was obtained within the research and constitutes the fundamental support to systematize the study. For the present investigation, the population is constituted by optics of the city of Riohacha in the department of La Guajira that are presented below: Óptica Riohacha Ltda, Óptica Central IPS SAS, Linares Óptica, Órisica Iris De La Guajira, and Óptica Guajira The table 2 shows the composition of the population.

Table 2. Composition of the population

Óptical	Location
Óptica Riohacha Ltda.	Cr15 10-14
Óptica Central I.P.S S.A.S.	Cr6 Cl 7 Esq.
Óptica Linares	Cl 2 6-43 L-103
Óptica Iris De La Guajira	Calle 14 10-27 Libertador
Óptica Guajira	Cl 7 9-36
Imagen Visual Moderna Óptica	Cr15 9-63 Av. Los Estudiantes
Toro Eiler Optómetras	Cr 11 14-39 Consultorio 214

4.2. Study show

In this sense, the sample is essentially a subgroup of elements that belong to that set defined in its characteristics, called population, contributing to the fact that the information collected is accurate to what would be obtained from the universe it comprises. That for this case are all the Opticians of the city of Riohacha in the Guajira (Colombia).

According [12], the use of sampling is advisable when the population is infinite or when there are large finite populations. In the present investigation the population is constituted only by (7) Opticians, which is considered a very small number; so it is used as sample shows the total number of the population.

Additionally, taking into account the nature of the variables investigated, the study considers as key informants the optometrists, and secretaries or advisors of the opticians of the city of Riohacha in the department of La Guajira. The above is summarized in Table 3.

Tabla 3. Target population

Óptical	Optometrists	Secretaries or advisors
Óptica Riohacha Ltda.	1	2
Óptica Central I.P.S S.A.S.	3	2
Óptica Linares	1	1
Óptica Iris De La Guajira	1	2
Óptica Guajira	1	2
Imagen Visual Moderna Óptica	1	1
Toro Eiler Optómetras	1	1

4.3. Techniques and instruments for data collection and research procedure

For the development of the theoretical framework, a bibliographic review related to the research variable was carried out, namely, Management system for the optics of the city of Riohacha in the Guajira, in order to support research and instruments of data collection used in it.

Regarding the instrument used to obtain the information, refer to the "tool used by the subject researcher to gather information about the event, event or phenomenon that investigates.

In this study, the interview was used as a technique to record the information and a questionnaire was applied as a tool, consisting of a series of questions with open-ended alternatives. In this specific case, the instrument is made up of 25 items, with an alternative of open response addressed to Optometrists and Secretaries or Advisors of the Opticians of the city of Riohacha. This interview allowed to have clarity about the processes, the actors, and the input and output information of the system and the relationships between them; it also made it possible to diagnose the current situation of the opticians in the city of Riohacha and to establish the requirements of the system in order to reach the final result.

In general, the interview answered the following questions:
What are the processes that take place in optics?
How does the process begin?
How does the process end?
Who are the actors in the process?
What standards and protocols related to the process?
What information enters and leaves the process?
What are the relationships of the process?

In the requirements analysis, a detailed description was made of the roles and tasks that each one of them must perform.

Subsequently, a specification of the scenarios was elaborated, which corresponds to the way in which the tasks for each of the established roles will be carried out; then use case diagrams were developed as a technique to specify the behavior of the system, continuing with the description of each one of the diagrams and after this analysis a conceptual design and model of the database was developed, later it was developed the information system, the respective tests were carried out before and after the implementation and finally, it was implanted in the Iris optics of La Guajira.

4.4. Development language
The information system for optics is developed in PHP and has a connection to a database in MySQL. When working based on PHP, the content and views of the system will be dynamic and will have an iteration with a database server designed for hosting, respecting the policies that exist for the access and development of the company that wishes to implement it. Tool for the storage and management of information.

4.5. Development tools
The development tools are open source, the information system design is made with the Laravel version 5.5 framework and the text sublime text editors in version 3, Visual Studio Code and Xampp local server version 7.2.1. The modeling of the system was carried out with the Balsamiq Mockups tool in its trial version.

5. Results
5.1 Requirements analysis
Based on the interviews carried out in the different perspectives, the identification of roles and tasks was carried out. The following describes the roles that will participate in each information exchange and the tasks that it will have to support according to the participation that each actor has on it.

5.1.1 Rol: administrator
• Authenticate as system administrator and have access to the privileges that this merits.
• See patients who are admitted to the system waiting to be seen.
• Attend patients and fill in the data corresponding to the procedure performed in the appointment.
• Consult patient data in the system corresponding to clinical history, previous consultations, and identification data.

IOP Conf. Series: Materials Science and Engineering **519** (2019) 012033 doi:10.1088/1757-899X/519/1/012033

• Formulate medications.
• Make glasses.
• Make and manage sales of products and services.
• View and manage inventory products.

5.1.2 Rol: advisor / advisor
• Log in with an advisor or receptionist and password through the authentication interface.
• Register new patients and update existing data in the system.
• Carry out the different queries regarding the information of the clinical history of patients through the information system.
• Print patient's medical records.
• Set appointments.
• Manage the status of appointments in the system.
• Enter patients to the system to meet appointments and make the corresponding payment to him or the services provided to the patient.
• Make sales of products or services.
• Enter new products into the inventory.

5.2. Specification of scenarios.
We will proceed to make a description of how the information system will be used from the tasks conceived for each role, in a specific way each of the scenarios that describe these tasks.

5.2.1. Administrator
• Authenticate as system administrator. To access the system as an administrator, you must Log in with a username and password, in this way you will have access to the privileges that this offers to insert, update, view and delete data that correspond to patient information, clinical history and sales.
• See patients who are admitted. The administrator can see the patients once they are entered into the system by means of a waiting list that indicates that they can already enter the office to receive the service for which they go to the Optician.
• Attend patients. The administrator will assist the patient and write the findings found in the patient's medical history.
• Consult patient data in the system corresponding to clinical history. After attending to a patient, the administrator will be able to make inquiries related to the patient's clinical history, last control, treatment, among others.
• Formulate medications. The system administrator after diagnosing the patient can order medications to treat their discomfort if necessary.
• Make glasses. If the patient requires it, the administrator will order medicated lenses according to the condition shown by the exams at the time of care.
• Make and manage sales of products and services. The administrator can make sales of the products in inventory and can also sell the different services offered by the optics.
• View and manage inventory products. The administrator can see the products with which the inventory counts, add new products, update and eliminate them if necessary.

5.2.2. Advisor / a receptionist
• Log in with an advisor or receptionist. To enter the system you must log in as an advisor or receptionist and the software will allow you to access the benefits that this type of user is assigned.
• Register new patients and update existing data in the system. An advisor or receptionist can create new patients in the system or update their data.
• Carry out the different queries regarding the information of the clinical history of patients through the information system. This user can consult the patient's medical records but only to see and print, you cannot modify or delete this information.
• Print patient's medical records. The consultant can make a consultation through which he will see the patient's clinical history and then print it.

• Set appointments. The counselor has the authorization to set an appointment once a patient arrives who requires it in the event that he or she comes in a particular way, this task will not be carried out.
• Manage the status of appointments in the system. When a patient needs to make some type of modification to their appointment, or failing to assist the advisor can make the relevant changes regarding this in the system.
• Enter patients to the system to meet appointments. When the patients arrive to meet their appointments, the advisor must make the respective entry and make the collection of the value of the service acquired by the patient if necessary, and then this will be placed on the waiting list to enter the office.
• Make sales of products or services. The advisor is authorized to make sales either of products or of the services offered by the optics to the public, or to the companies respectively according to the customer's need.
• Enter new products into the inventory. When new products arrive, the advisor will register them in the system.

5.3. Design and Modeling

5.3.1 Diagrams of cases of use. This point consists of modeling the system from the use case diagrams that are the description of the actions of a system from the point of view of the user. It is a valuable tool given that it is a technique of successes and errors to obtain the requirements of the system, just from the point of view of the user.

The use case diagrams model the functionality of the system using actors and use cases. The use cases are services or functions provided by the system for its users.

According to the above, below will be seen in images the use case diagrams that specify the behavior of the system, which were designed with the modeling tool in the UML standards, StarUML which is a graphic language to visualize, specify, build and document a system. The version used was the test one, See the figure (1, 2, 3, 4, 5).

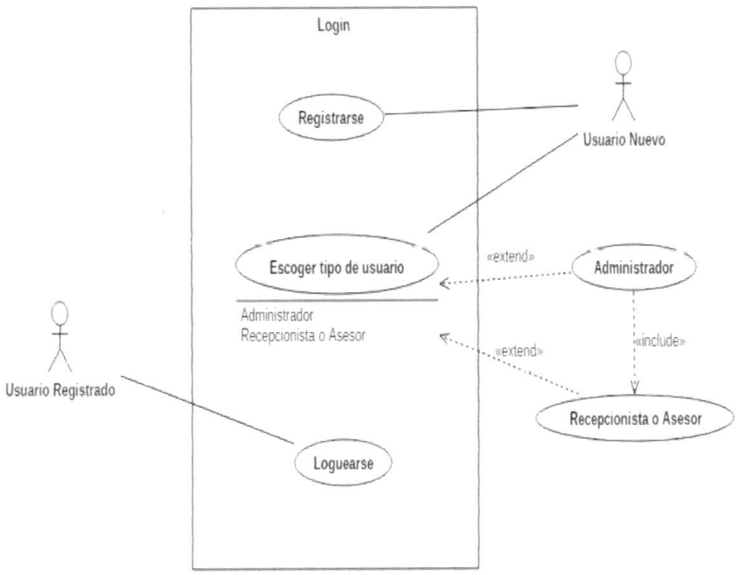

Figure 1. Case of Use of User Entries to the System

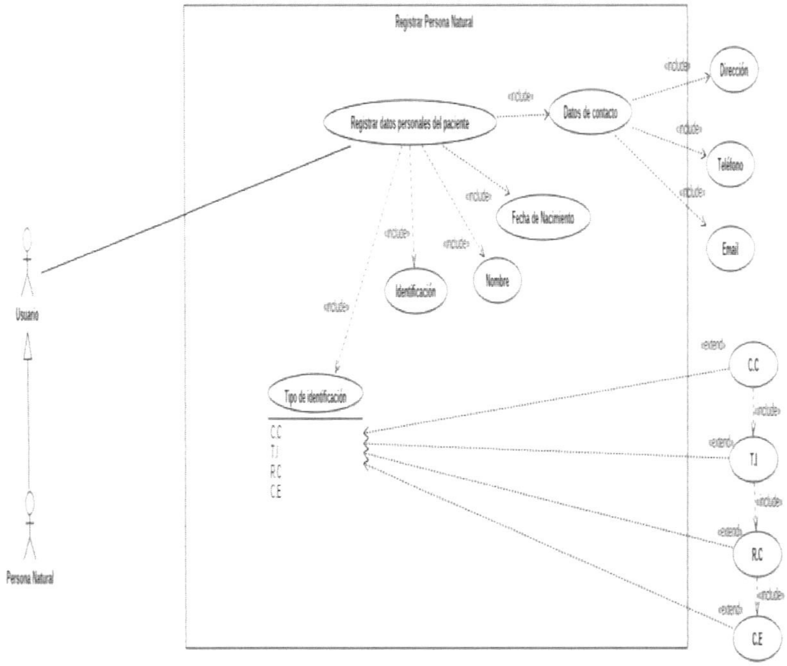

Figure 2. Patient Registration Use Case

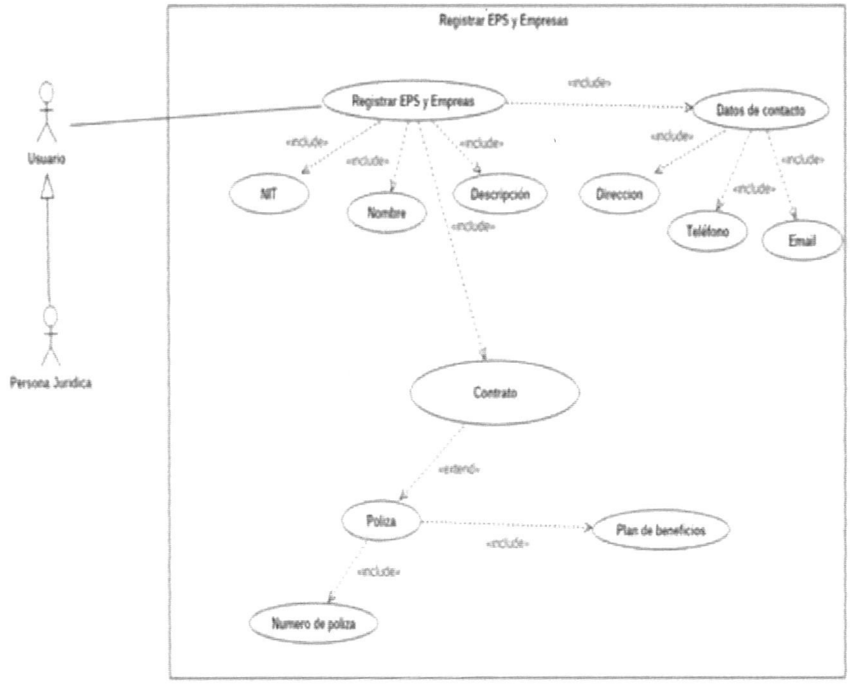

Figure 3. Use Case Register EPS and Companies

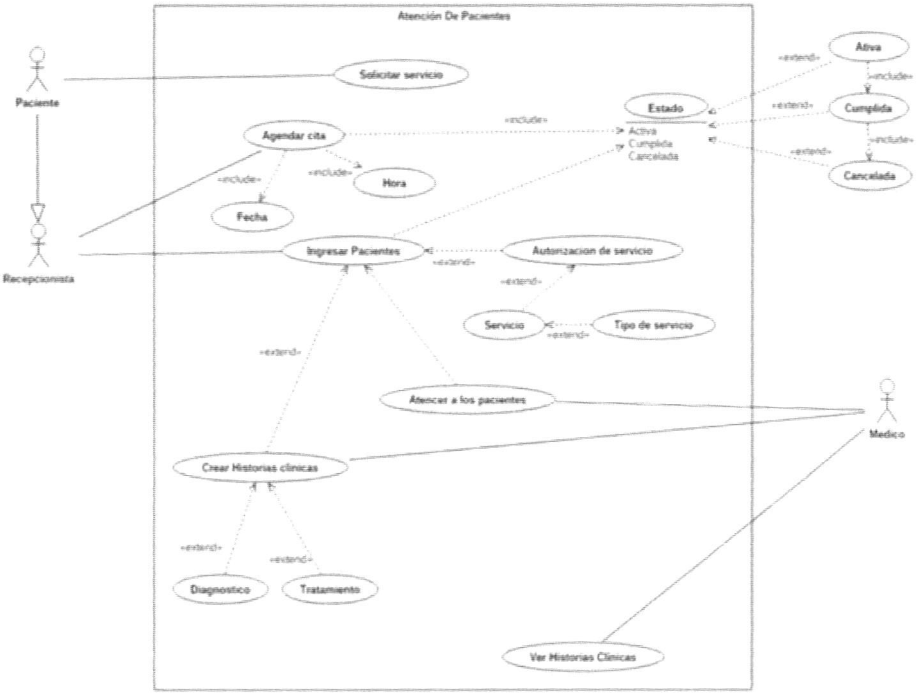

Figure 4. Case Use Patient Affiliation

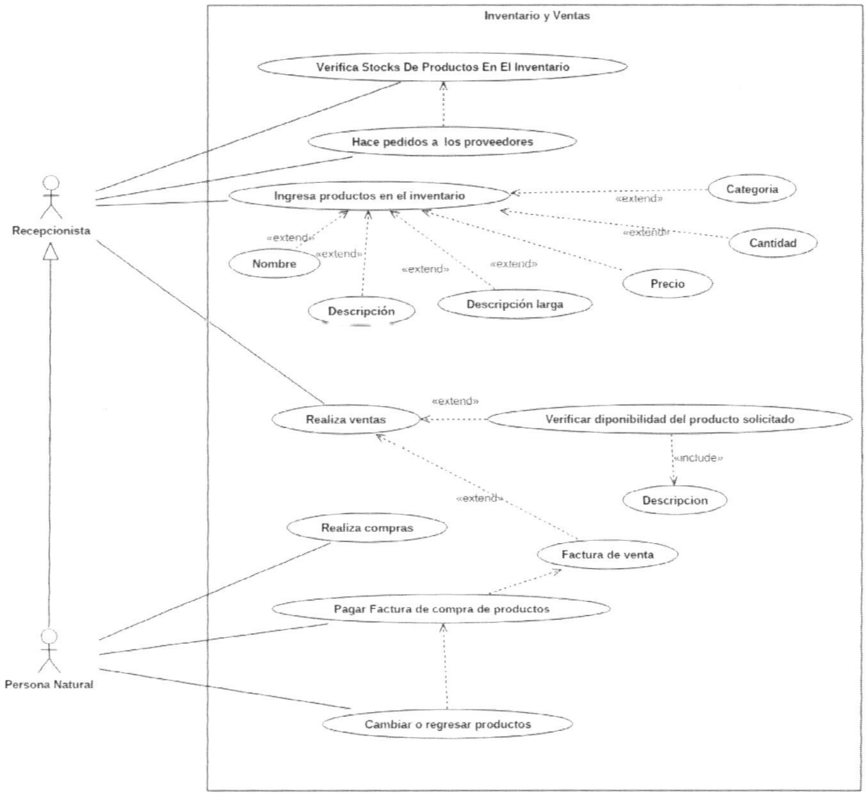

Figure 5. Use Case Sales and Inventory

IOP Conf. Series: Materials Science and Engineering **519** (2019) 012033 doi:10.1088/1757-899X/519/1/012033

5.3.2 Design of the topic base. The information system for opticians, aimed at the Opticians that are located in the city of Riohacha in the Guajira, uses a database developed with the MySQL engine in version 5.4 because it is a database management system relational developed under a dual license GPL / Commercial License by Oracle Corporation and is an open source database; which does not report costs or additional expenses for the company that wishes to implement the system. The use of MySQL allows the system not to have too many demands and does not generate additional costs since it is possible to use it in basic computer equipment that are currently the majority of the companies. On the other hand, the MySQL database engine is very complete and can store a large amount of information without causing disruption in the correct functioning of the information system. See figure 6.

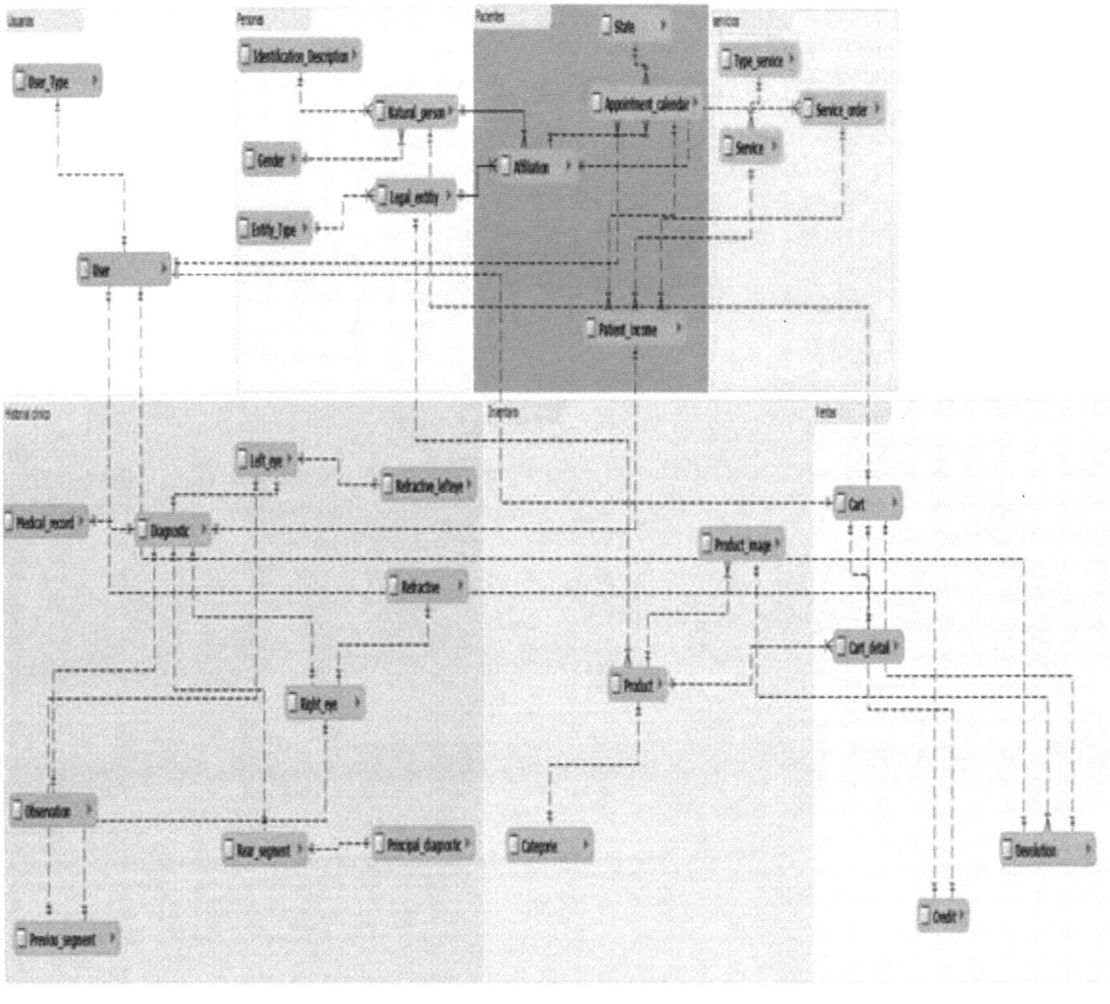

Figure 6. database model

6. Conclusions

Throughout the investigation, it was observed that in the District of Riohacha there are seven Opticians, which provide diverse services to its users; some of them handle software tools, but do not have complete information systems, which handle the clinical part and the administrative part, which allow them to have a good treatment of the information. As a main problem we determine the need that exists in these companies to have an integrated software that allows the reliable and timely management for the effective functioning as entities providing visual health services in Riohacha.

It also made it possible to diagnose the current situation of the opticians in the city of Riohacha and to establish the requirements of the system in order to reach the final result.

In the requirements analysis, a detailed description was made of the roles and tasks that each one of them must perform.

Subsequently, a specification of the scenarios was elaborated, which corresponds to the way in which the tasks for each of the established roles will be carried out; then use case diagrams were developed as a technique to specify the behavior of the system, continuing with the description of each one of the diagrams and after this analysis a conceptual design and model of the database was developed, later it was developed the information system, the respective tests were carried out before and after the implementation and finally, it was implanted in the Iris optics of La Guajira.

With a stable model of the database, we determined that the design of the system interfaces was made, for which the development language Php and the Laravel framework in its version 5.5 were defined as a development environment. to develop what we will see as the Information System for management in the optics of the city of Riohacha la Guajira, where we will find a patient module where all the necessary actions can be taken to give attention to any person or user of the services of some optician in this city, an inventory and sales module was also carried out where you can keep track of the products that are available for sale and the sales that are made.

At the end of the design and development phase of the system, the product was tested in an Iris de la Guajira optic in the city of Riohacha and training was carried out for 15 days so that the users of the system were familiarized with their functions and characteristics. in such a way that they could also say if it meets their needs or not, at this stage it could be observed that the system was accepted, since it meets the necessary characteristics to provide care to patients and also make sales using the same software. A solution was given to the needs that we initially observed.

At the end of the testing phases of the system and training to the end users, and with the approval of them for the implementation of the software in their company, we can conclude that this project had the impact that we expected when solving the problems presented by the users. to fulfill their work tasks on a day-to-day basis; It is also clear that the implementation in this company is intended to motivate others to try and appropriate the same system so that over time the impact of this is greater in the city and can meet the needs of more optical.

References

[1] Castell 2006 *The information age vol. 3 economy, society and culture*. End of the millennium Editorial Alliance, Spain.

[2] Romero, Sandy, and Jaider Quintero 2014 *Programming and information systems: level of knowledge in programming and its impact on the development of automated information systems in the systems engineering students of the University of La Guajira* International Review of technology, knowledge and society 3 (2): 55-74. Doi: 10.18848 / 2474-588x / cgp / v03i02 / 55-74.

[3] Guzman C. 2002. *Análisis y Desarrollo de Sistemas de Información.*Chile.
Source: http://www.observatorio-iberoamericano.org/.

[4] O'brien, james. 2003. *Management information systems*. Fourth edition. Irwin-mcgraw hill. Colombia.

[5] O'brien, and marakas. *Management information systems*. Mcgraw-hill / irwin, 2006.colombia.

[6] Senn 2005. *Analysis and design of information systems* Madrid Spain: pearson prentice hall.

[7] Kendall and Kendall. *Analysis and design of systems* . Pearson. 2005 Edition.

[8] Joyanes, l 2000 *Cibersociety Social challenges in a new digital worl*d. First edition. Editorial mc. Graw hill. Bogota Colombia.

[9] Bunge M. 2009. *The science. His method and his philosophy.* Editorial New Image. Buenos Aires.

[10] Hurtado J 2012. *The Research Project.* Ediciones Quirom. Caracas Venezuela .

[11] Hernandez, fernandez and batitsta. 2010 *Research methodology* Fourth edition Ediciones Mc. Graw hill, mexico.

Application of the Wilcoxon Test to correlate the results of the Saber 11 and Saber T&T Test

L K Ahumada[1] and I D Sanchez[1]

[1]Antonio de Arevalo Technological Foundation TECNAR

Luz.ahumada@tecnar.edu.co

Abstract. In Colombia there is a model of monitoring the Quality of Education in which different tests are applied throughout the academic lifecycle of a student, in basic education, secondary education, until finally reaching higher education. Among these tests are: the Saber Test 11 and the Saber T & T Test, among others. The first, applies to students of the last grade of secondary education and the second to students of the last semester of the Technical and Technological program. It is possible to think that a person who obtained good results in the test saber 11, after preparing in a Higher Education program and presenting the corresponding test, should present a notable improvement in their results, if compared with the first exam. The foregoing is not necessarily true, in order to make this statement it is essential to carry out a study that supports this theory. The present research makes use of Test Wilcoxon to compare the means of 2 distributions, for this case particularly: the Saber 11 and Saber T & T tests of the same student. From the Saber 11 test, the generic Mathematical and language competences are chosen, and for the Saber T & T tests, quantitative reasoning and critical reading were selected. The selected population are the students of the Technology in Systems program who took the test in 2012, belonging to the Faculty of Design and Engineering of the Technological Foundation Antonio de Arevalo - TECNAR, they are students, product of the first accreditation of the program.

1. Introduction

In Colombia, education is defined as a process of permanent, personal, cultural and social formation that is based on an integral conception of the human person, their dignity, their rights and their duties. [1]

The Colombian educational system is made up of: initial education, pre-school education, basic education (in primary, five grades; and in secondary, four grades), secondary education (two degrees and culminates with a bachelor's degree) and higher education [2]. The latter was selected to carry out the investigation.

Quality, in the context of education, and especially applied to institutions of higher education, refers to an attribute of the public service of education in general and, in particular, to the way in which that service is provided, according to the type of institution concerned [3]. Quality, well understood, entails the continuous effort that the state and its institutions make to fulfill with responsibility the specific demands of each one of its functions [4].

One of the mission objectives of the Ministry of National Education-MEN is to improve the quality of education regardless of what level we are talking about.

Content from this work may be used under the terms of the Creative Commons Attribution 3.0 licence. Any further distribution of this work must maintain attribution to the author(s) and the title of the work, journal citation and DOI.
Published under licence by IOP Publishing Ltd

IOP Conf. Series: Materials Science and Engineering **519** (2019) 012034 doi:10.1088/1757-899X/519/1/012034

In order to measure quality, the MEN devised an instrument called Saber Tests, which is applied according to the degree of training in the following way:

Saber Test. Applies to 5th and 9th grade students. It focuses on areas related to Language, Mathematics, Natural Sciences and Citizen Competencies.

Saber Tests 11. Applies to students who finish grade 11 and focuses on areas such as language, math, physics, chemistry, biology, geography, history, philosophy, foreign language

Saber Tests T & T. It is applied to students who finish technical and professional programs in Higher Education Institutions. The learning and the competences of the academic program that is being studied are evaluated. The present investigation focuses on these last two tests, it should be mentioned that the presentation of these exams is obligatory and sometimes they are an essential requirement to obtain the degree to which they aspire.

2. Institutions of Higher Education in Colombia

The Institutions of Higher Education (IES) are entities that count, according to the legal norms, with official recognition as providers of the public service of higher education in the Colombian territory [5].

The academic character is the main feature that since the constitution (creation) of a higher education institution defines and gives identity to the competence (field of action) that in academics allows you to offer and develop higher education programs, in a another academic modality [6]

- Higher education in Colombia is divided according to its academic character into 4 levels:
- Professional Technical Institutions: These institutions that are characterized by offering short-term technical programs.
- Universities: those that are able to offer all academic programs including masters and doctorates.
- University Institutes: Are institutions that offer university degrees, graduates get titles as "professionals"
- Finally, the technological institutions: These are institutions that offer technological programs at the professional level in a medium term.

3. Characterization of the Program

Precisely, in this last category is where the Antonio de Arévalo Technological Foundation is located. It was created at the beginning of the 80s as a technological foundation in the city of Cartagena. [7]

Currently, TECNAR has three faculties: Faculty of Economics, Faculty of Social Sciences and Faculty of Design and Engineering. This latter has 14 academic programs, among which is the Systems Technology Program. Accredited on three successive occasions as a High Quality program, fulfilling the mission objectives established by the Ministry of National Education.

The Technology-in-Systems Program has been characterized by training students in the areas of software knowledge. in order that graduates can perform with a technological and humanistic approach in functions of: Development of computer programs, deploying skills and abilities for the use of technological means in an open and permanent manner also involves other areas of knowledge such as: exact sciences, communication skills among others.

To date, the program has approximately 340 graduates. The communication with them is given through the division of Social Projection, which is a specialized area that tends to maintain close and permanent relations with them.

Currently, the institution does not have a mechanism to measure the knowledge that graduates acquired during their permanence at the institution. Therefore, the academic units can not reflect or assess themselves about the relevance of the microcurriculum with strong bases. Due to the above, it is necessary to analyze the performance of the graduates of the Technology-in-Systems program. it is then suggested to make a comparison of the results related to the generic competence of the Saber Test 11 (The students in what state they enter the institution to study? academically speaking), Saber T & T (The students in what state they finish their studies in the institution? academically speaking)

IOP Conf. Series: Materials Science and Engineering **519** (2019) 012034 doi:10.1088/1757-899X/519/1/012034

4. Methodology

In order to obtain answers to the previous interrogations, we chose as a unit of analysis those high school graduates from the different schools of the city of Cartagena - Colombia, who later attended and later graduated from the Technology in Systems program. The databases used were downloaded from the Colombian Institute for Education Evaluation and the Institutional Information System portal. Subsequently, mathematical and statistical models were used to correlate variables and give a light in their behavior, with the Wilcoxon test being the best option to achieve it.

It must be taken into account that the Saber 11 test differs in the year of presentation and therefore in the structure and evaluation of it. To remedy that it is necessary to make a transformation in the scores of the Saber T & T Tests because the Score Range is 0.200 and you want to make the comparison with the Saber Tests 11 whose Range is 0.100. For this circumstance in the present investigation, the division of the test score will be done by T&T by 2 and thus parity will be made between the results of the 2 tests.

The Wilcoxon test is the tool used to perform the correlation, it is a statistical test that contrasts two samples of the qualitative variable relationship. [8] For the present study (Saber Tests) and another quantitative variable (results obtained by the students) and which makes a comparison contrast of the central tendency, the median.

Testing is used when there are very small sample spaces. Wilcoxon contrast can be used when a random sample of linked pairs is available [9]. In the present study they are related because the same students are evaluated in two different moments, a first moment with the Saber Test11 and a second moment with the Saber Test T & T.

With this test, we will compare the differences of the averages of the Groups. Between the average of the Results of the Saber Test T & T, and Saber Test 11, and answer the Hypothesis if the differences between this median is zero, this is Null Hypothesis (Ho: Difference between averages = 0) or if the differences between the averages is different from zero, this is the Alternative Hippothesis (Ha: Difference between averages ≠ 0).

If the statistical result yields a result of **Ho** then it would be concluded that the results of the Saber T & T Tests and the Saber Tests 11 do not have a significant change after the process of Formation of the students in the Technology Systems Program. Otherwise. If the statistical result yields a result of **Ha** then it would be concluded that the results of the Saber T&T Tests and the Saber 11 Tests have a significant change after the process of training of the students in the Program. In the latter case, it would be defined if the change were positive or negative depending on which of the two averages is greater.

The sample used corresponds to 30, who made their Saber T&T Test in 2012. The Wilcoxon test is applied for data greater than 25 datas, where n> 25, and is calculated as follows:

1) The differences in each element of the sample are calculated for the two variables to be studied (Saber T&T, and Saber 11).
2) Subsequently the differences are ordered by discarding the signs (+ or -).
3) Then an order range is assigned to them. (The values that are repeated are not given different order ranges, we assign an average value).
4) The ranges are added according to the signs that have the differences and the estimators are obtained:
a. W (+) : Sum of Ranges corresponding to positive differences.
b. W (-) : Sum of Ranges corresponding to negative differences.
5) The Contrast Statistician is defined, which is used to make the acceptance decisions of the Ho or **Ha**, that is, the minimum value between the two sums of ranges: W = Min [W (+), W (-)]

Steps to follow when the sample is greater than 25 data, n> 25: Follow the same steps as before to point five.

6) As the statistician fits a normal distribution, the W statistic is transformed into a new value designated as Z with the formula (1):

$$Z = \frac{W - \frac{n(n+1)}{4}}{\frac{\sqrt{n(n+1)(2n+1)}}{2a}} = N(0,1) \qquad (1)$$

8. This Z Value is compared to a Normal value $Z\propto$, which we know for a 95% confidence level $Z\propto$ = **1,96**. If the experimental value $| Z | < Z\propto$ would accept the **Ho** and if experimental value $| Z | > Z\propto$ then it reject the **Ho** and the **Ha** would be accepted.

5. Results

In Table 1 where there are related calculations with the Wilcoxon Test. Two-tailed test: Quantitative Reasoning of the Saber test T&T and Mathematics competence in Saper 11, it is observed that the result of the experimental value transformed "Z" is greater than the critical value of the Wilcoxon Table. Therefore, there would be a risk if the null hypothesis is rejected. For this reason, we can conclude that, since the differences between the medians are null, no statistical evidence has been found in the set that indicates that between the results of the Critical Reading tests and the Reading Comprehension tests there is no Change significant after the process of training students in the program for the academic period under study. Although it is observed that the transformed experimental value Z is lower than the critical value of the Wilcoxon Table, therefore, there is a high acceptance rate of the alternative hypothesis, bearing in mind that the change is highly significant for this case.

5.1 Quantitative Reasoning

This table shows that Quantitative Reasoning (QR) has a significant change of "positive" character in the training process that is to say that students after having lived their formative process reached to overcome with respect to the first test performed (Saber 11) , it is observed that when evaluating the student in the Saber T & T Tests compared to the initial ones, in the evaluated competence that for this case was QR Quantitative Reasoning, 66.6% of the students obtained a considerable improvement. Improvement in their performance in Saber tests T & T.

$$Z\ QR > Wo\ QR$$

5.2 Critical Reading

To the right of Table 2 the analytical summary of the results of the method for generic competence is observed. Critical reading, although there is no statistical evidence to reject the Ho null hypothesis, this does not indicate that there are negative changes in the results. It should be noted that 60% of the population obtained notable improvements in their performance despite not having achieved the desired results.

$$Z\ CR < Wo\ CR$$

Table 1. Calculations with Wilcoxon Test Two-tailed test - Saber T&T Quantitative Reasoning and Saber 11 in Mathematics

DQR	DCR	IDIFI RC	IDifI LC	QRR	CRR	RRQR	RRCR	RSQR	RSRC	W+ RC	W- RC	W+ LC	W- LC
-13.66	0.65	13.66	0.65	26	1	26	1	-26	1	-26			1
-1.05	-1.17	1.05	1.17	3	2	3	2	-3	-2	-3		-2	
4.17	1.51	4.17	1.51	15	3	15	3	15	3		15		3
3.53	2.23	3.53	2.23	13	4	13	4	13	4		13		4
-3.64	-2.78	3.64	2.78	14	5	14	5	-14	-5	-14		-5	
1.67	-2.99	1.67	2.99	4	6	4	6	4	-6		4	-6	
2.17	-3.11	2.17	3.11	7	7	7	7	7	-7		7	-7	
-7.54	3.41	7.54	3.41	18	8	18	8	-18	8	-18			8
-8.73	-3.57	8.73	3.57	20	9	20	9	-20	-9	-20		-9	
-11.12	4.22	11.12	4.22	23	10	23	10	-23	10	-23			10
-10.5	-4.5	10.5	4.5	21	11	21	11	-21	-11	-21		-11	
-14,7	-4.9	14.7	4.9	27	12	27	12	-27	-12	-27		-12	
-4.55	-6.15	4.55	6.15	16	13	16	13	-16	-13	-16		-13	
3	6.95	3	6.95	11	14	11.5	14	11	14		11		14
-2.5	-7.5	2.5	7.5	8	15	8.5	15.5	-8	-15.5	-8		-15.5	
-3	-7.5	3	7.5	12	16	11.5	15.5	-12	-15.5	-12		-15.5	
-10.77	-7.58	10.77	7.58	22	17	22	17	-22	-17	-22		-17	
-7.85	-7.86	7.85	7.86	19	18	19	18	-19	-18	-19		-18	
2.5	8.5	2.5	8.5	9	19	8.5	19	9	19		9		19
-18.12	-9.24	18.12	9.24	29	20	29	20	-29	-20	-29		-20	
-18.23	-10.59	18.23	10.59	30	21	30	21	-30	-21	-30		-21	
0.19	10.73	0.19	10.73	1	22	1.5	22	1	22		1		22
-2.97	-10.9	2.97	10.9	10	23	10	23	-10	-23	-10		-23	
-12.62	11.75	12.62	11.75	25	24	25	24	-25	24	-25			24
0.19	12.3	0.19	12.3	2	25	1.5	25	2	25		2		25
-17.44	-12.4	17.44	12.4	28	26	28	26	-28	-26	-28		-26	
5.64	13.14	5.64	13.14	17	27	17	27	17	27		17		27
-12.5	-14.5	12.5	14.5	24	28	24	28	-24	-28	-24		-28	
2.05	17.77	2.05	17.77	6	29	6	29	6	29		6		29
-1.99	-19.72	1.99	19.72	5	30	5	30	-5	-30	-5		-30	
										-380	**85**	**-279**	**186**

Where: DQR = Difference in quantitative reasoning, DCR = Difference Critical Reading, QRR = Quantitative reasoning range, CRR = Critical reading range, RRQR = Range real of quantitative reasoning, RRCR = Real range of critical reading, RSQR = Range with sign quantitative reasoning, RSRC = Range with sign reading critical.

Table 2 show the analytical summary of the results of the method:

Table 2 Analytical summary of the results of the method

QR				CR			
W RC	N	Z RC=	Wo RC	W LC	N	Z RC=	Wo LC
85	30	3,03	1,64	186	30	0,956	1,64

Table 3 Analytical summary of the results of the method

QR			CR		
T	N	Z	T	N	Z
85	30	-3,03	186	30	-0,96

The results shown in this research are susceptible to improvement. It is suggested to make curricular changes and trainings that tend to strengthen each of the generic competences in the level of teaching and the student body. For which it is important to take into account the following topics:

In Quantitative Reasoning: Interpretation of Tables and Charts in which it involves numerical information. Justification of procedures and strategies in problem solving. Basic topics such as Percentages and probability of elementary events and events.

In Critical Reading: Reading of texts that generate reflection, understanding and criticism. Find the general meaning of a text and make inferences from it.

It is important to take a diagnostic entrance exam, so that students know the status of generic competences.

Finally, carry out an institutional program that starts from the first semester to strengthen these Competitions

It is important to perform a diagnostic entrance exam to the students to know the status in the generic competences. Carry out an Institutional program that starts from the first semester to strengthen these competences.

6. Conclusions

The Saber T & T tests are external evaluations, carried out by the Colombian Institute for the Evaluation of Education (Icfes), to students who are about to finish their higher education studies and although many students assume it as a test that responds to the interests of the entities of higher education. Higher education and the state, and not the interests of students. These exams are important because they provide information on the degree of development of the skills of students who are close to completing an academic program, but it should also be quite significant for the same student because it gives him the opportunity to self-evaluate and obtain great benefits such as the condoning of credits with Icetex, among others. It must be important for institutions to review the results of their students because it allows them to make adjustments and make medium-term decisions about their curriculum.

Statistics is a powerful and fundamental tool that allows us to infer the behavior of certain events, it also allows us to obtain accurate information about a group of quantitative data and puts them at the service of decision making.

The Wilcoxon test is non-parametric that contrasts two samples) of the Relationship between a qualitative variable (in the present investigation the Saber Tests) and another quantitative variable (the Results) and that it makes a comparison contrast of the central tendency, in this case the median. This tool was used to correlate the results of Saber 11 Test and Saber T & T in the Institutions of Higher Education IES. Thus becoming a tool that can help the Faculties evaluate their academic programs in relation to the quality of education that provide and the curriculum they offer and in this way make decisions based on solid foundations generating improvement actions to strengthen the skills that students need to meet the objectives of any institution and achieve educational quality.

References

[1] Ministry of National Education 2016 *Document of Self-evaluation for the purposes of renewal of the international certification of the national accreditation council* - CNA colombia Bogotá October 2016 pag. 9
[2] *Dynamics of inclusive education in colombian higher education: process, principles and challenges*
[3] *Self-evaluation Document for the Purpose of Renewal of the International Certification of the national Accreditation Council* - CNA colombia, p. 13
[5] https://www.cna.gov.co/1741/article-187264.html last access Dec 15 2018
[6] Minimum Quality Conditions of Academic Programs 2004 Bogotá, colombia, Ministry of National Education Vice Ministry of Higher Education Quality Department
[7] http://www.tecnar.edu.co/tecnar/historia_last access Dec 15 2018
[8] Anderson D R, Dennys J S and Williams 2008 *Statistics for Administration and Economics* 10th Ed Editorial Asv
[9] Corso S., Jimmy A 2005 *Non-parametric statistics* Editorial UN

IOP Conf. Series: Materials Science and Engineering **519** (2019) 012035 doi:10.1088/1757-899X/519/1/012035

Design and Simulation of a Slot Waveguide Array Antenna (SWAA) for Satellite Communications

L Ripoll[1] and L Valdez[2]

[1]Universidad del Norte - UNINORTE, Barranquilla, Colombia.
[2]Fundación Tecnológica Antonio de Arévalo-TECNAR, Cartagena, Colombia.

libis.valdez@tecnar.edu.co

Abstract. One of the most used antennas today, is that of slots in waveguides. These antennas are an important type within the microwave range with various applications in electronic communications systems, radar, and satellite applications, due to its high power capacity, high efficiency, low weight, compact structure and few lateral lobes, all added together to its proven resistance to radiation, high temperatures and vibration, characteristics without which they can not be used in aerospace applications. The following lines show the design and simulation of a Slot Waveguide Array Antenna (SWAA) for satellite communications in the 30 GHz band. The antenna will be approximately 170x170 mm, constituted by 24x24 slots. It will be fed by waveguides in a lower layer connected to the radiating guides through slots. The modeling and design of the different elements that make up the power grid as T-junctions and elbows in the E and H planes were carried out.

1. Introduction

At present, different types of arrays have been developed [1, 2, 3, 4], all with some restrictions, mainly in terms of efficiency and bandwidth [5]. The increase in operating bandwidth of slit antennae in waveguides is currently stimulated by the development and applications of synthetic aperture radars (SAR) [5, 6, 7]. According to Takeshima and Isogai [8], the larger the number of slots in an array, the lower its bandwidth. Therefore in [9,10] it was found that by reducing the number of slots in an array and using subarrays fed by waveguides the bandwidth can increase. In [11], it was found that by managing conventional longitudinal slots and subarray techniques it is possible to achieve bandwidths of approximately 15%. In [12] to extend the bandwidth of the array, it went away in two subarrays fed by a convex waveguide power splitter. In [13] a geometry of slots for large arrays that offer an increase in the bandwidth of the slot on the conventional geometry is exhibited, this new model was compared with the conventional one and its advantages and disadvantages were discussed in large arrays applications of slots. While most researches are based on narrow grooves, the coupling between grooves is very important, since this coupling can optimize bandwidth and power handling capacity, Sinha [14] and Sangster [15] have investigated the coupling of wide grooves. Distribution networks constructed with various power dividers are widely used to power the array's radiating elements with the correct amplitudes and phases [16, 17, 18]. They have the advantage of having low losses, high power handling and are used to feed arrays of microstrip antennas [19] and slots [20], but their use has been restricted by having large relative electrical dimensions [21].

Content from this work may be used under the terms of the Creative Commons Attribution 3.0 licence. Any further distribution of this work must maintain attribution to the author(s) and the title of the work, journal citation and DOI.
Published under licence by IOP Publishing Ltd

IOP Conf. Series: Materials Science and Engineering **519** (2019) 012035 doi:10.1088/1757-899X/519/1/012035

The aim is to design and simulate a Slot Waveguide Array Antenna (SWAA) for satellite communications, which will operate in the 29.5-31 GHz band. The power coefficients will be separable in such a way that the use of linear current distributions to independently generate the radiation pattern with different beam width values in the main planes of the antenna. A distribution of currents associated with a Taylor diagram of 20 dB will be used. The antenna will be 170x170 mm, consisting of 24x24 slots and taking into account that the separation between slots is 7,031 mm, 48 linear subarrays of 12 slots each will be required, for a total of 576 slots. The way of feeding has been raised by waveguides in a lower layer coupled to the radiating guides by means of slot couplings.

In the following lines, we seek to develop a waveguide slot antenna with vertical polarization and broadside radiation, to operate in the 29.5-31 GHz band at an operating frequency of 30 GHz, with a gain greater than 30 dBi, a beam width of 3 degrees in both planes, a fractional bandwidth of 5%, a Lobe Level (NLPS) of 20 dB and an ROE less than 1.4: 1. A Slot Waveguide Array Antenna (SWAA) was selected. A distribution of currents associated with a Taylor diagram of 20 dB will be used in this design. An efficiency greater than 55% is expected.

2. Approach of the solution (design of the complete array) and results.

2.1. Antenna size

To determine the size of the antenna, the antenna beam width specifications are evaluated, as well as their gain and efficiency, taking the least restrictive selection criteria.

2.2. Beam width (BW)

Taking into account that a beam width (BW) of 3° (0.05236 rad) is required in both planes, we can determine approximately the length in the planes of the array, using the equation $L = f_B \lambda \frac{0.88}{BW}$

Where λ is the wavelength in the vacuum and f_B is the spreading factor of a Taylor array given by:

$$f_B = 1 + 0.632(\frac{2}{R}\cosh\sqrt{(\cosh^{-1} R)^2 + \pi^2})^2 \qquad (1)$$

For a lobe level of 20dB, R = 10 and the spreading factor is practically unity. For the frequency of operation the wavelength is 10 mm, therefore.

$$L = (10mm)\frac{0.88}{(0.05236\,rad)} = 168.1mm$$

An antenna array of 170 x 170 mm is required

2.3. Gain and Efficiency

Taking into account that at least one gain (G) of 30 dBi (1000) and an efficiency (η) of 55% is required, we can determine approximately the length in the planes of the array, using: $A = G\frac{\lambda^2}{4\pi\eta}$

$$A = 1000\frac{(10mm)^2}{4\pi(0.55)} = 14.468.6mm^2$$

An array of at least 120 x 120 mm is required. Taking these results into account, the beam width criterion is taken as a reference, given that it has a higher value. Therefore the size of the antenna would be at least 170x170 mm. And taking into account that the separation between slots is 7.031 mm, ((λg/2) the antenna will be constituted by 24x24 slots, for a total of 576 slots.

2.4. Design subarray
This was based on the specifications of the antenna. The criterion established to define the maximum number of slots in a subarray was to have a bandwidth from 29.5 to 31 GHz, with a fractional bandwidth of 5%. In the first stage of the design a formula was found that determines the maximum number of slots that are required to comply with the bandwidth specification, which will allow us to determine the number of slots of a subarray.

Taking into account that it is desired to operate in the frequency band of 29.5 - 31 GHz, a fractional bandwidth of 5% is required, therefore, using the equation: $\frac{\Delta f}{f} < \frac{1}{2m}$.

Where $\Delta f / f$ is the fractional bandwidth, confirming that increasing the length of the antenna reduces the operating bandwidth.

We can determine that the separation between the first and last radiation slots of the subarray, must be less than 5 and therefore each subarray, will consist of six linear arrays, each of 12 radiation slots. Therefore, a total of 8 subarrays divided into four quadrants are required, for a total of 48 linear arrays of 12 slots each.

2.5. Generation of excitation coefficients and displacement of the array slots (x_n)
For the generation of the coefficients the Taylor method with n = 4 is used and for obtaining the displacements of the slots of the array they were based on their electrical equivalent. The resonant arrays are designed for broadside operation and to obtain said radiation all slots must be excited in phase, this is achieved by spacing the slots λg/2 alternating their location around the central axis of the guide, as shown in figure 1.

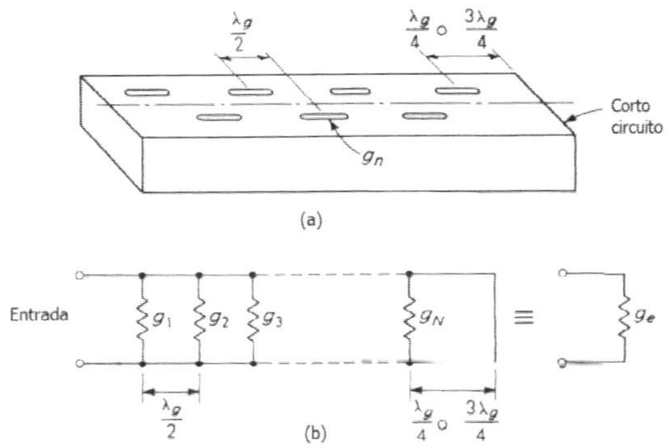

Figure 1. Longitudinal resonant arrays and their electrical equivalent [22]

The equivalent circuit of the array consists of N conductances connected through a transmission line with spacings of λg/2, a short circuit is located at a distance λg/4 or 3λg/4, from the last slot, for minimize the reflected wave and this produces that the equivalent input conductance of the array ge is simply the sum of all the individual conductances; Then, you have to:

$$g_e = \sum_{n=1}^{N} g_n \quad (2)$$

If it is the desired excitation coefficient for the e-nth slot in the array. It can be shown that the relative excitation coefficient of said e-nth slot is proportional to $\sqrt{g_n}$ and therefore, it would be given by:

$$g_n = Ka_n^2.$$

Where K is a constant to be determined and to radiate all the available input power, it must be fulfilled that the equivalent input conductance of the array g_e equals 1, having that: $K \sum_{n=1}^{N} a_n^2 = 1$.

Therefore, the constant K is determined, knowing that the excitation coefficients are already known. To later use the previous equation and define g_n and obtain the individual conductances of each slot and finally, using the equation that equals 1, the equivalent input conductance of the array g_e, determine the displacement x_l of said slots. The slots are designed to be resonant in their frequency of operation, for this reason they are of lengths close to a λ/2, allowing to maximize the bandwidth of the array [23].

Table 1 shows in its second and fourth column the coefficients for the 24 slots (N = 24) of the first row of the array for an NLPS of -20 dB, and in its third and fifth columns, the respective displacements of the slots with respect to the axis of the guide.

Ranuras		(mm)	Ranuras		(mm)
1	1	0.8347	13	1.67798493	0.9665
2	0.99926318	0.8340	14	1.6618726	0.9566
3	1.00855231	0.8422	15	1.62339987	0.9331
4	1.04291921	0.8723	16	1.55533467	0.8918
5	1.11241761	0.9339	17	1.45665268	0.8324
6	1.21522678	1.0263	18	1.33699436	0.7612
7	1.33699436	0.7612	19	1.21522678	1.0263
8	1.45665268	0.8324	20	1.11241761	0.9339
9	1.55533467	0.8918	21	1.04291921	0.8723
10	1.62339987	0.9331	22	1.00855231	0.8422
11	1.66187260	0.9566	23	0.99926318	0.8340
12	1.67798493	0.9665	24	1	0.8347

Table 1. Normalized excitation coefficients and slots displacements

2.6. Theoretical Radiation Diagram

The figure 2 shows the array factor of the electric field radiated by a rectangular flat grouping of 24x24 elements equispaced to λg/2, which is given by:

$$FA(\theta, \emptyset) = \sum_{m=1}^{M} \sum_{n=1}^{N} a_{mn}\, e^{jm.k.\frac{\lambda g}{2} sen\theta cos\emptyset}\, e^{jn.k\frac{\lambda g}{2} sen\theta sen\emptyset} \quad (3)$$

Where are the excitation coefficients, that for the separable feed ($a_{mn} = a_m a_n$), k is the wave number of the free space and λg is the wavelength in the guide.

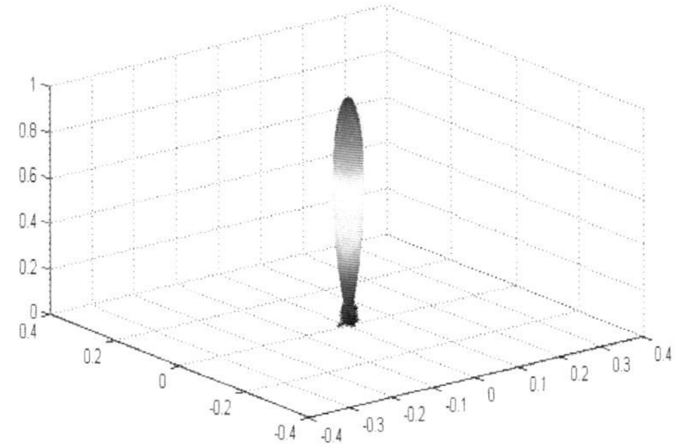

Figure 2. Array factor of a rectangular flat array of 24x24 elements

In figure 3, it is shown the total electric field radiated by the array, which includes the radiation of the elements, which in this case is a slot of λ /2.

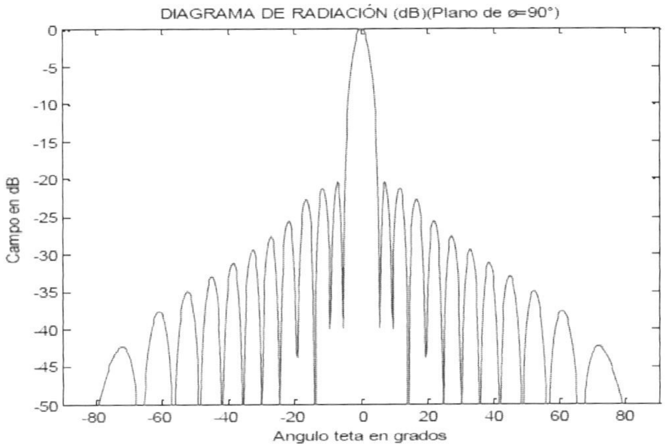

Figure 3. Total electric field radiated by the array, includes the radiation of the elements (∅=90°)

2.7. Power Distribution in the Array
The power input to the network of the array is equally distributed to the waveguides of the four sectors of the array (Top right, Top left, Bottom right and bottom left), using a corporate type power network and from each sector is distributed the power using power dividers type T and elbows H and E to each subarray, which are those that contain the radiation slots according to the chosen excitation law, which in our case follows a Taylor distribution.

2.8. Power Distribution in the Subarray (Parameters S)
The input power of each subarray can be estimated as the sum of the powers of each linear array and taking into account that the power of each linear array is proportional to the square of the excitation coefficients of the law, in the following way $P1 \sim b1^2$; $P2 \sim b2^2$; $P3 \sim b3^2$,.., $Pn \sim bn^2$

Where b1, b2, b3,..., bn are the excitation coefficients. Therefore, the total input power that enters to the three linear arrays is: Total power to the subarray; $PT = P1+P2+P3= b1^2 + b2^2 + b3^2$

IOP Conf. Series: Materials Science and Engineering **519** (2019) 012035 doi:10.1088/1757-899X/519/1/012035

a. For the second linear array: If we consider that in the first linear array there is a unit power.

$$S_{21}^2 = \frac{P1}{P1+P2} = \frac{b_1^2}{b_1^2+b_2^2} \tag{4}$$

$$S_{21dB} = 10\ Log\left(\frac{P1}{P1+P2}\right) = 10Log\left(\frac{b_1^2}{b_1^2+b_2^2}\right) = 10\ Log\left(\frac{0.999^2}{1^2+0.999^2}\right) = -3dB$$

b. For the third linear array:

$$S_{21}^2 = \frac{P1+P2}{P1+P2+P3} = \frac{b_1^2+b_2^2}{b_1^2+b_2^2+b_3^2} \tag{5}$$

$$S_{21}^2 = 10Log\left(\frac{1^2+0.999^2}{1^2+0.999^2+1.009^2}\right) = -1.79\ dB$$

Based on these calculations and the use of the CST Microwave Studio software, the displacements of each coupling slot for each subarray are determined.

2.9. Determination of the phases of the radiating grooves and S11

Taking into account that a broadside radiation is required, it is necessary that all the slots have the same phase, therefore with the help of the CST Microwave Studio the phases of the slots of the subarrays are optimized to achieve this objective. The simulation obtained shows the phases obtained in one of the subarrays of the antenna, where it was observed that at the frequency of 30 GHz the phases of the slots are similar. The simulation also showed through the reflection coefficient, that for the S11 obtained in the subarray, it is observed that although they have acceptable values they are not good enough and this is due to the great length of the linear arrays that are part of the subarray.

2.10. Symmetric, Non-Symmetric Power Shunts, and Elbows

The CST software is used for the design of both symmetric and non-symmetrical shunt models.

2.11. Symmetric shunt

As it is known for this type of union, its objective is to divide the signal that is excited at the entrance to each of the outputs with the same amplitude and phase. For this design a discontinuity known as a septum was used and was made using curvatures around a metal plate [24] located exactly in the center of the joint. In addition to the entrance of the transition, an inductive symmetric iris was added, made with two progressive narrowings [25, 26]. In figure 4, the results of the simulations performed for the power and phase parameters for the modules S11, 12 and 13 are shown.

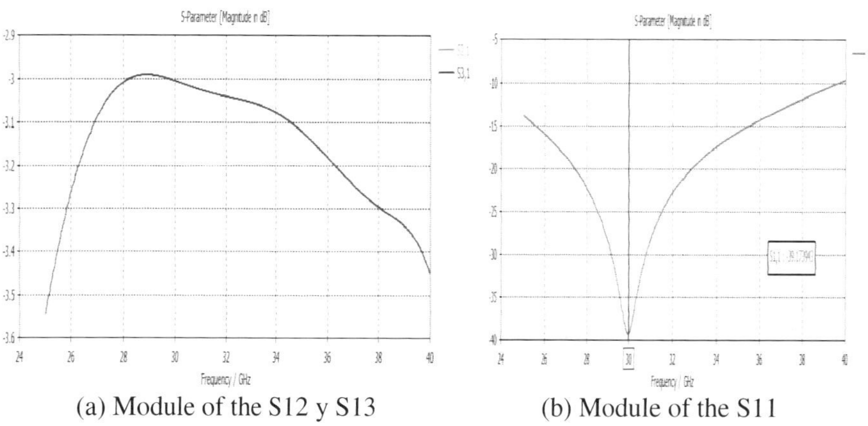

(a) Module of the S12 y S13 (b) Module of the S11

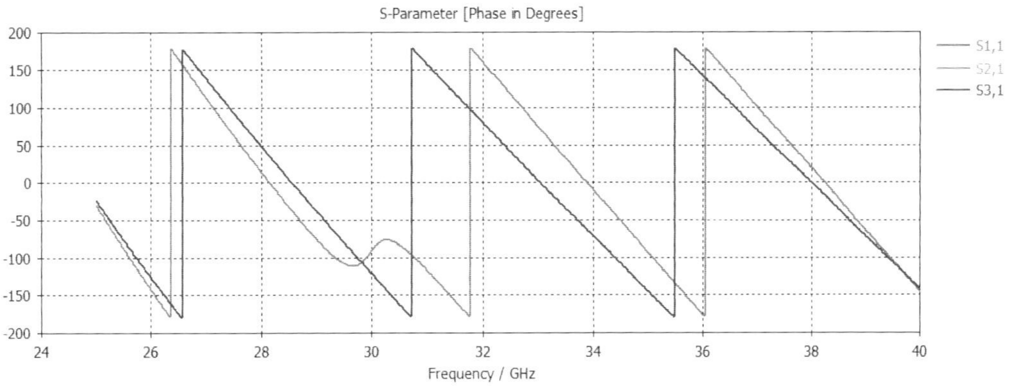

(c) Fase del S13, S12 y S11

Figure 4. Simulations of the symmetric power shunt

2.12. Non-symmetric shunt

For this design, as was done with the symmetric shunt, curvatures were used around a metal plate [24]. Additionally, two progressive narrowings were added at the entrance of the transition. Figure 10 shows one of the results obtained for a desired theoretical S13 of -3.61 dB and a desired theoretical S12 of -2.48 dB at the frequency of 30 GHz. When performing the simulations of the non-symmetric power derivative, corresponding to the module from S21 and S31 and the phase from S21 and S31, it is observed that acceptable operating values were obtained, for both parameters.

2.13. Elbows H and E

The elbows H and E were designed and simulated, introducing discontinuities at the 90 degree junction and for this a circular curvature was used [27, 28] with the appropriate values to obtain the lowest S11 possible.

In the design it was possible to observe the elbow in H, with a cut in its interior and the results of the simulations are shown where satisfactory values were obtained. With respect to the elbow in E, with a cut in its interior the circular curvature is evident, obtaining also satisfactory values.

2.14. Distribution network of 8 and 16 ports

Figure 5 shows the 8-port network, which includes symmetric shunts, elbows E and H.

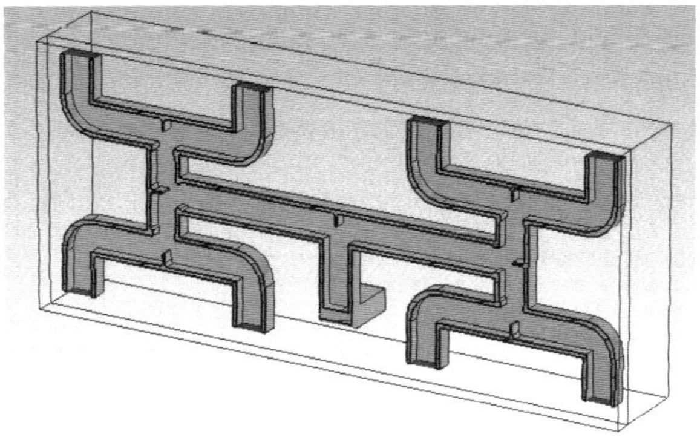

Figure 5. Cutting the plant view of the 8-port power network

With respect to the power network, according to the simulation, the bandwidth is approximately 6.2 GHz (26.2-32.4 GHz) for a SWR = 2 and 2.8 GHz for a SWR = 1.5

Figure 6 shows the values of the modules of the transmission coefficients, observing acceptable values of equitable distribution of power.

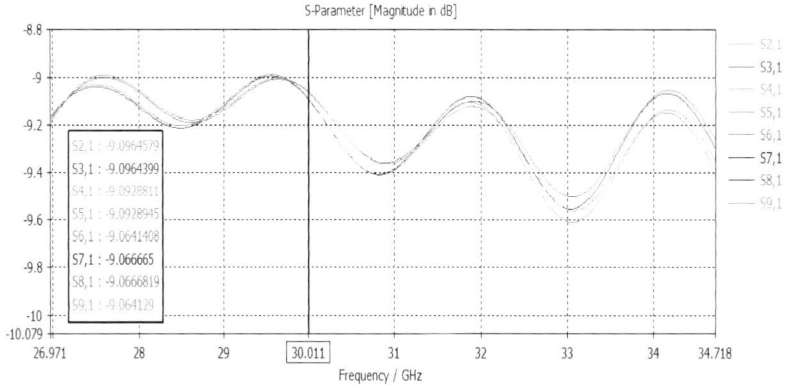

Figure 6. Cutting the plant view of the power network

Figure 7 shows the 16-port network, which includes non-symmetric shunts.

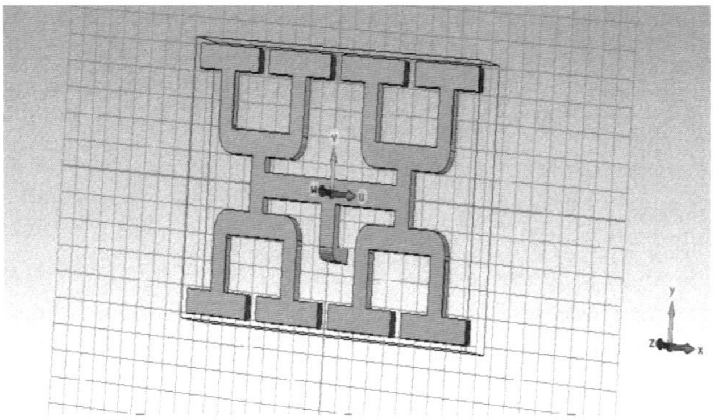

Figure 7. S11 of the 16-port power network

The simulation for S11 of the 16-port power network shows that the S11 of the power network has a bandwidth of approximately 3.9 GHz (26.5-30.4 GHz) for a SWR = 2 and 1.21 GHz for a SWR = 1.5.

2.15. Design of the complete Radiation array

After the design of the subarrays, they are joined in the appropriate positions. Figure 8 shows a plan view of the complete antenna and a bottom view where you can see the ports of each subarray, which will allow us to simulate a study of mutual links between subarrays and passive reflection in each of them.

2.16. Mutual coupling between elements

It is considered as a mutual coupling between elements such as the signal level, in module and phase, which is coupled between the i-th and j-th subarrays. Since the simulation presented here has a port for each radiant subarray, this parameter will correspond, in module and phase, with 4.5. The results show the mutual coupling, taking into account that there are 8 subarrays, in figure 9 the worst and best mutual coupling that each subarray receives from its neighboring subarrays is shown [29].

$$sij \mid i \neq j, \forall j \qquad (6)$$

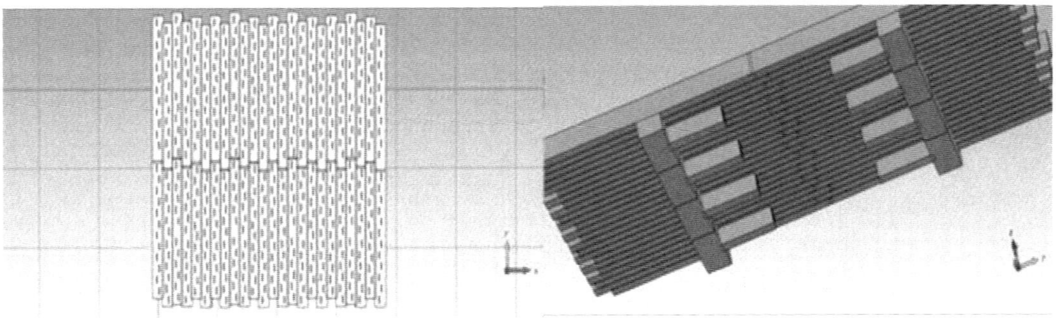

Figure 8. Plan view of the complete array and bottom view of the array for the study of couplings

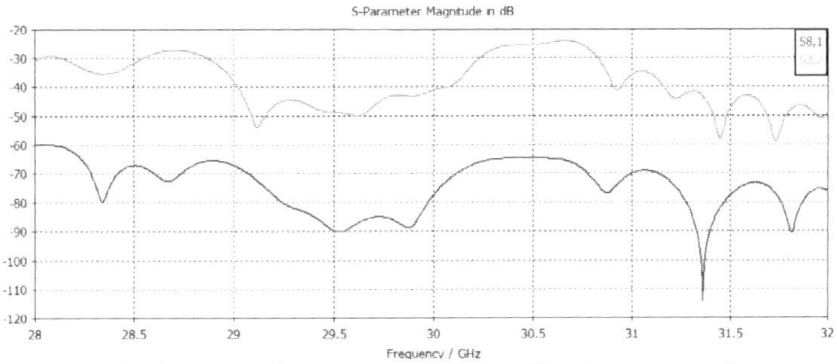

Figure 9. Better and worse mutual coupling between subarrays

As expected, the mutual coupling between sub-arrays at t

The operating frequency of 30 GHz does not affect the radiation pattern.

2.17. Passive reflection parameter

This parameter informs about the reflection that occurs at the entrance of each of the radiating elements of the array when the rest of the elements are present, but not fed. We want to emphasize that this parameter is not the same as the reflection parameter of the isolated element, given that the fact that the rest of the elements of the array are present, even though they are not being fed, modifies the impedance of the element (in a non-active way). It corresponds in simulation for the i-th subarray with parameter 4.6. The results show the values for the passive reflection of each subarray of the array [29].

$$Si, \forall i \qquad (7)$$

In the simulation and analysis of the passive reflection for each subarray of the array, it is observed that each of the ports presents an adequate reflection value.

2.18. Active reflection parameter

It informs about the reflection that occurs at the input of each of the elements of the array, but this time taking into account that the rest of the elements of the array are being fed as well [29]. Figure 10 shows this parameter for each of the sub-arrays of the antenna.

The small difference between the active and the passive values is observed.

IOP Conf. Series: Materials Science and Engineering **519** (2019) 012035 doi:10.1088/1757-899X/519/1/012035

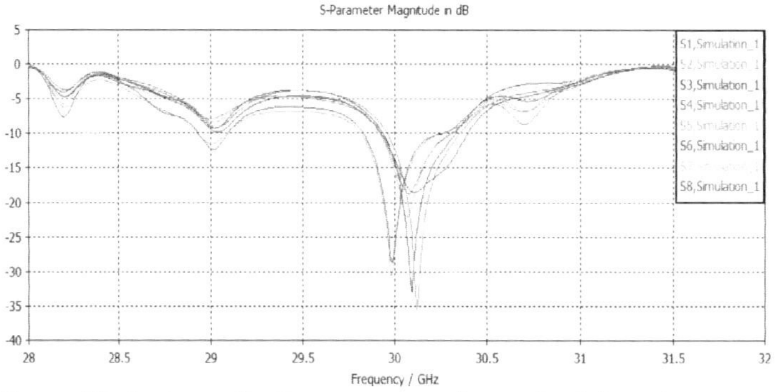

Figure 10. Active reflection parameter for each of the subarrays

2.19. Obtaining the Radiation diagram and the S11 of the Array

Figure 11 shows the bottom profile view of the complete array, where the power network with its input port is observed

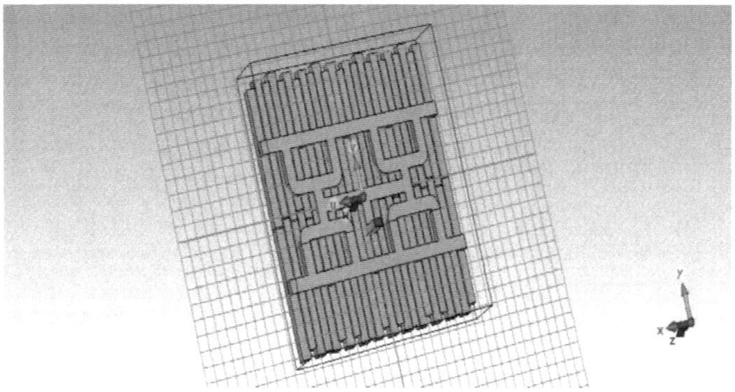

Figure 11. Bottom perspective view of the complete array

2.20. Radiation diagram

Figure 12 shows the 3D radiation diagram of the complete array, for a frequency of 30 GHz, where it is observed that it complies with specifications such as the gain and beam width at the frequency of 30 GHz, but does not comply with the desired NLPS of 20 dB and this is due to the fact that the mutual coupling between slots was not taken into account.

Figure 12. 3D radiation diagram

Figures 13 and 14 show the diagram in cartesian coordinates, where the polar and cross-polar radiation is observed for the planes ø=0° and ø=90° respectively. Finding that the target of an NLPS of 20 dB is not met and this may be possible due to the fact that the effects of the mutual coupling between slots were not contemplated

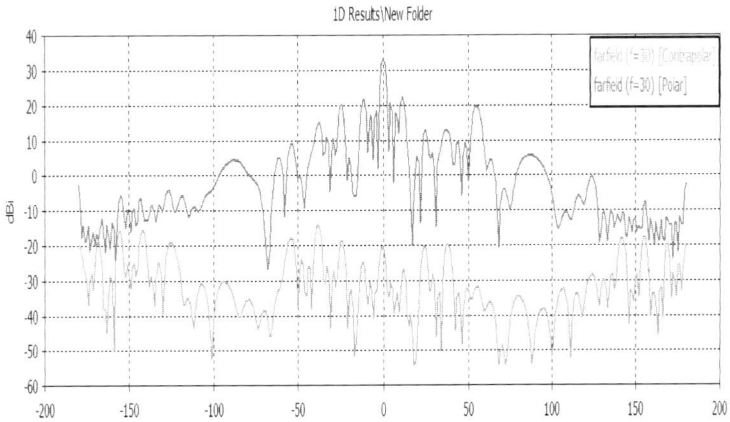

Figure 13. Radiation diagram in Cartesian coordinates (plane ø = 0°)

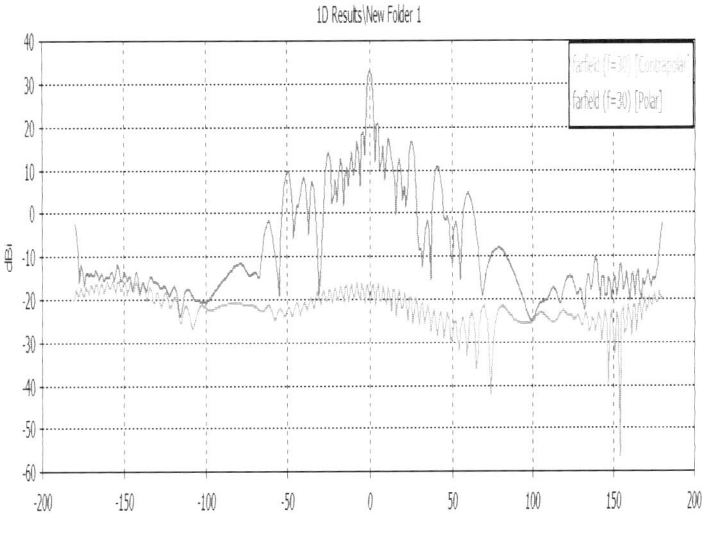

Figure 14. Radiation diagram in Cartesian coordinates (plane ø = 90°)

When plotting the S11 of the power network, it is observed that the bandwidth is approximately 0.5 GHz (29.8-30.3 GHz) for a SWR = 2 and 0.26 GHz for a SWR = 1.5, this is because at the moment to characterize the slot the equation was used:

$$g = 2.09\frac{\lambda_g}{\lambda}\frac{a}{b}cos^2\frac{\pi\lambda}{2\lambda_g}sin^2\frac{\pi\lambda_1}{a}$$

Where λ is the wavelength in the free space, λ_g is the wavelength in the guide, x_1 is the displacement from the axis of the guide and *a* and *b* are the width and height of the rectangular waveguide, as described in [10, 11], where the effects of the mutual coupling between slots are not contemplated, but which gives satisfactory results in linear arrays but not in flat arrays, additionally and taking into account that the subarrays were designed of 12x6 slots, presenting arrays very long linear (12 slots), this influenced the reduction of bandwidth.

3. Conclusions

With the realization of this design and its corresponding simulation, it is confirmed that in the construction of slot antennas on waveguide, one of the most important problems is the construction of the guide and the precision of the slots. A Slot Waveguide Array Antenna (SWAA) was designed and simulated for satellite communications, which was performed using the CST Microwave Studio software. It was found that the parameters of the antenna comply with some of the specifications of the initially proposed design, such as the gain and beam width, but it does not comply with the bandwidth and the NLPS and this is because the subarrays very long, 12x6 slots were made, reducing the bandwidth, however it is verified that the subarray method is appropriate for increasing the bandwidth in this type of arrays. It was found that the Taylor design criterion is appropriate for this type of flat arrays because it allows a more equitable distribution of power in each subarray allowing for easier design. It is necessary for future work to perform the modeling of elements in a more precise and detailed manner, an important aspect for the improvement of the power network of the flat array and the joints and bends in both modeled planes.

References

[1] Sakakibara K, Kimura Y, Akiyama A, Hirokawa J, Ando M and Goto N 1997 Alternating phase-fed waveguide slot arrays with a single-layer multiple-way power divider, *IEE Proceedings on Microwaves, Antennas and Propagation* **144** 425-430.

[2] Levine E, Malamud G, Shtrikiman S and Treves D 1989 A study of microstrip patch array design *IEEE Transactions on Antennas and Propagation* **AP-37** 426-434.

[3] Rammos E 1982 New wideband high-gain stripline planar array for 12 GHz satellite TV *Electronic Letters* **18** 252-253.

[4] Masa-Campos J L, Sanchez-Sevilleja S, Dominguez-Grano-De-Oro C, Sierra Perez M and Jose Luis Fernandez-Jambrina 2008 Monopulse beam-scanning planar array antenna in L band *Microwave Opt Technol Lett* **50** 1812–1819.

[5] Sehm T, Lehto A and Raisanen A V 1998 A 38 GHz Horn Antenna Array *EuCAP, Amsterdam,* **1** 184- 189.

[6] Wang W, Jin J, Lu J-G and Zhong S-S 2005 Waveguide slotted antenna array with broadband, dual-polarization and low cross-polarization for X-band SAR applications *IEEE International Radar Conference,.*

[7] Stangle M, Werninghaus R and Zahn R 2003 The TerraSAR-X Active Phased Array Antenna *IEEE Int.Sym. On Phased Array Systems and Technology* **2003** 70-75

[8] Takeshima T and Isogai Y 1969 Frequency bandwidth of slotted array aerial *system Electron. Eng.* 201-204.

[9] Coetzee J C, Joubert J and Tan W L 1999 Frequency performance enhancement of resonant slotted waveguide arrays through the use of wideband radiators or subarraying *Microwave Opt. Technol. Lett* **22** 35–39.

[10] Sekretarov S S, Vavriv D M and Shevtsova L V 2010 Slotted Waveguide Antenna Arrays for Airborne Radars *18th International Conference on Microwave Radar and Wireless Communications (MIKON)* **14-16** 1 – 4.

[11] Sekretarov S and Vavriv D M 2010 A wideband slotted waveguide antenna array for SAR systems *Progress In Electromagnetics Research M* **11** 165-176.

[12] Richardson P N and Yee H Y 1988 Design and analysis of slotted waveguide antenna arrays *Microw. J.* **31** 109–125.

[13] S.R. Rengarajan, M S. Zawadzki, and R E. Hodges, "Bandwidth Enhancement of Large Planar Slot Arrays", Antennas and Propagation Society International Symposium, 2007 IEEE, 9-15 June 2007, pp 4405 – 4408.

[14] Sinha S N 1987 A generalized network formulation for a class of waveguide coupling problems *Proc. IEEE* **134** 502– 508.

[15] Sangster A J and McCormick A H I 1970 Resonance characteristics of waveguide slot radiators with moderate width-to-length ratios *Proc. IEEE* **135** 319–322.

[16] Wang W, Zhong S S, Zhang Y M and Liang X L 2006 A Broadband Slotted Ridge Waveguide Antenna Array *IEEE transactions on antennas and propagation*, **54**, NO. 8, AUGUST 2006. P 2416-2420.

[17] Sehm T, Lehto A and Raisanen A V 1999 A highgain 58-GHz box-horn array antenna with suppressed grating lobes *IEEE Trans. Microw. Theory Techn.*, vol. **47** 1125- 1130.

[18] Xiao B, Hong J and Wang B 2011 A novel UWB out-of-phase four-way power divider *Applied Computational Electromagnetics Society (ACES) Journal* **26** 863-867.

[19] Demir V, Elsherbeni D, Kajfez D and Elsherbeni A Z 2006 Efficient wideband power divider for planar antenna arrays *Applied Computational Electromagnetics Society (ACES) Journal* **21** 318-324.

[20] Yu B, Wu D and Seo K 2008 Array of Waveguide-fed Microstrip Antennas *11th IEEE Singapore International Symposium on Communication Systems (ICCS)* 797 – 800.

[21] Homayoon A, Mahmoud N T 2008 Design and Optimization of Linear and Planar Slot Arrays on Rectangular Waveguides *EuMic* **2008** 1304 – 1307.

[22] Collin R E and Zucker F J 1961 Antenna Theory, Part I (New York: McGraw-Hill) 587-601.

[23] E. Collin, Robert 1985 Antenas and radiowave propagation (1ª. Ed. McGraw Hill, Singapore) 265-268.

[24] Holy A, Krihaly M and Matzner H 2011 A 1 : 8 Waveguide Feeding Network for Antenna Arrays **1** 8–10.

[25] Sehm T, Lehto A and Ra'isanen A 1997 Matching of a rectangular waveguide T-junction with unequal power division *Microwave and Optical Technology Letters* **14** 141- 143.

[26] Bang J H, Hwang S M, Lee S G and Ahn B C 2003 Design formulas for the H-plane septum power divider in a rectangular waveguide *Microwave Opt Technol Lett* **37** 390-393.

[27] Joubert J and Rengarajan S R 1996 Design of unequal H-plane waveguide power dividers for array applications *IEEE Antennas Propagat Soc Int Symp* 1636 – 1639.

[28] Zhewang M, Yamane T and Yamashita E 1997 Analysis and design of Hplane waveguide bends with compact size, wide-band and low return loss characteristics in *IEEE MTT-S Int. Microw. Symp. Dig.* **2** 417–420

[29] Jui-Pang Hsu, Anada T and Hiraoka T 1997 Analysis of H-plane circular bend of rectangular waveguide based on equivalent multi-transmission line network representation *Microwave Conference Proceedings 1997. APMC '97 1997 Asia-Pacific* **2** 773-776.

IOP Conf. Series: Materials Science and Engineering **519** (2019) 012036 doi:10.1088/1757-899X/519/1/012036

Zinc adsorption capacity with humic acids extracted from bituminous coal

M Vásquez-Osorio[1], F Colpas-Castillo[2], J R Castro-Suarez[1] and Y Vásquez-Osorio[3]

[1]Facultad de Diseño e Ingeniería, Fundación Tecnológica Antonio de Arévalo, Cartagena-Bolívar, Colombia
[2]Facultad de Ciencias Exactas y Naturales, Departamento de Química, Universidad de Cartagena, Campus San Pablo, Cartagena-Bolívar, Colombia
[3]Laboratorio V.O S.A.S, Cartagena-Bolívar, Colombia

mayerlis.vasquez@tecnar.edu.co

Abstract. The contamination of bodies of water by heavy metals represents not only a danger for the environment, but also for human health; because they do not have the capacity to biodegrade, accumulate in the environment and in the tissues of the living organisms that inhabit it. In order to minimize these impacts, Zinc adsorption (Zn) was studied in aqueous solutions using humic acids extracted from bituminous coal from the Cerrejón mine (Guajira). This investigation was divided into two phases: The extraction of Humic Acids and the adsorption study. In the first phase, an aqueous oxidation of coal samples with different particle sizes (125mm, 250mm, and 425mm) was performed, with hydrogen peroxide as the oxidizing agent, at a concentration of 30%. In the second phase Zn^{2+} removal using Humic Acids extracted from bituminous coals was studied in aqueous media as a function of pH, highest percentage of adsorption obtained was of 66.8% at a pH of 5.

1. Introduction

Contamination of bodies of water by heavy metals, which can reach dangerous levels for humans and ecosystems. It is due to the activity of different industrial sectors such as mining, the cement industry, dyes, tannery, electroplating, steel production, photographic material, corrosive paints, energy production, textile manufacturing, wood preservation, among others [1]. But one of the main sources of contamination, is the extraction of oil, because it generates a high environmental impact, reflected in the amount of water used as fracture water, as it is discharged, reaching groundwater, streams, aquifers or streams. About 632 million barrels of water can be consumed to produce 441 million barrels of oil. The resulting wastewater contains demulsifiers, defoamers, antioxidants, among others, providing a considerable amount of heavy metals such as Arsenic, Cadmium, Cobalt, Chromium, Mercury, Molybdenum, Nickel and Lead [2].

In order to remove these heavy metals from liquid waste, various methods have been used such as chemical precipitation, oxidation, reduction, electrochemical treatment, recovery by evaporation, filtration, ion exchange and membrane technologies [3]. But all these methods are very expensive and of low efficiency, when the concentration of metals in solution is very low, being a great disadvantage for these processes [1,3]. This has led to research aimed at obtaining new technologies that allow the reduction of heavy metals in the treatment of wastewater at environmentally acceptable concentrations

Content from this work may be used under the terms of the Creative Commons Attribution 3.0 licence. Any further distribution of this work must maintain attribution to the author(s) and the title of the work, journal citation and DOI.
Published under licence by IOP Publishing Ltd

and at a low cost. Adsorption has been postulated as a promising technology in this respect, it has the potential to remove the dangerous metal efficiently, faster and at a lower cost compared to the other methods [4]. Humic acids have been studied and used as adsorbents for the removal of mercury, lead and other heavy metals in wastewater. On the other hand, coal in its natural structure has some capacity for ion exchange, however, this can increase drastically, when humic substances are extracted from them, as they have a greater capacity for adsorption [5].

The removal of heavy metals cadmium (Cd), lead (Pb) and mercury (Hg) was evaluated, using fulvic acids (humic substance) extracted from lignite coal as a chelating substance. For the removal of the metals, they prepared different solutions from previously lyophilized fulvic acids, with concentrations of 0.8% and 2% each. As a result, they determined that the chelating activity of fulvic acids, allowed a satisfactory removal of heavy metals, since the cadmium was between 80% and 100% removal, for Lead 23% and 99%, for Arsenic 44% and 97% of removal and finally for Mercury, 13% and 89%. With the statistical analysis they concluded that the removal of heavy metals does not present a normal distribution (which explains the difference in the percentages), since it depends on the metal and the acidification process, therefore it was analyzed non-parametrically [2].

Similarly, the adsorption of heavy metals cadmium (Cd), copper (Cu), nickel (Ni), lead (Pb) and zinc (Zn) with humic acids (AH) extracted from leonardite carbon samples has been studied. The results of adsorption of the metals were adjusted to the Freundlich model showed a different behavior of the adsorbents in relation to the metals studied, being the maximum adsorption capacity (K) and the retention force (n) of the different metals ($P < 0.05$). According to these parameters, the metal adsorption selectivity sequences presented the following order of preference, for K in humic acids: Cd> Pb> Cu> Ni> Zn [6].

Also, the adsorption of metal ions in humic acids extracted from lignite coal has been investigated, demonstrating that the strong bonding capacities of metals with humic acids from different origins are similar under the same loading conditions. The results obtained showed that humic acids have higher affinity for the lead ions (Pb), silver (Ag), mercury (Hg), copper (Cu), barium (Ba) and cadmium (Cd), of the 17 metals studied and that the union is carried out mainly through carboxylic and phenolic groups. In addition, they were able to demonstrate that Pb, Ag, Hg and Cu ions can interact with phenolic groups, even at low pH [7].

Taking into account the environmental problems generated by heavy metal pollution, it is proposed to use adsorption as a more economical and efficient treatment [4], using the humic acids extracted from bituminous coal as adsorbents, thanks to its ion exchange capacity . In this research, it evaluates the adsorption capacity of zinc heavy metal (Zn) in aqueous solutions, using humic acids (AH) extracted from samples of bituminous coals, from a mine in the Cerrejón region in the department of La Guajira (Colombia).

2. Experimental

2.1. Extraction of Humic Acids (AH): Aqueous oxidation

The bituminous coal sample with favorable thermal characteristics, with low ash level and low nitrogen oxides emissions, from the Del cerrejón mine (Guajira-Colombia). It was crushed in a ball mill, to reduce particle size; it was sifted to select the different sizes. No.40 mesh (420mm), No.60 mesh (250mm), and No.120 mesh (125mm) were used, as noted in figure 1. Subsequently, 100 g of the sample were immersed for 1 h at room temperature with hydrochloric acid (HCL) 0.5 N, to be demineralized. To remove the bitumen, an ethanol-benzene mixture was used in a ratio of 1: 1 v / v under reflux for 24 hours. The solvent was removed by distillation and vacuum filtration with distilled water. The excess moisture was removed by heating at 80 ° C for 12 hours [8].

Then, 5 grams of carbon were taken at the different sizes, and oxidized with 15mL of a hydrogen peroxide solution (H_2O_2) at a concentration of 30%. The resulting mixture was heated to a temperature of 60 ° C and under continuous stirring for 6 hours. To extract the humic acids from the mixture, to 5 g of the oxidized samples, 100 mL of 0.1 M Sodium Hydroxide (NaOH) was added [8], maintaining this system at 60 ° C for 1 hour in continuous agitation. The resulting solution was filtered under vacuum, and 100 mL of HCl (0.1 M) was added to the filtrate for the precipitation of the humic acids. This system was allowed to stand for 24 h, then it was centrifuged at 3600 rpm for 10 minutes. The colloid obtained was then washed with ethanol (reactive quality) and finally heated at 100 ° C for 1 hour [8].

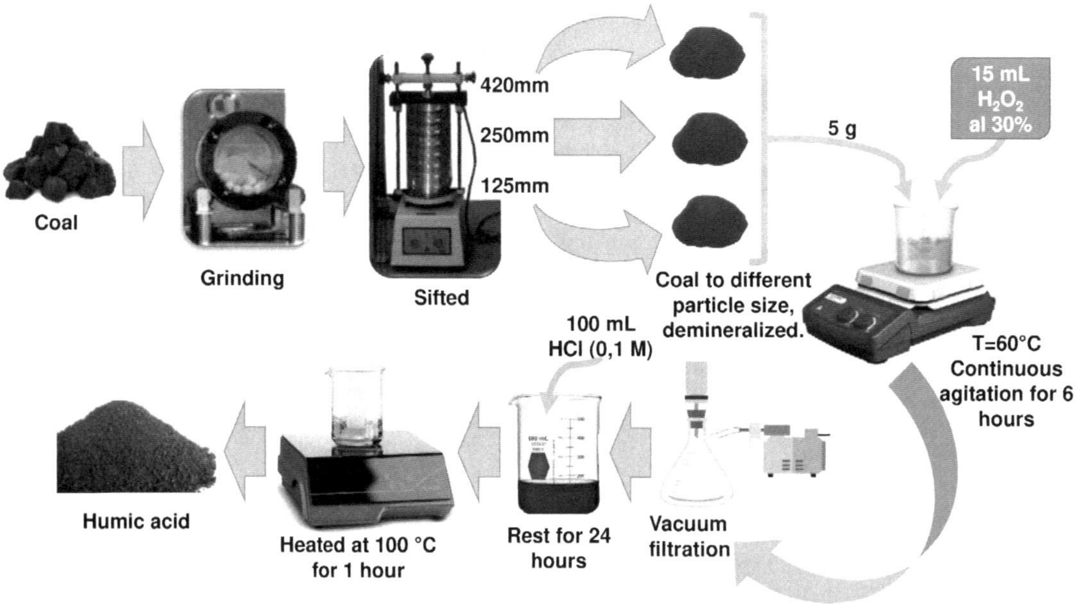

Figure 1. Aqueous oxidation of carbon.

The percentage of AH extracted was calculated taking into account the relationship between the mass extracted with the initial weight of the sample, according to equation 1[8].

$$\frac{Humic\ acid\ mass}{Coal\ mass} * 100 \qquad (1)$$

The samples of carbon and humic acid were characterized by infrared spectrometry (FTIR) using a Perkin Elmer model 1600.

2.2. Adsorption Study

To determine the cation exchange capacity, the effect of pH on Zn^{2+} adsorption was first studied, from samples of 50mg of HA with 10mL of 100 ppm solution of the metal under study, in a 250 mL Erlenmeyer, at 25°C and in constant agitation for 2 h. The analysis was carried out at different pH values, which varied between 3 and 6, adjusting by adding 0.1M HNO_3 solution and 0.1M NaOH. Then, 100 mL of a 0.1 M HCl solution was added, to precipitate the AH and to filter the solution 0.4 mL of HNO_3 was added, to preserve the medium. The above, allowed to find an optimal pH value, where the highest percentage of adsorption of the metal is obtained [9].

Later, the adsorption capacity using AH extracted from carbon samples of different sizes was studied. 50mg of AH was mixed with 10mL of 100 ppm solution of the metal under study, in a 250 mL Erlenmeyer, at 25 ° C and in constant agitation for 2 h at an optimum pH value (previously determined) [9]. The amount of metal adsorbed was determined using equation 2:

$$\%\ adsorción = \frac{(C_O - C_e)}{C_O} * 100 \qquad (2)$$

Where: C_0, is the initial concentration Zinc of in mgL^{-1}; C_e, the concentration in equilibrium of Zinc in mgL^{-1}.

The concentration was analyzed by atomic absorption spectrophotometry.

3. Results and Discussion

3.1 Extraction of Humicós Acids (AH)

Humic acids were extracted from carbon samples at different particle sizes, the percentages of extraction are recorded in Table 1. The highest amount of AH extracted (60 ± 0.1%) was from the sample with particle size 125 mm, oxidized with H_2O_2, at a concentration of 30%, and an extraction time of six (6) hours. These treatment conditions lead to the decomposition of oxygenated functional groups of coal, which are favored by the presence of structures similar to humic acids, such as fulvic acids, among others [8].

Table 1. Percentages of extraction of humic acids with different particle sizes.

Particle size	**% extraction of AH**
420mm	35.7 ± 0.01
250mm	47 ± 0.03
125mm	60 ± 0.01

In Table 1, it is observed that an oxidation during 6 hours causes a decrease in the amount of humic acids in the structure of the coals, when the particle size of the carbon sample increases. In all cases statistically significant differences were found ($p < 0.05$).

The FTIR spectra of the humic acids obtained after oxidation with 30% hydrogen peroxide (see Figure 2), present between 3500-3200 cm^{-1} a broad and strong band of absorption of -O-H groups, distinctive of the carboxylic groups. Likewise, the increase of narrow bands between 3000-2800 cm^{-1} was observed, which are due to the C-H vibrations in aliphatic structures, which explains that during the oxidation process with H_2O_2, there was a breakdown of macrostructures that caused the presence of groups aliphatics that can be as substituents in aromatic structures or as bridges between them [10].

Other bands were observed between 1800-1600 cm^{-1} generated by the stretching vibrations of C = O, in 1270 cm^{-1} by the presence of C = C or Ar-OR, and between 1300-1000 cm^{-1} by the C-O-C bonds. These results correspond to the molecular structure of humic acids reported in the literature [11-13].

3.2 Adsorption capacity of humic acids

The union of Zn^{2+} cations to Humic Acids extracted from bituminous coals, was studied in aqueous media as a function of pH. In Figure 3, it is observed that this is a factor that affects the adsorption of the heavy metal. At low pH values, a low Zn^{2+} adsorption is observed, due to the electrostatic repulsion force of the H^+ on the surface of the AH with the metal ion, making it difficult for them to interact with the adsorbent [14]. But when increasing the pH, this repulsive force becomes weak, benefiting the diffusion of the Zn^{2+} ions towards the surface of the adsorbent, where the adsorption increases.

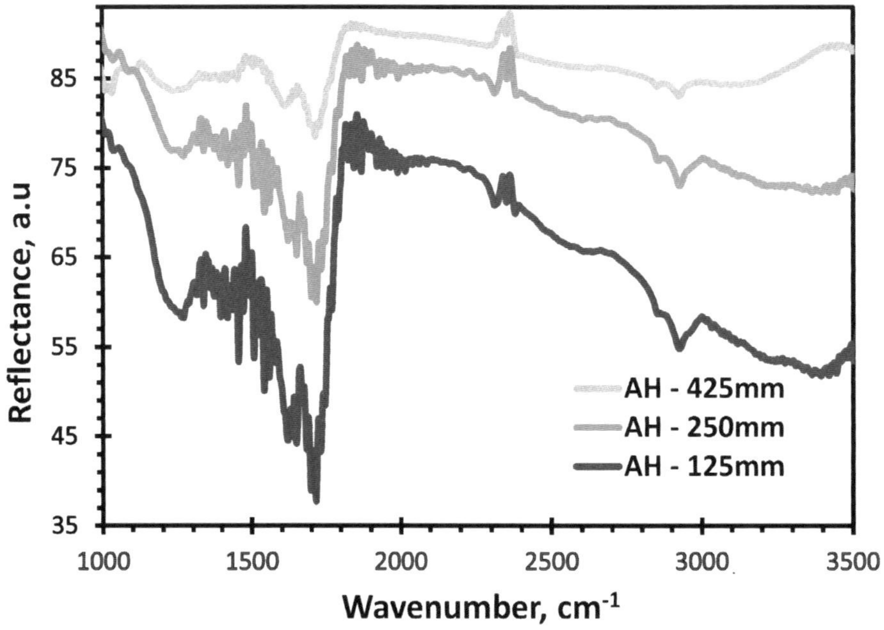

Figure 2. FTIR Spectra of Humic Acids at different particle sizes.

From figure 3, it can be deduced that the highest percentage of adsorption is obtained at a pH of 5, with a value of 66.8%. These results agree with that reported by Pehlivan & Arslan [14], comparing the adsorption capacity of Zn^{2+} and Cd^{2+} cations using relatively young brown coals YBC (lignite), humic acids, and commercial humic acid; Observing that effective metal removal was demonstrated at pH values between 5-5.7 [14].

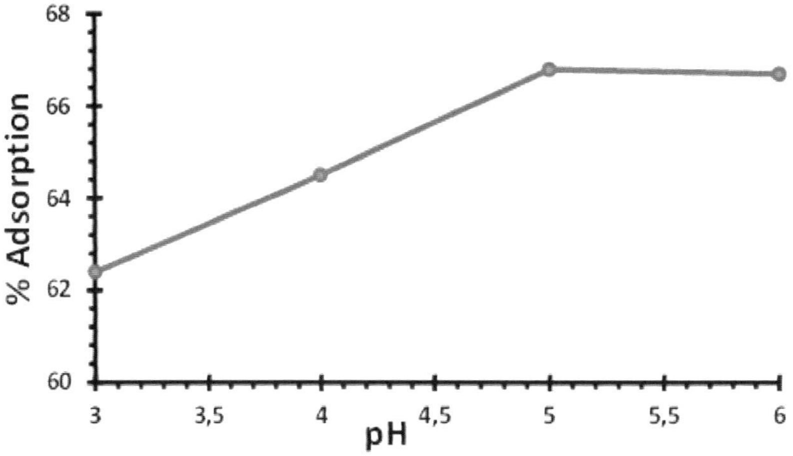

Figure 3. Effect of pH on the adsorption of Zn^{2+} with AH.

Similarly, the adsorption capacity of humic acids extracted from carbon samples at different particle sizes was studied. The percentages of adsoricon in relation to the size of the coal samples are shown in Table 2.

Table 2. Percentages of Zn^{2+} Adsocricón with AH extracted from carbon samples of different sizes.

Sample	Particle size (mm)	Adsorption (%)	Average Adsorption (%)
M1	425	62,716	
M2	425	62,698	63,002
M3	425	63,591	
M1	250	63,084	
M2	250	63,135	63,293
M3	250	63,66	
M1	125	66,999	
M2	125	67,361	67,245
M3	125	67,376	

Where it is observed that the highest percentage of Zinc adsorption, was achieved with the AH extracted from carbon sample with particle size of 125mm. However, in all cases the adsorption percentages were efficient, due to the low concentrations of the contaminant in the water used. Where generally the common methods are economically unfavorable or technically complicated [14, 15].

4. Conclusion

The highest percentage of humic acid extraction was obtained with the 125mm carbon sample, oxidized with hydrogen peroxide at a concentration of 30% and for 6 hours of extraction. The FTIR spectra of the humic acids, obtained by treating the carbon with the oxidizing agent corroborate that the material extracted in the oxidations are humic acids, according to the macromolecular structure defined for this type of substance. In addition, it was possible to identify that the particle size of the coal influences in the amount of humic acids that can be extracted from the coal sample. Also, the bituminous coal from the Cerrejón mine in the department of La Guajira (Colombia) can be a potential raw material to obtain humic acids, useful as heavy metal absorbers. The results show that the pH has an important role in the determination of the Zinc adsorption capacity when AH is used. The high Zinc-adsorption values make the method useful for wastewater treatment.

Conflict of Interest.

The authors report there are no conflicts of interest.

References

[1] Tejada C, Villabona A and Garcés L 2014 Adsorción de metales pesados en aguas residuales usando materiales de origen biológico *Tecno Lógicas* **18** 109-123.
[2] Dueñas J and Mora R 2015 Evaluación de la actividad quelatante para los ácidos fúlvicos en la remoción de metales pesados y cloruros en aguas congénitas *(Bachelor's thesis) Universidad de Ciencias Aplicadas y Ambientales Bogotá, Colombia.*
[3] Cañizares R. 2000 Biosorción de metales pesados mediante el uso de biomasa microbiana. *Revista Latinoamericana de Microbiología* **42** 31-143
[4] Torres T, Roa G, Cheikh F, Barrera C, Ureña F, and Pavón T 2013 Improving lead sorption through chemical modification of de-oiled allspice husk by xanthate *Fuel* **110** 4-11.
[5] Salati S, Papa G and Adani F 2011 Perspective on the use of humic acids from biomass as natural surfactants for industrial applications *Biotechnology Advances* **29** 913-922.
[6] Cortes L, Bravo I, Martín F and Menjivar J 2014 Adsorción de metales pesados en andisoles, vertisoles y ácidos húmicos *Acta Agronómica* **64** 61-71.

[7] Klucakova M 2012 Comparative Study of Binding Behaviour of Cu (II) with Humic Acid and Simple Organic Compounds by Ultrasound Spectrometry *The Open Colloid Science Journal* **5** 5-12.

[8] Anillo R, Colpas F and Meza E 2013 Aumento del contenido de ácidos húmicos en un carbón de bajo rango a través de la oxidación con aire y con peróxido de hidrogeno o ácido nítrico *Quím. Nova* **36** 387-392.

[9] Colpas F, Tarón A and Vásquez Y 2017 Capacidad de adsorción de plomo con intercambiadores catiónicos de carbón xantado *Rev. U.D.C.A Act. & Div. Cient.* **20** 425-433.

[10] Quintero J and Bonilla L. 1991. Obtención de ácidos húmicos por oxidación de carbón con permanganato de potasio y su efecto sobre la disponibilidad de fósforo proveniente de roca fosfórica *Revista Colombiana de Química* **20** 9-14.

[11] López R, González G,Vázquez R, Olivares E, Vidale J, Carranza R and Ortega M 2014 Metodología para obtener ácidos húmicos y fúlvicos y su caracterización mediante espectrofotometría infrarroja *Revista Mexicana de Ciencias Agrícolas* **8** 1397-1407.

[12] Izquierdo M, Antelo J, Fio S, Gondar D, Iglesias A, López R and Arce F 2004 Caracterización de ácidos fúlvicos y ácidos húmicos extraídos de una turbera minerotrófica *Revista EDAFOLOGIA* **11** 329-339.

[13] Manzak A, Kurşun, C and Yildiz, Y 2017 Characterization of humic acid extracted from aqueous solutions with polymer inclusion membranes *Journal of the Taiwan Institute of Chemical Engineers* **81** 14-20

[14] Pehlivan E and Arslan G. 2006. Comparison of adsorption capacity of young brown coals and humic acids prepared from different coal mines in Anatolia *Journal of Hazardous Materials* **138** 401-408.

[15] Jiang Y, Pang H and Liao B 2009 Removal of copper(II) ions from aqueous solution by modified bagasse *Journal of Hazardous Material* **164** 1-9.